Working Knowledge

T0225947

Karl Hess

Working Knowledge

STEM Essentials for the 21st Century

 Springer

Karl Hess
Electrical and Computer Engineering
  Physics
Center for Advanced Study
Beckman Institute
Professor Emeritus of the
  University of Illinois
Urbana, IL, USA

ISBN 978-1-4614-3274-6   ISBN 978-1-4614-3275-3 (eBook)
DOI 10.1007/978-1-4614-3275-3
Springer New York Heidelberg Dordrecht London

Library of Congress Control Number: 2012948344

Printed on acid-free paper

Springer is part of Springer Science+Business Media (www.springer.com)

Dedicated to the STEM education of my granddaughters Audrey and Natalia Calef and my grandsons Lukas and Tomas Hess.

Acknowledgements

Thanks are due to Josh E. Brunner, who read the first chapters while attending high school and assured me that the material was understandable and interesting to him.

I am deeply indebted to Prof. Joseph Lyding for his friendship and his detailed, precise and important corrections. Theodore L. Brown supplied many valuable suggestions and his book on chemistry (the central science), which is an example of a great educational text for a STEM discipline. The STEM gang of Hawaii has influenced the book content and given me inspiration and courage: Sylvia Hess (my loving wife), Liz and David (MD) Sonne, Barbara (PhD, J.D.) and Ron (PhD, J.D.) Winters, and the astronomy crowd led by Taft Armandroff (PhD), currently director of the Keck observatory, and his associate Debbie Goodwin. Comments and suggestions of Prof. David Ferry and Dr. Anirban Basu are also greatly appreciated.

Illustrations by Michael Aschenbach, VisionBuild Author services.

Preface

Introduction

Working Knowledge: STEM Essentials for the 21st Century presents my list of fundamental, age-old (and very-recently-developed) concepts that explain the workings of the natural world and underpin tomorrow's expectations related to STEM: science, technology, engineering, and mathematics. At a time that's witnessed a huge surge of interest in STEM Education—evident in everything from government programs and funding, calls to action from political leaders, and urgent appeals from academic, corporate, and industry representatives—the questions arise. What can and should students expect to learn and master at the start of the twenty-first century? What insights can they expect to gain for their lives, intrinsically and practically, from a working knowledge of how present-day technology operates? How does this relate to historic discoveries and inventions by the likes of Newton, Maxwell, Einstein, Edison; and Bardeen, Brattain, and Shockley? How does mathematical analysis, even that based on elementary quantitative description, illustrate our understanding of the science and engineering? And, finally, what set of facts might parents, or the inquisitive reader, outside a formal STEM academic setting, explore to appreciate more fully the massively STEM-dependent world in which we live?

My own expectation in writing this book is to impart what I believe is the essence of what a student of a STEM curriculum should be able to grasp and what teachers of STEM should understand in great detail.

Aim of the Book

For whom is this book written? First and foremost, I wrote this book to inspire high-school students who have a STEM interest and for STEM teachers and those who wish to become teachers. I wrote many sections of the book directly for the students:

Sects. 1.1 and 1.2, and the introductory sections of all chapters contain information for students as young as eighth grade ready to decide whether they would want to pursue a STEM education in high school and later in college. And all sections, even some of the more advanced, are written for students who wish to find out whether they have the interest and talent to pursue STEM-related jobs after school.

Last but not least, I wrote this book for parents of students interested in STEM and for all who wish to participate in discussions that involve modern STEM education. All chapters, except for the last, can be understood by use of elementary algebra and geometry, and the last chapter, Advanced STEM Problems, requires only an introductory understanding of calculus (such as is taught in high school). All of the necessary mathematics is presented at the beginning of the book, in spite of the fact that the M in STEM is last! (One cannot understand basic science and engineering without mathematics.)

A novelty of my presentation is the proposed use of mathematical software. Calculators and computers are commonplace in high schools. Pocket calculators are, in principle, enough to do all the problems that are discussed in this book. I have added boxes that explain how the software MATHEMATICA can be used to solve the problems. My hope is that students and teachers will have access to MATHEMATICA or other software packages. As can be seen in the text, the use of these packages is not too complicated and permits solutions to many real-world problems with negligible effort. I show the reader how to solve equations with several unknowns, an expertise often needed in science and engineering. Many of these solutions, when performed by hand, involve so much labor that students may become frustrated and disinterested. Students are also sometimes turned off by problems that can be done easily by hand, but are oversimplified and artificial and do not seem to serve any "real" purpose. My goal in introducing the use of mathematical software is to rectify this problem.

The contents of this book are mainly dedicated to mathematics, physics, chemistry, engineering, and technology. Another book needs to be written for the expectations in the life science/biological areas. You might ask if I covered all important STEM topics, including all those pertaining to important technological advances of recent years? Of course, I have not. It is amazing, however, that most of the recent developments in technology are based on just a few engineering and scientific principles, and therefore can be explained in condensed form. It is my goal to combine the well-known principles, often dating back to ancient knowledge, with recent developments in science, engineering, and technology. In this way, I wish to present what we know already, where we are, as well as where we might improve and go in the future. The last chapter contains a description of some more advanced problems of science and engineering and introduces the rudimentary calculus of differentiation and integration. This chapter is added for those who truly love STEM and, in particular, mathematics and desire to understand topics such as Einstein's relativity theory, quarks, and superconductivity.

My hope for this book is that it will inspire students and teachers, give them a sense of the extent of their STEM aptitude, and serve as an example to parents and other of what we should know about STEM in all walks of life. As an illustration of how important basic STEM knowledge is, even for top executives, I offer the following true story.

My friend, Dr. Adler, was director of the research laboratory of a very well-known firm I will call "Acme Electronics Corporation." Acme is a substitute name for a real company, which was a venerable American manufacturer of consumer electronics for nearly all of the Twentieth century, most famous for producing television sets. Adler visited me one day and was very upset. "Can you imagine," he said, "one of the top executives of Acme told me that the company does not need a research laboratory anymore, because everything to know about TVs is already known." This incident happened in 1978 when there existed only very limited and primitive cable TV options, and no satellite, digital, or high-definition televisions. Nor were there screens based on LCDs and LEDs, plasma screens, DVDs, etc., not even a widely used remote-control box. And certainly no 3-D TV!

With this in mind, the statements of the Acme executive, and his following actions to dismantle much of Acme's research, were preposterous and, indeed, very damaging to the company. If the CEO had any significant understanding or appreciation of STEM, he would have also had some feeling for where TVs could go with respect to (potential) functional developments in any number of other allied technologies. (All he had to do was to look at what was already known, such as the picture quality of color photographs taken with the best cameras of this time, and compare this to the lousy pictures of the late 1970s television sets.) He may very likely have then, instead, asked his researchers: How can we reach the type of resolution and perfection that already exists in photography? They would have told him that the answer to his question would require research and a number of specific advances. Perhaps then Acme would not have dismantled its research and development activities, and as such the company might still today be a leader in the field.

It's not a stretch to my mind to assert that a broad working knowledge of STEM across the entire population, beyond simply the group of people who choose careers in science and engineering, is essential and that approaching problems armed with this way of thinking is not only critical to the survival of companies, but it is also critical to our survival as a species.

Prelude at Hapuna Beach

The Hapuna Beach of the Big Island of Hawaii is one of the nicest beaches in the world. It is located on the slopes of the Mauna Kea, an extinct volcano more than 4000 m (13,000 feet) high. High on top of that mountain, one can see the brilliant white observatories that include the two powerful Keck telescopes that penetrate the universe toward its most remote places. Right above the beach is the Hapuna Beach Hotel, a very modern and beautiful landmark. And on the beach, we meet the creatures of the ocean, beautiful fish and turtles.

So come with me, walk with me on the beach. Regard the world as a scientist or engineer would, and see what nature has to offer, and what we can find out about its intricate workings.

Let's first imagine we are looking at the beach like the Greek scientist and philosopher Democrit. We approach the beach and see the beautiful white sand. The grains of sand are, at first, bigger, more like little pebbles. They become smaller and finer as we approach the beach. In the distance, we see the waves moving in from the open ocean and rising higher as they approach the shore; sometimes they reach 5 m (15 feet) high. As the waves move toward the shallower waters, they crush into the sand and turn the water from its blue-green color to a sandy brown, grinding the sand into smaller pieces. This is why the sand becomes finer and finer as one approaches the beach front. What might Democrit have thought when he observed the beaches of his homeland Greece? How fine can the sand get? Can one grind it to an ever smaller size, or is there a limit? Democrit thought that there was a limit below which the grain size could not be further reduced, no matter how much we grind. He thought there was a smallest, fundamental size, and he called this "atomos" and gave thus the name to what we today call the atom. What an important concept that was! Scientists have been thinking about atoms ever since. Two scientists, Mach and Boltzmann, both from Austria, argued whether atoms really exist. Boltzmann was for atoms, Mach against. In the end, Boltzmann won the day. Tragically, however, he started to doubt his own important work and committed suicide before the existence of atoms was proven.

We now know, without any doubt, that atoms exist. One can even make them directly visible with modern microscopes. We also know that Democrit was not quite right when he thought that atoms were the smallest possible pieces. Atoms can be split into still smaller entities, and we will learn more about that in Chap. 5. The important idea, of course, is how Democrit's thought process illustrates the essence of the scientific way of thinking. Democrit expanded the analogy of smaller and smaller sand grains as an example for the smaller building blocks of all of our surroundings. Then he took the ultimate limit to arrive at his hypothesis: the existence of a smallest possible, elementary entity, namely, the atom. In the centuries that followed, careful, replicable experimental tests were used to support the hypothesis and bring forth the theory of atomic structure.

Euclid was another great scientist and mathematician who drew inspiration from analogies and limits. He made drawings of circles and straight lines and dots in the sand or on paper. He used very simple tools, a piece of string and a straight rod with measurement markings (such as centimeter and inch markings on modern rulers), and with these everyday articles laid down all the basic laws that one needs to know in geometry, whether one wishes to measure the area of a garden or wants to determine the height of a tower. Or, sitting on Hapuna Beach, we could estimate the height of the Mauna Kea with a few measurements as will be discussed in the Chap. 1. Euclid's findings were so important and fundamental that, thousands of years later, they formed the basis for the development of the global positioning system (GPS) used to find locations when driving a car or determining the position of a ship or airplane.

Everywhere we look around at the beach we can see some phenomena or facts that relate to STEM. We can see, for example, that the water has receded at the shore line, and some rocks that were submersed at first are now out of the water. This is related to the tides. Tides arise because of the forces that the moon and the sun exert on the earth and its oceans. Scientists and engineers understand these forces with great precision and have even been able to use them to create energy, such as electrical power from the tides.

In the distance we see a big freight ship that transports cars plus other cargo between the islands. How many cars can such a ship hold? Archimedes, another Greek scientist, figured that out: every ship displaces a certain amount of water because part of every ship is under water. The weight of the water that is displaced amounts, exactly, to the weight of the ship itself and all the freight it has on board. This is the principle of Archimedes, and it is as important today as it was thousands of years ago. The principle is not only valid for ships in water, but it is also valid for balloons in air. A balloon floats in air if its total weight—consisting of the balloon material, the gas in the balloon, and the freight—is equal to or smaller than the weight of the air that the balloon displaces (one could also say replaces). To give an example, there was a balloon that looked like a flying saucer hovering over Colorado and made news in 2009, in part, because people thought it might be carrying a boy who got onto the balloon by some accident and flew away. Anyone who saw this story on the news could see that the balloon had a diameter of around 4 m and a height of about 2 m. From the section on geometry, in Chap. 1, we see that the approximately pyramid-shaped balloon displaces about 8 cubic meters of air. We know from physics (and from weighing things) that 8 cubic meters of air weigh about 20 pounds. We can therefore easily conclude that balloon could not have carried its own weight and the boy, who probably weighed at least 30 or 40 pounds or more. Even if our size estimate was too small, the boy could not have been on board. The whole nation and news media were upset over nothing. A careful estimate afforded by a working knowledge of STEM would have provided critical insight into whether or not they should worry.

Today, of course, we can also find things on the beach that Archimedes, or any of the ancient scientists, could not possibly have even imagined, cell phones, iPods, personal computers with Internet connections, etc. How could they have even dreamed that you could sit at the beach with a small device that could connect you with anyone, or any piece of knowledge, around the world? We will look at the World Wide Web and other technologic achievements in detail in Chap. 3.

By now evening has approached and the sun is setting. It was a beautiful day and the sun, a bright-red disc, is disappearing into the ocean. Just in the last moment when there is still a last sliver of sun visible, its color changes to green and the sun ends the day with a green spot that rapidly vanishes at the horizon. This is what the Hawaiians call a green flash. Experts of optical science can explain it by similar reasoning as they explain the colorful effects of glass prisms in sunlight. But there is another interesting feature of the sunset. The sun has just totally set at the beach and it has become dark. However, the sun is still shining its bright light onto the mountaintop. Particularly one can see the observatories on top of the mountain, radiating white and orange in the evening sun. What is the explanation of this fact? Nowadays everyone knows: the earth is spherical, a round rotating object in space. The rotation of the earth causes the sun to set first on the beach and later on the mountaintop. If one thinks about this effect carefully, one can see that, even without telescopes or modern satellites, there is significant evidence, which has always been there, that the earth is round. Some of the ancient scientists realized this. However, this fact became clear only much later because of the work of astronomers like Kepler, Copernicus, and Galileo. Astronomers are still at work with giant telescopes. The Keck telescopes work with two mirrors, each 10 m in diameter, and are situated about 80 m apart. This twin telescope on top of Mauna Kea reaches the most distant places in the universe and has discovered not only new stars and new clusters of stars (galaxies) but also many planets that move around stars far away from our star, the sun. Keck may someday reveal another earth-type planet with water and an atmosphere containing oxygen—suggesting, yet again, that we might not be alone.

I hope this prelude on the beach has helped illustrate how important STEM is for our life and that it is worthwhile (and fun!) to learn about it even if one does not wish to pursue a professional career in science or engineering.

This book will give you an idea about STEM knowledge that we should have and that a high-school education should provide. Of course, as with any such book of but a few hundred pages, the story is not complete. I offer, therefore, these chosen topics as one scientist–engineer's collection of important and interesting examples. I hope you will find this working knowledge about our natural world as fascinating and essential as I do.

Hapuna Beach, HI, USA   Karl Hess

Contents

Chapter 1
Mathematics: The Study of Quantity, Structure, Space, and Change

1.1 Just Numbers

We do not know when numbers were first used, but we may assume that the use of numbers is as old as mankind. In the beginning of all intelligent life there was language. Part of any language was certainly a method of counting: counting people, counting oranges, and counting weapons. This correspondence of elements of our world to the mathematical abstraction of a number is at the root of the beginning of mathematics and in a way also of science, for it is clear that a theory is developed here about nature. If one develops symbols for numbers and rules such as adding these symbols, then one also develops rules for things in nature, for oranges, apples, or livestock which are, although in principle all different, abstracted and regarded as mere numbers.

This abstraction is basic for the development of a theory: the theory says that, for certain purposes, one can deal with all of these things, apples, bananas, and livestock, just by using numbers. The second and most important step of a scientific approach, the experimental investigation and confirmation, is straightforward in the case of numbers and is performed by everyone not only by scientists. It is also clear that these numbers can help with any form, even the most primitive form, of engineering and technology. For example, how many palm leaves does one need to put a roof on a hut and how many wooden planks do we need to construct a raft? Thus numbers represent an "elementary force" when it comes to understanding science, technology, engineering, and mathematics (STEM), and as a consequence it is obvious that we must demand a detailed understanding of the various systems of numbers, and uses thereof, from anyone who has finished a high school education. The importance of numbers must not be underestimated, and the necessity for any student to be good with numbers cannot be stressed enough. Everyone needs proficiency with numbers, even the greatest artist. For example, Michelangelo needed to project to the Pope the expense of his frescos, he needed to know the ratios of various paints and fresco material, he needed to pay his assistants, and he needed to rent a studio for material storage.

K. Hess, *Working Knowledge: STEM Essentials for the 21st Century*,
DOI 10.1007/978-1-4614-3275-3_1, © Springer Science+Business Media New York 2013

Fig. 1.1 The number system of the Mayas consists of *lines* and *dots*. The *line* corresponds to five dots. This is therefore not a decimal system. In a decimal system a *line* would correspond to ten dots. The figure describes the equation $18 - 6 = 12$ in Maya numbers

The aspects of proficiency with numbers did not at all come easy to mankind and did not develop quickly. Even the choice and development of the most efficient symbols took hundreds and thousands of years, and the advanced theory of numbers is still not a finished area of research. Mathematicians still continue to report important progress in number theory. Therefore we pull up our sleeves and work on this chapter with a pencil.

1.1.1 The Natural Numbers and Integers

Natural numbers are the numbers we use when counting. We start with considering simple symbols of natural numbers as they were used by the Mayas. The Mayas used a dot to symbolize a "one," two dots for a "two," three for "three," and four for "four" but then a line for "five." Interestingly enough, they also had developed a symbol for "zero," a shell (probably an empty shell). This symbol represents, of course, a higher form of abstraction then the numbers that we usually use when counting. A shell means that we have nothing here. As we will see, the "zero," the shell, or the 0 as we write it now, has indeed a special meaning and use and needs to be included in the system of natural numbers. In the mathematics of all numbers, it also has a special standing and needs to be excluded from mathematical operations and statements that involve division (see below). It is no small achievement for any culture to have included a zero into their system of numbers.

With numbers we can explore the basic operations of adding and subtracting. If a general had a large number of warriors and wanted to leave some of them home for protection of his castle and therefore "subtracted" these from the total, how many did he have left. If we deal with such questions, i.e., with relations of numbers, then we deal basically with "equations" and we use symbol combinations that involve =, the equal sign. An example is given in Maya notation in Fig. 1.1.

This equation solves the problem of the general: he has 18 warriors and wishes to leave 6 at home for protection. Then he has 12 left for the battle. Thus writing the symbols of Fig. 1.1 in our modern way we have

$$18 - 6 = 12. \tag{1.1}$$

It is neat how the Maya notation permits you in an easy way to "solve" the generals "puzzle" or equation. The subtraction is performed by simply taking the symbol that is subtracted (one dot, one line) literally out of the original total to arrive at two dots and two lines. We would do the same for oranges, of course, except that we have now a "shorthand" for five of them (the line). The fact that the line represents five is arbitrary. Had we considered the line to stand for a "ten" and had we been willing to put up to nine points above the line, then we would have had a decimal (stands for ten) number system. The equation of Fig. 1.1 would then describe a different problem: the general would have 33 warriors at his disposal, would leave 11 at home, and would have 22 for the battle.

The decimal system is very convenient as was proven over a history of thousands of years. Of course, the decimal system is not in any way special. As we know, most important instruments such as clocks or modern computer chips do not use it. Clocks count up to twelve, and computer-chips use the binary number system that we will discuss later. Nowadays, everyone is familiar with the decimal number system and with the modern notation that differs from that of the Mayas. The notation we use comes from the Arabian culture. We use the ten symbols or "digits" $0, 1, 2, 3, 4, 5, 6, 7, 8, 9$ instead of the points and the sequence of digits 10 instead of a decimal line, 20 instead of two lines, 30 instead of three lines, and so on. For "eleven" we use the two symbols 11 indicating that we have a "line" that is valued as 10 and one dot. Then we can write the equation of Fig. 1.1 as

$$33 - 11 = 22. \tag{1.2}$$

We can still think of the operation in this equation in the same way the Mayas thought about it: we start with three lines and three dots, take away one line and one dot, and end up with two lines and two dots. I do not believe that it is necessary for me to give here further examples of subtraction and addition.

The more advanced concept of multiplication can be thought of as follows: if we multiply the number 12 by 5, that simply means that we add 12 five times. Of course, multiplying large numbers can only be done quickly when certain procedures are well known. I believe that even in our time of computer and information technology, it is of utmost importance to teach all students how to handle numbers small and large. Only if a "sense" for numbers is well developed, can we hope to have the students prepared for life. I have seen on numerous occasions how calculators have totally confused students because they mistyped numbers. For example, one student had typed one digit too many, 222 instead of 22, and then obtained for $5 \cdot 22$ the result of $1,110$ instead of 110 and accepted this result uncritically! I still remember the multiplication exercises with my high school teacher who had a Ph.D. in mathematics. She was emphasizing simple operations with numbers for years. She taught us frequently little tricks and then went through many minutes of exercises. This was time well spent. For example, if one multiplies a number that ends with 5 such as 35 by itself, all one needs to do is multiply the first symbol or digit which is in our case the 3 by the next higher number (the 4) to obtain 12 and then attach 25 to obtain $1,225$. This is the result, i.e., $35 \cdot 35 = 1,225$ where "·" is the symbol for multiplication. One can do this type of multiplying a number by itself faster than by use of a pocket calculator, and it always works when the number ends with a 5. The reader may try to find out why.

Natural numbers follow by their "design" some very natural rules. We can express these rules in a very general form because they are valid for all natural numbers. The way to do this is to write a letter symbol for a natural number, for example, the letter a. What does this mean? It means nothing else than: a stands for some natural number, could be for 3 or for 3000 or whatever. Similarly we can say that the letter b also stands for a natural number. It could be that a and b are equal which we express as $a = b$ or they are not equal which we express as $a \neq b$. So what are these general number symbols good for? Well we can, for example, state the rule: for every natural number a one can find a higher natural number b. One such higher natural number is $b = a + 1$. You can play games with kindergartners and tell them that for any number they know you know a higher number. In this way one can foster some understanding for the fact that no natural number, and no number of any kind, can be called the highest number.

Another very important rule for natural numbers, and again for all kind of numbers, is the following:

$$a + b = b + a. \tag{1.3}$$

It is pretty clear that this rule, that is called the commutative law, must be true . Consider, for example, that we have $a = 10$ oranges and $b = 5$ oranges. If we add the two, we obtain 15 oranges no matter in which sequence we add. Another "obvious" rule would be

$$a \cdot b = b \cdot a \tag{1.4}$$

the commutative law of multiplication, meaning that if, for example, I multiply three things by two or two things by three I will get six both times. Still another obvious rule is the so called distributive law:

$$a \cdot (b + c) = a \cdot b + a \cdot c. \tag{1.5}$$

Here the parentheses means simply that the expression inside of the parenthesis stands for a single number. You can verify the rule by substituting certain numbers of oranges for the symbols b and c and some number for a.

These rules give us advantages in the handling of numbers as we will see over and over. While we can derive these rules from our comparison with handling things of nature such as oranges and apples, we can also take them as the basic rules to establish numbers, meaning that these rules define what sets of objects can be numbers. In that case we just assume that these rules are valid a priori, that is, without any further justification, and call them the "axioms" for our particular system of numbers. The axioms for the natural numbers were found by the Italian mathematician Giuseppe Peano, who showed that the natural numbers can be completely defined and described by very few rules such as we have given them above. The interested reader can find a lot of details about this by looking up "Peano's Axioms" using a search engine on the Internet.

It is worthwhile to think about axioms a little more. Axioms accomplish the following. We are dealing with a system of numbers that has been abstracted from experience with real things such as oranges and apples. By establishing a

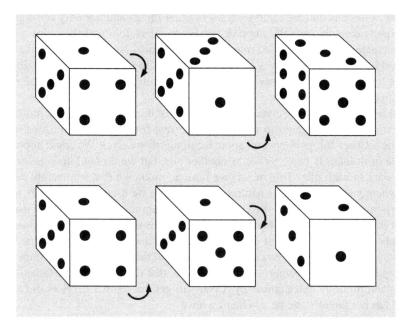

Fig. 1.2 Rotations of a die that show that the commutative law is, in general, not valid for rotations. The first *arrow* at the *upper left* indicates rotation in the vertical plane of the cube face with five dots. The *second arrow* in the *middle* indicates rotation in the horizontal plane (that of the cube face with three dots). The *two arrows below* show first a rotation in a horizontal and second in a vertical plane

small number of rules, or axioms, that are necessary and also sufficient to describe everything that can be done with those numbers, we have completed the process of abstraction. Once one has the axioms, one does not need to think about the real things connected to our symbols. We just deal with the axioms and can derive from them by logic all rules (often called theorems) that are derivable for the number system that follows these axioms. We can then go back and apply these rules, no matter how complicated, to real things, and the results must be correct as long as the axioms hold and are consistent with these real things. Can the axioms be inconsistent with real things? Of course they can be. Consider, for example, the law of commutation. Clearly this law is valid for the addition of oranges. But now, try to apply the law to the process of driving to the house of your friend by taking several turns: it is certainly in general not the same if you turn right at the first intersection and then turn left at the next one, or if you first turn left and then turn right. Other examples are easy to find. Take a die and put it on the desk in front of you. Now turn the die first one turn toward yourself and then one turn toward the right. If you first turn the die toward the right and then toward yourself, you will have a different outcome as shown in Fig. 1.2.

Here we see the wonderful interconnection of mathematics and nature. Starting from counting oranges we have created numbers and then generalized these numbers by rules or axioms that are valid for a lot of other things and not only for oranges. The axioms actually give the numbers an independent life, and they also can be used to create more complicated rules and mathematical laws by logical deduction. It should be remembered, however, that not all things in nature will be following the axioms for numbers. As we saw in Fig. 1.2, the rotations of a die do not obey the commutative law.

To finish this section, we consider applications of the axioms to certain problems. The first example is related to the fact that in real-life situations we often know relations between things but not all about the things themselves. We could know that one pile of oranges is twice as big as another pile, but we do not know how many oranges are in each pile. This means we have an unknown that we usually denote by x, where x stands now for a natural number just as we used the letters a, b, and c above for numbers. Then, we often have some clues about x. These clues are usually written in form of equations, because the clue tells us that two given combinations of numbers are equal. We will deal with equations and important problems later. Here we just solve a simple puzzle by using some of the axioms of real numbers.

The puzzle is the following: A friend tells you that if you add to her father's age 5 years and multiply that number by 3 you will get the result 120. How old is her father? Let her father's age be x. Then we have

$$3 \cdot (x + 5) = 120.$$

Now we apply the distributive law to obtain

$$3 \cdot x + 3 \cdot 5 = 120.$$

At this point we use two rules for equations that we will discuss later in more detail: one is permitted to deduct from both sides of any equation the same number and still obtains a valid equation. This is very easy to understand: the equation is like a scale, and one can say that one has equal "weight" on each side. If one deducts equal amounts on each side then we still have equal weight on each side and we still have a valid equation or a balanced scale. Therefore, if we deduct 15 from both sides above, we get

$$3 \cdot x = 105,$$

and now we can easily see that we must have

$$x = 35,$$

because $3 \cdot 35 = 105$. Thus we have solved the puzzle and the father is 35 years old.

Historically speaking, we have covered here thousands of years of mathematical development, starting from the process of counting and understanding natural numbers to finally arrive at the axioms that permit us to solve the equations. Operating with known numbers is called arithmetic (a Greek word). Working out the values of unknown numbers from clues is called algebra (an Arabic word). We have used both arithmetic and algebra to solve a simple puzzle, and we will see that we can solve much more complicated puzzles that give us solutions to problems of engineering and technology, as well as puzzles that give us solutions for questions related to the number system itself. Not all such questions are or can be answered, but the ones that could be resolved in science, technology, and engineering have enriched our lives and will continue to do so.

We finish the section with some remarks on the system of natural numbers itself. There are basic problems related to natural numbers that can be derived from their original use of counting apples or oranges. Less basic problems use more advanced concepts, for example, the concept of factorization. We know that some numbers can be factorized into smaller numbers, i.e., written as a product of smaller numbers. For example, we have $15 = 3 \cdot 5$. But then there are other numbers (such as 13 and 5) that cannot be written as a product of smaller numbers. These numbers are called prime numbers. You see that for the definition of prime numbers we have already used two concepts: that of smaller numbers and that of factorization. Because of the use of these additional concepts in the definition of prime numbers, it is naturally more difficult to derive truthful statements about prime numbers as compared to just amounts of apples and oranges. To illustrate this fact, we add two more complicated concepts related to prime numbers. First we define neighboring pairs of prime numbers or "twin primes" as pairs that are next to each other exactly as 3 and 5 are; that is they differ from another prime number by 2. Other such pairs would be 5, 7; 11, 13; and so on. Then we ask the question: is there an infinite number of such twin primes? By this question, we have introduced another advanced concept, that of "infinity" with which we will deal often later. For the time being, infinite or infinity just refers to numbers bigger than any given natural number, no matter how big. Thus we have asked a very simple question that everyone can understand. However, the mathematicians still have not resolved this question. We do not know whether or not there exists an infinite number of twin primes. And there are, of course, more advanced concepts and more difficult questions, and the mathematics of the theory of natural numbers is still an active area of research. We refer the reader to technical books and to the Internet on this topic.

We turn now to extensions of the natural number system. A rather simple extension represents the system of integers. If we deduct eight oranges from five oranges, we obtain by the method shown in Eq. (1.1) minus three oranges. Now, the expression "minus oranges" does not necessarily make any sense. However, we can see the "minus" as something that we have borrowed, that we owe to someone. Of course, at the end, we can work with the abstraction of negative numbers without recourse to things like oranges. We just need to give meaning to the addition, subtraction, and multiplication of negative numbers which can be done in the following way. We let

$$a + (-b) = a - b \tag{1.6}$$

and

$$a - (-b) = a + b, \tag{1.7}$$

as well as

$$a \cdot (-b) = -a \cdot b \tag{1.8}$$

Fig. 1.3 A square partitioned
into four parts; one-fourth of
it *shaded*

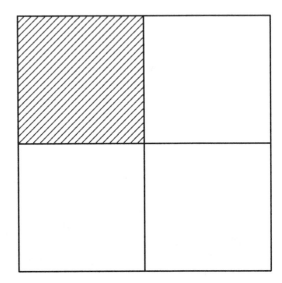

and

$$(-a) \cdot (-b) = +a \cdot b. \tag{1.9}$$

Here we have tacitly understood that a symbol in parentheses represents just one
integer number, i.e., a natural number but now with positive or negative sign.
The student readers are encouraged to make themselves familiar with the above
equations and to "explain" them by logic and language. For example, we know
from language that two negatives give a positive. It will become clear then that
integers, i.e., the numbers $0, \pm 1, \pm 2 \ldots$, form a significant conceptual extension of
the natural numbers, and we understand therefore that historically integers appeared
long after natural numbers were well established.

1.1.2   Rational Numbers

Another significant extension of the natural number system arises from the partitioning of objects into smaller entities. Take a square as shown in Fig. 1.3 and partition it
into precisely four equal parts. The shaded region represents one-fourth of the area
of the whole square. Correspondingly a rational number is a number which can be
expressed by the ratio, also called the quotient or fraction of two numbers. For the
case of our square, the shaded area represents $\frac{1}{4}$, in words one-fourth, of the whole
area. The remainder of the square, the unshaded area, forms therefore $\frac{3}{4}$ of the whole
area.

   In general, every rational number can be written as

$$\text{rational number} = \frac{a}{b}, \tag{1.10}$$

where a and b are integers and b must not be zero. Here we have a crucially important rule for the 0: it cannot be used in the denominator of a fraction. All other integers can be used and make sense, i.e., there is some interpretation of partitioning possible, however, with a zero that does not work. Division by zero does not lead to any number but only to contradictions.

Any rational number is said to be in simplest form when a and b cannot be divided by any common number except 1. For example, $\frac{1}{4}$ is in simplest form while $\frac{2}{8}$ is not because both 2 and 8 can be divided by 2. From Fig. 1.3 we can also see that the two different fractions $\frac{1}{4}$ and $\frac{2}{8}$ represent the same ratio and therefore the same rational number. By partitioning squares or other entities, we can derive the rules for dealing with rational numbers, i.e., we can derive the arithmetic of rational numbers.

These are the rules that we need to remember. The student is encouraged to derive these rules for simple cases of partitioning geometrical objects.

• Two rational numbers $\frac{a}{b}$ and $\frac{c}{d}$ are equal if and only if

$$a \cdot d = b \cdot c. \tag{1.11}$$

• Two fractions are added as follows:

$$\frac{a}{b} + \frac{c}{d} = \frac{a \cdot d + b \cdot c}{b \cdot d}. \tag{1.12}$$

• The rule for multiplication of fractions is given by

$$\frac{a}{b} \cdot \frac{c}{d} = \frac{a \cdot c}{b \cdot d}. \tag{1.13}$$

Note that the symbol $\cdot$ for multiplication is often dropped if no confusion can arise. If one adds or subtracts numbers, one simply uses the signs $+$ or $-$, respectively. If one multiplies numbers, one just uses often no multiplication sign. Then the previous equation reads

$$\frac{a}{b} \frac{c}{d} = \frac{ac}{bd}. \tag{1.14}$$

A lot of exercising is necessary in order to become proficient with these rules. In particular, the rules can be applied repeatedly. For example, a and b can be fractions themselves. This book is not for providing you with exercises for such proficiency. However, it cannot be stressed enough that handling fractions correctly and with ease is the precondition for any capabilities in algebra. I would expect any mathematically talented student to handle fractions and fractions of fractions in their sleep. The teacher should be alerted by the above and following discussions that many conceptual steps are involved in the process of mastering the arithmetic of fractions. Many centuries had to pass for the experts to handle fractions efficiently and to understand rational numbers. Naturally it takes patience for and persistence of the students to get familiar with rational numbers.

Rational numbers cannot only be represented by fractions. Another very important representation of rational numbers is the decimal representation. We have already used the decimal system for integer numbers and seen that each higher digit corresponds to a multiplication by 10. For example we have

$$423 = 3 + 2 \cdot 10 + 4 \cdot 100, \tag{1.15}$$

where we have indicated multiplication by a dot $\cdot$ that is raised to the middle of the line. The dot at the bottom of the line is used for the sentence period and also for the so-called "decimal point." We define a rational number with a decimal point similarly to the number of Eq. (1.15), but now we also include multiplications by fractions $\frac{1}{10}$, $\frac{1}{100}$, etc. for the digits after the decimal point. For example:

$$423.67 = 7 \cdot \frac{1}{100} + 6 \cdot \frac{1}{10} + 3 + 2 \cdot 10 + 4 \cdot 100. \tag{1.16}$$

It can be shown, and this fact is discussed in detail on the Internet, that any fraction of two integers can be written as a number with a decimal point as shown in Eq. (1.16). This number is usually obtained by what is taught in elementary school as "long division" of the two integers of the fraction. The result of "long division" of two integers is either a number with a finite amount of digits behind the decimal point or a finite number followed by periodic sequences of digits as, for example, 3.72123123123 . . . where we have the digits 123 repeated over and over forever. Students and teachers are encouraged to search the Internet for essays on this topic to see how complex knowledge of all facts of digital representation of fractions really is.

We proceed with further generalizations of the system of natural numbers and ask ourselves the question whether there are other numbers that cannot be written as fractions. Naturally we can see from the above that a decimal number that does neither have a finite number of digits nor a periodically repeated sequence cannot correspond to a fraction, but what can it then correspond to?

1.1.3   Irrational Numbers, Imaginary Numbers

The term irrational number has, nowadays, not the meaning of being "crazy" or not logical. However, when it first was found that numbers exist that cannot be formulated as the ratio of integers, this fact certainly must have sounded crazy. The history of irrational numbers can be traced back to India 2,700 years ago, to the Greek culture, to the middle ages, and even to more modern mathematics. All kinds of logical proofs have been presented for the existence of irrational numbers, and we give here one that is basic to the mathematical thinking: the proof by contradiction.

In this type of proof, one assumes a certain statement to be true; in the present case one assumes that a certain number is rational, and then one shows that this leads to a logical contradiction. Hence, the assumed statement must be false, the number must be irrational. Naturally such proof involves a number that is not known and, therefore, because we are dealing with the unknown, this involves algebra. Let us think about the number a that multiplied by itself results in 2:

$$a \cdot a := a^2 = 2. \tag{1.17}$$

Here we have used the symbol := which means we deal with a definition. We just define or write the product $a \cdot a$ by the symbol a^2 (spoken as a-squared or a to the power of 2 or a to the second). We assume now that a is rational and therefore can be written as the fraction of two integers m and n, i.e., $a = \frac{m}{n}$. Let us further assume that this fraction is in the simplest form. This latter assumption can always be achieved, because if m and n could be divided by a certain common integer number other than one, we just divide by that number and arrive at the simplest form. Thus we have from Eq. (1.17)

$$\frac{m}{n} \cdot \frac{m}{n} = 2. \tag{1.18}$$

From Eq. (1.13) and the definition of the square it follows that

$$\frac{m^2}{n^2} = 2 \text{ and thus } m^2 = 2n^2. \tag{1.19}$$

From this equation we can conclude that m^2 is an even number, a number that can be divided by 2. But then m itself must also be an even number because if it would be odd, its square would also be odd (the attentive student may wish to check this out). Thus we can write $m = 2p$ and consequently $m^2 = 4p^2$. Then Eq. (1.18) reads

$$4p^2 = 2n^2 \text{ and hence } 2p^2 = n^2, \tag{1.20}$$

which shows us that also n^2 is even, and therefore n is even. But this means now that both m and n can be divided by 2, and here we have a contradiction because we assumed that $\frac{m}{n}$ is in simplest form. In other words there is no simplest form of fraction that equals a and therefore there is no fraction of any integer numbers that equals a. Consequently a is irrational. a is usually defined as the square root of 2 written as

$$a := \sqrt{2}, \tag{1.21}$$

and we know now that $\sqrt{2}$ is an irrational number.

The method of extending the natural number system is thus quite clear. We just use symbols like a and b for more general numbers, and we apply the rules that we know from using the natural numbers, such as the distributive law or the commutative law, and we perform some algebraic manipulation like squaring and assigning a known number to the square. Subsequently we check whether a and b can still be integers or fractions or whether we deal now with some other type of numbers. This process can go extremely far and lead us to numbers that are quite different from the ones we are used to from counting. For the case above, we were looking at a number a with the property that $a^2 = 2$.

Now let us consider a number i with the property

$$i^2 = -1. \tag{1.22}$$

Obviously it is impossible to find such a number among the integers. If an integer is positive its square is also positive, and if an integer is negative then according to Eq. (1.9) its square is again positive. The same is true for fractions and rational numbers, and it is even true for irrationals. Thus we need to admit a new type of number. This number i was even for its inventors so strange that they called it an "imaginary number" and defined it by

$$i =: \sqrt{-1} \text{ and } i^2 = -1. \tag{1.23}$$

The system of imaginary numbers is very useful and behaves in all respects like numbers should behave. Naturally it is much more elaborate than the numbers of counting, and one can use it to describe natural phenomena and relations that are much more complicated than just counting oranges. We will briefly return to this advanced number systems in connection with quantum mechanics. To really become an expert though, the reader has to consult the special literature. For now, it is important to remember that we have various number systems, and these systems can be linked to different natural phenomena and objects such as whole squares, fractions of squares as well as more complicated entities.

The question of how many numbers of a certain type exist has been the subject of much research, and the answer to this question shows us how different irrationals and imaginary numbers really are when compared to the natural and rational numbers. We cannot discuss this area of mathematics in detail, but we give a brief description of what it is about.

1.1.4 How Many Numbers: Various Kinds of Infinity

As mentioned above, one can play games, sometimes already with kindergartners, about how big a number can be. Give me any number and I know a bigger one. So the child says: One trillion and one gives a bigger number: one trillion plus one. Thus the natural numbers cannot, in their extension, be described by anything finite or any given number. One says there exists an infinity of natural numbers. But what does infinite really mean? If I have something infinite, like the natural numbers, can I have something bigger? The answer is that, in fact, one can. The natural numbers are representing a certain measure for infinity, and one calls this "countable infinity," because natural numbers cover all counting. So if we have any collection or set of things and we can number them with natural numbers, then this

Fig. 1.4 A "gated" electronic element with an electrical current "off" and "on" corresponding to the bits "0" and "1" respectively

set is said to be at most "countable infinite." This gives us some sense of infinity. There are very surprising facts connected to countable infinity. For example, the rational numbers are countable infinite. One certainly would not think that one can count all fractions just by using the natural numbers, because fractions are made out of two integers: one in the numerator and one in the denominator. However, there is a way to show that all fractions can be counted, and the interested reader is encouraged to go to the Internet and search for "countability of rational numbers." There you will find proofs from high school to university level that the rationals can be counted. It is also proven, however, in a less elementary way, that the irrational numbers are not countable and that there are so many more of them around that it is impossible to find a one to one correspondence of the irrational numbers to the natural or rational numbers. Thus the irrational numbers are of a much higher kind of infinity. They are simply in a different category, or as one also says of higher "cardinality". The research in this area is connected to the theory of sets (collections of arbitrary things mathematically abstracted) and was performed by many famous modern mathematicians including Georg Cantor and Henri Lebesgue who proved many beautiful results related to the measure of infinity of number systems. We can just scratch the surface here and turn to a different theme, one that is even more modern and brings us to research of the presence.

1.1.5 Computers and Numbers

The personal computer has invaded modern life in all of its aspects. Although the computer chip handles nowadays our photos, trade on the Internet, taxes, cars, dishwashers, and almost everything we can think of, its basic capability is to deal with numbers. We will hear more about computers in the next section and later chapters. Here we discuss just the basic relation of computers and numbers. How are numbers represented by computers, and why are they so represented?

The basic element that makes a computer tick is an electrical switch that can be turned on and off electrically by a "gate." The principle of such a switching device is shown in Fig. 1.4. The switching from conducting (an electric current I flows) to not conducting, can be done extremely fast. Such a gated element of the computer (e.g., a transistor, as described in Sect. 3.2) represents two pieces or "bits" of information: one corresponding to the on position of the switch (that we identify with the digit 1) and one corresponding to the off position (that we identify with the digit 0). The expression bit is a contraction of the two-word binary digit. As just explained, a single bit can stand for a one or a zero that, in turn, correspond to the "on" or "off" position of a transistor that conducts or does not conduct an electric current, respectively. Thus we can express, with a suitable electronic device, two digits or the natural numbers zero (0) and one (1).

How can we express all numbers that we need for calculations if we have only two basic numbers available? We can do this, similar to what we did before in the decimal number system, by using powers of two instead of powers of ten. We call this new system of numbers the binary number system. For example, we have in analogy to Eq. (1.15) for the binary number 110:

$$110 = 0 + 1 \cdot 2 + 1 \cdot 4, \tag{1.24}$$

which corresponds in the decimal system to the number 6. We can express in this way, of course, any natural number as we have done in the decimal system. We can also deal with rational numbers as we have in Eq. (1.16):

$$110.01 = 1 \cdot \frac{1}{4} + 0 \cdot \frac{1}{2} + 0 + 1 \cdot 2 + 1 \cdot 4. \tag{1.25}$$

The decimal point, which now is called the binary point, can be shifted easily. We know that in the decimal system $21.03 = 2.103 \cdot 10$. Similarly we have in the binary system $11.01 = 1.101 \cdot 2$. Thus we can shift the decimal point or binary point by multiplying by 10 or 2, respectively. We can shift the decimal point to the left twice and then multiply by 100 to arrive at the same number. 100 is, of course, 10^2. Inversely we can shift the decimal point to the right by one digit and divide by 10 to get the same number and we can do so in the binary system dividing by 2 instead of 10. We can formulate this rule in a very general way and create arbitrary number systems using the number b instead of 2 or 10 and defining the powers of b as usual: $b \cdot b = b^2, b \cdot b \cdot b = b^3 \ldots$ and $\frac{1}{b} = b^{-1}, \frac{1}{b} \cdot \frac{1}{b} = b^{-2} \ldots$. Then Eq. (1.25) becomes

$$110.01 = 1 \cdot \frac{1}{b^2} + 0 \cdot \frac{1}{b} + 0 + 1 \cdot b + 1 \cdot b^2. \tag{1.26}$$

In this way we arrive at the generalization of the decimal point or binary point to arrive at the concept of a "floating point number."

To represent arbitrary, very large or very small, rational numbers by binary numbers, it is convenient to express the rational number as a product of a pre-factor and an exponent. The pre-factor may have several digits and a binary point and is called the "significand" (sometimes also called mantissa) denoted by s. The significand can also have both a positive or a negative sign. In addition we have a second factor which is a power of 2 for the binary system (10 in the decimal system or b in a general system). We denote this power by $b^{\pm ex}$, and we thus admit also

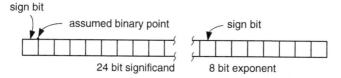

sign bit

assumed binary point   sign bit

24 bit significand   8 bit exponent

Fig. 1.5 A 32 bit floating point number. Each box (there are 32 but not all are shown) corresponds to a "gated" electronic element as shown in Fig. 1.4

negative powers. b is, of course, 2 for the binary system, and e is called the exponent. Thus we have the following general form of a floating point number F:

$$F = \pm s \cdot b^{\pm ex}. \tag{1.27}$$

In the example of Eq. (1.25) we can express the binary number 110.01 by using the binary significand $s = 1.1001$, the base $b = 2$, and the exponent $ex = 2$.

The binary floating point number representation is advantageous for computers because of the following facts. Remember that computers use switches (transistors) with two positions: on and off, as shown in Fig. 1.4. Each digit of a number is represented by a different switch (transistor). Thus a digital computer must use many switches to represent a rational number. Figure 1.5 shows the 32 bit floating point representation of a rational number. The number occupies 32 bits (32 gated electronic elements) that represent a 0 or a 1. We can use, as is typically done for computers, 24 bits for the significand and the remaining 8 bits to store the exponent. There are two bits reserved for the sign, one for the sign of the significand and the other for the sign of the exponent, respectively. The base is automatically 2. It is also automatically assumed that the binary point of the significand is after the first bit. These facts are summarized and illustrated in Fig. 1.5.

Because of the use of the exponent, we can make these (so represented) numbers very large or very small if the exponent is negative. This is why we have to reserve one digit for the sign of the exponent. Without the exponent, the largest number that we could represent by the 32 electronic elements would be a binary number with 31 ones, i.e., a decimal number close to 2 billion. Try to get the exact number as an exercise! If we have an exponent of 7 digits (8 digits with one reserved for the sign) then the largest number that we can represent is obtained by using $ex = 1111111$ in binary representation which is 127 in decimals. This gives $2^{127} = 1.7 \cdot 10^{38}$, a number much larger than 2 billion. The way one arrives at the exact highest number that a computer can deal with is a little more complicated and depends on the precise representation of the floating point number in the computer that is often not exactly as shown in Eq. (1.27) and may contain clever engineering tricks. For example, since every binary number starts with 1, one can take this as granted and save one bit. The interested student may look this up on the Internet where precise examples are given. The point to remember is that digital computers, no matter how sophisticated and powerful, have a highest and smallest number that they can deal with.

Why do we need to know about all of this? The computer knows it and does everything by itself! Yes, this is true, but if we do not know the computers limitations, then we may run into problems. Take the following example. Assume that a computer cannot store a number that is greater than 10^{39}. Now, let us try to perform some simple mathematical operations with the computer. We like to compute the value of the following arithmetic expression:

$$3 \cdot (4 \cdot 10^{38} - 2 \cdot 10^{38}) = ? \tag{1.28}$$

Then, if we let the computer first perform the difference in parenthesis and afterwards the multiplication, the computer will arrive at the exact result of $6 \cdot 10^{38}$. However, if we use the distributive law of Eq. (1.5) and multiply the first number in parenthesis by 3, this number equals $1.2 \cdot 10^{39}$ which is higher than the highest number the computer can do, and we will then get no result or worse, the computer will show the highest number 10^{39}. Then we multiply the second number to obtain $6 \cdot 10^{38}$ and deduct it from 10^{39} to obtain $4 \cdot 10^{38}$ which is false. You can try similar examples with your pocket calculator as soon as you know the highest number that the pocket calculator can do (often 10^{99}). Therefore, you need to know some things about your calculator or computer if you do not wish to be fooled! As soon as you know the number of bits (e.g., 32 or 64) that are reserved for a floating point number, then you can estimate, as done above, the highest number that the computer handles or look it up in the computers manual.

You can see also that it is possible that the distributive law is not valid for computer numbers, just as demonstrated in the above example. It is also important to remember that a computer can only deal with numbers that have a finite number of digits, which means with rational numbers. Of course we can find some way of letting the computer calculate $\sqrt{2}$ to as many digits as we can imagine. Fast computers, and our new computers are very fast, can calculate thousands of digits in seconds or less and you can print them out. Nevertheless, all computer numbers have a finite number of digits, and a computer can never represent an irrational number completely. Fortunately, for most of our needs in science and engineering, this is also not necessary. However, you can see why questions of mathematics cannot always be answered by a computer.

What I wanted to stress in this chapter is that numbers are of basic importance. However, they are not as simple to understand as often believed. They involve numerous conceptual hurdles that have only been mastered over thousands of years. Lots of exercising is needed to become proficient with numbers, and teachers need to repeat both concepts and exercises over the course of the years. It simply is not acceptable that students know how to automatically solve equations but has forgotten the distributive law and stumbles when they need to apply it. It is also not acceptable when a STEM student has absolutely no idea what the binary number system is and how it relates to the decimal system. Here we deal with the foundations of mathematical understanding, and these need be taught and exercised with great care. Practice your numbers!

1.1.6   Computers Crunching Numbers, Large Sums

We have learned in the previous section that we can represent numbers by the transistors or switches of a computer. Modern transistors or electronic switches can switch extremely fast. A single transistor can switch from on to off in a trillionths of a second. When many transistors are involved, for example, millions of transistors on the processing chip of a personal computer (PC), then switching operations take a little longer, more like a billionths instead of a trillionths of a second. From this we can conjecture that a computer may be able to digest numbers extremely quickly, and it is probably true to say that computers can deal with most problems related to rational numbers faster than any human can. How does a computer do it? Here are some of the major steps that may be taken in the course of a computation on a personal computer. The main inventors of these steps include Charles Babbage, Alan Turing , John von Neumann, and others that you can explore on the Internet. More than 150 years ago, Babbage constructed mechanical machines that could add and multiply with high precision. In a modern computer, such arithmetic and other tasks are performed by the so-called processor (a microchip). Turing and von Neumann developed the abilities of computational machines by considering sequences of commands to perform arithmetic and other tasks. These sequences are stored in some form of memory of the computer (e.g., on a memory microchip). The actual way of storage and processing is explained in Chap. 3. For now we need only remember that a sequence of commands to the computer can be stored and the commands can be executed by a processor.

When supplied with two binary floating point numbers the processor can add, subtract, or multiply these numbers and then return the results and store them again as binary numbers. For example, we may have a 32 bit floating point number that is stored at a certain location of the memory chip denoted by the symbol S1, as well as another floating point number stored at another memory location denoted by the symbol S2. These two numbers are sent to the processor chip by a command that chip can execute. For the command written as $S3 = S1 + S2$, the processor adds S1 and S2 and stores the result in memory location S3. We can then view or print the result by "looking" at S3 or letting the computer display or print S3. Note that what just looks like an equation $S3 = S1 + S2$ is now a command that makes the computer do certain tasks.

A processor can also perform other tasks, not just arithmetic ones. An important task of this kind is the following. One can label certain places in a sequence of commands. The labels or markers can be denoted, for example, by $Label[1], Label[2], Label[3] \ldots$. You can then write the command "go to Label[1]," and the processor of the computer will perform this command and actually go to Label[1] of the memory. Thus, we can write, for example, the following sequence of commands that the computer will execute:

begin
$S1 = 15.6$
$S2 = 0.0$
Label[1]
$S2 = S1 + S2$
go to Label[1]
end

Here the words "begin" and "end" just signal the beginning and end of our sequence of commands (also called a computer program). If the processing chip of the computer is activated to go through this sequence, starting with begin, what does it do? It first stores the decimal number 15.6 in binary form at the memory location denoted by $S1$; then it stores the number 0.0 at $S2$ and then just recognizes that there is the label *Label*[1]. The next step is typical for a computer, and you should consider it very carefully. As mentioned, this step does not represent a mathematical equation as we have described it previously. $S2 = S1 + S2$ as a mathematical equation would simply mean that $S1 = 0$ because $S2$ can be canceled out by subtraction from both the left side and the right side of the equation. It is of utmost importance to remember that this expression written for a computer represents a command: send $S1 = 15.6$ and $S2 = 0$ to the addition "machine" and return the result and store it in $S2$. So what this means is that $S2$ changes from 0 to 15.6. The next command sends the processing chip back to *Label*[1] from where it proceeds and stores next 31.2 in $S2$; then the processor goes to *Label*[1] and puts $S2 = 46.8$. In this way, the computer continues to add 15.6 to $S2$ for ever and ever. Obviously, that way, the computer processor would just be busy with adding and never stop and do anything else for us. Therefore we need to have some way to stop the process and print out or display the result. You still might think that even then the computer does not do much for us. But wait, we will show you that it can accomplish quite a bit. We just need to describe first how the process is stopped. Before we do that, we add here a remark on the way we talk to the computer.

The sequence of commands that we have described above represent our way to talk to the computer. This way to talk is therefore called a computer language. Many computer languages have been developed over the course of time. Frequently used languages are Fortran, C, and most recently Java. These languages have great similarities. All of them permit you to send the electronic machinery to certain places or labels in an electronic memory, as was accomplished above with the command that we have written as *goto*. All languages permit you to perform arithmetic operations. Different words and symbols are used in different computer languages, resulting then in essentially the same procedure that the computer performs. Our way of writing things here should be seen only as an indication for what is going on and not how commands are actually written in any particular computer language. You could say that we speak some "pidgin" computer language.

To stop a process, one needs a command that sends the machine to a different label or marker, as soon as a certain number of iterations through the command sequence has been performed. For example, we can monitor the iterations by increasing in each iteration a number m by one. Then, conditional to m reaching a certain value (like 3), we send the machine to another marker, for example, to $Label[2]$. Such a conditional command can be written as: If $m = 3$ go to $Label[2]$ else go to $Label[1]$. The above string of commands changes then to:

```
begin
S1 = 15.6
S2 = 0.0
m = 1
M = 3
Label[1]
S2 = S1 + S2
m = m + 1
If m = M go to Label[2] Else go to Label[1]
Label[2]
Display or Print S2
end
```

Now the computer will display or print $S2 = 31.2$ after the string of commands is executed.

To see the usefulness of such routines or procedures or, as the computer scientists say, of such an algorithm, just replace the addition by multiplication and the starting value of $S2$ by 1. Then we have

```
begin
S1 = 15.6
S2 = 1.0
m = 1
M = 3
Label[1]
S2 = S1 · S2
m = m + 1
If m = M go to Label[2] Else go to Label[1]
Label[2]
Display or Print S2
end
```

This algorithm multiplies 15.6, or any number that we put in, $M - 1$ times. For $M = 3$ one obtains $S2^2$ which is in words $S2$ to the power of 2. One can obtain any other power by just modifying M. To show how useful this is, we extend the algorithm to give approximate solutions to several important mathematical problems. First, however, we describe the story of these problems.

Mathematicians have puzzled about the value of big, or even infinite, sums of numbers for thousands of years. For example, they wanted to know the value A of the following sum:

$$A = 1 + \frac{1}{16} + \frac{1}{81} + \frac{1}{256} + \dots \qquad (1.29)$$

In words, we add fractions, with the denominator being the fourth power of the numbers $1, 2, 3 \dots$. Mathematicians write such a sum in a general way by using the sum-symbol $\sum$:

$$A = \sum_{k=1}^{\infty} \frac{1}{k^4}. \qquad (1.30)$$

Here ∞ stands for infinity and the sum goes over all k starting at 1 and never ending. One of the greatest mathematicians of all times, Leonhard Euler found the result for A in the eighteenth century. He needed to involve the complicated concept of "differentiation" that we will discuss later; you also can find a description of differentiation and Eulers findings on the Internet. Euler's result was $A = \frac{\pi^4}{90}$. Here the Greek letter π is related to the circumference of a circle. A circle of diameter d has a circumference of $d \cdot \pi$ where $\pi = 3.14 \dots$ is an irrational number. This means that there is no regularity or end of digits after the decimal point of this number. We can easily write down the symbol π to denote this number; however, we can never fully compute this number, because we cannot compute an infinite number of digits. Before the computer was invented, the mathematician Ludolph van Ceulen tried to calculate π to as many digits as he could. It took Ludolph all his life to calculate 35 decimals of π, and he requested that the 35 decimals be engraved on his tombstone.

Using the above computer algorithm we can easily approximate A to many digits, and a good personal computer (PC) can do this just in minutes or even seconds. All we have to do is let the processor go through the following sequence:

```
begin
M = 1000
m = 1
A = 0.0
Label[1]
r = m · m · m · m
s = 1/r
A = A + s
m = m + 1
If m = M go to Label[2] Else go to Label[1]
Label[2]
Display or Print A
end
```

Here we have again written the numbers as decimals that the computer turns (automatically) into binary numbers. We have used the relatively small number of thousand summations, by choosing $M = 1,000$. Any PC can do 1,000 additions in fractions of a millisecond (one thousands of a second). The accuracy of the result A is then only a few decimals. To obtain 35 decimal accuracy, as Ludolph did hundreds of years ago by hand, we would have to choose a much larger M, and we would also have to do some additional work to actually print the 35 decimals. Ordinary computers do not deal with so many decimals. However, computer software packages such as MATHEMATICA can accomplish to print even larger numbers of decimals in fractions of a second.

The term software refers to collections of command sequences such as those listed above, starting with "begin" and finishing with "end." A typical software package may contain from thousands to millions of commands and command sequences that any suitably constructed computer can execute (run). Thus, you can run different software packages (sometimes referred to as applications or Apps) on the same computer or PC. The software is run (executed) by the electrical switches, devices, and "chips" of the computer, the so-called hardware of the computer or PC.

We have mentioned MATHEMATICA and have used, for the above strings of commands, a notation which is similar to that used for running MATHEMATICA. For example, we have capitalized the word "Label" and used the symbol [1] to number the label. In a different computer language, the corresponding words may not be capitalized and different parenthesis, for example, (), may be used. However, the computational steps are necessarily similar to the ones discussed above, because the same mathematical goal needs to be achieved.

The computer does not react, or reacts badly, if one uses just slightly different words or spellings than those prescribed in a particular software package. For example, if one types *label* instead of *Label*, the computer may not understand. Minor and major differences in the computer commands present often an annoying obstacle for mastering available software. However, the differences of software packages and generally of computer languages have sometimes very good reasons. For example, the Java language does not contain any command corresponding to *go to*. It does use commands such as

$$if(m < n)\{takeaction1\}\ else\ \{takeaction2\},$$

that accomplishes similar purposes (the symbol < stands for smaller). The reason behind this is that the *go to* command was frequently used by programmers even if it was not really necessary. For large software packages, this abuse of *go to* commands was leading to mistakes, because the programmers did not remember to which labels the computer had actually been sent. We are only interested in learning the basics and how a computer works, and for that purpose the *go to* statement is useful.

MATHEMATICA has already a built-in routine that performs the whole summation of our above example. If you type Sum[expression,m, 1, M], MATHEMATICA will perform the summation. Here "expression" stands for any mathematical expression that you wish to sum up. The MATHEMATICA box shows how to obtain the sum over $\frac{1}{m \cdot m \cdot m \cdot m} = m^{-4}$.

MATHEMATICA

s = Sum[m^{-4}, {m, 1, 20}]

If you type this on your MATHEMATICA notes-pad and hit the shift- enter keys you obtain the exact result as a fraction of large numbers. MATHEMATICA does more for you than the ordinary computer program would do, which gives you only a certain number of digits. You can get a given number of decimals also from mathematica by writing:

N[s, 10] and hit shift-enter to obtain the result to 10 digits:

1.082284588

MATHEMATICA can even do more than that. By writing:

s = Sum[m^{-4}, {m, 1, Infinity}]

and hitting shift-return you obtain

$\frac{\pi^4}{90}$, which is the exact result that Euler derived by using very complicated mathematical reasoning.

While Euler found the exact value of the above sum in terms of π, he could not find any value for sums with an odd exponent. For example, neither he nor anyone else has found the value B of the sum:

$$B = \sum_{k=1}^{\infty} \frac{1}{k^5}. \tag{1.31}$$

Yet, we can easily find B to 35 digits with a PC and a few seconds of computation. This illustrates the number-crunching power of computers. All computer power is, in the last analysis, based on the computer's speed to go through algorithms (sequences of commands) and to crunch out numbers quickly.

Processor chips that perform all the tasks described above and many more have been created by a number of well-known companies such as INTEL, AMD, and others. The reader is encouraged to search the Internet for more information. We will also return to computers and their capabilities in many other sections of this book.

1.2 Equations and Algebra for Solving Problems

The word algebra is related to the mathematical calculation of quantities that are at first not known. One only knows certain relations between these quantities. Consider as an example of these quantities the age of a husband and a wife and assume that we know that their age is equal. How do we formulate that mathematically? We cannot write any specific numbers such as 23 and 21. for their ages, because we do not know the actual ages yet. Therefore, one uses letters such as $x, y, \ldots$ instead of the numbers with the understanding that the letter x symbolizes the age of the husband (that could be 23 or any other reasonable age) and y stands for the age of the wife (that also could be 23 or any other reasonable age). The dots $\ldots$ mean just "and so on". The fact that husband and wife are, in this example, of equal age is then expressed by what one calls a mathematical equation:

$$x = y.$$

If we now know the age of one of the persons, say $y = 23$, we know immediately the age of the other because then we must have $x = 23$ also. As another example, we may know that the husband is 3 years older than the wife and we do know the age of the wife and wish to know the age of the husband. We then have

$$x = y + 3.$$

If the age of the wife is 23 years, then we have $y = 23$ and $x = 23 + 3$ and therefore $x = 26$. The advantage of the use of such equations is that they work for any number and they represent the mathematical "clue" that lets us calculate what we wish to know. The advantage of such a mathematical formulation becomes very significant if we have a larger number of such clues. Consider, for example, a garden architect who wants to design a water sprinkler system with 50 different outlets that need to supply different quantities of water. The architect deals then with 50 unknowns and needs as many clues to solve his problem. If we deal with so many unknowns we cannot just use letters $x, y, \ldots$, because we might run out of letters, and it is then better to denote the unknowns by $x_1, x_2, x_3 \ldots$. How do we use all of these clues to get an answer for all the unknowns? This is the subject of this section. The mathematicians of the past have found a general way to solve equations with unknowns. This way can be followed by hand or programmed into a computer and helps us solve many important problems in science, engineering, and even in our daily life. Before we discuss the general methods for solving equations, we present two more examples that represented important steps in the history of algebra.

Solving equations for unknown quantities was already known by the mathematicians of ancient cultures. Of particular fame is Diophantus of Alexandria (Egypt), who is one of the fathers of algebra. He solved, for example, equations of the form

$$2x - y = 3, \tag{1.32}$$

and he was mostly interested in solutions involving integers (and not rationals or even irrationals). Equations that are solved considering integers only are named after him "Diophantine equations." Equations are called linear if they contain only terms that include x and y but no products such as xy, or squares x^2, or square roots $\sqrt{x}$. The name linear is related to the fact that such equations have the geometric interpretation of representing a straight line, as we will explain in the section on geometry.

Eq. (1.32) (meaning the equation with the number 1.32) has infinitely many solutions. We can find these infinitely many solutions by rewriting the equation as follows:

$$2x - 3 = y,$$

Fig. 1.6 Diagram of currents
of water in three connected
pipes or electrical currents in
three connected wires

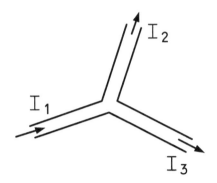

which is obtained by adding $y - 3$ to each side of the equation. Remember, an equation can be compared to a scale. If we add equal weights to each side, the scale remains balanced and the equation remains true. We can also rewrite this equation by exchanging the left with the right and the right with the left side:

$$y = 2x - 3. \tag{1.33}$$

Now you can insert for x any integer $0, \pm1, \pm2, \ldots$ (or, in fact, any number) to obtain the corresponding solutions for y. For example, if $x = 0$, then $y = -3$, and if $x = +3$, then $y = +3$. Diophantus was actually not quite as comfortable with negative numbers as we are nowadays. In a practical application, this equation could give the solution of the following puzzle: A brother is older than his sister. If you multiply her age by two and deduct then 3 years, you obtain his age. He is not yet 7 but above 3, how old is he? Well, you can work with the above equation and you will see that he must be 5 and she is 4. This example may look somewhat artificial. Why on earth would we really need to solve such puzzles. However, in science, technology, and engineering we often know only relations between quantities and not the quantities themselves, and we then need to solve mathematical equations in order to solve the tasks of science and engineering. Solutions of equations in terms of integer numbers have great importance in the mathematical theory of numbers and some importance in science. For engineering, we almost always wish to find solutions in terms of rational numbers, not just integers.

The following example is typical for many equation-related engineering type of problems. Figure 1.6 shows three currents of water flowing along three pipes. We may also think of these currents as electrical currents I_1, I_2, and I_3 in three connected power lines (wires). Let us say we deal with electrical currents and assume that we know that $I_1 = 12$ A. Here A stands for Amperes which is a unit for the electrical current named after Ampere who contributed significantly to the knowledge about electrical currents. Assume further that we know from some facts related to the wire thickness that $I_2 = 2I_3$. For example, the line that carries I_2 could have a lower resistance to the electrical current and therefore carries twice as much current than the line that carries I_3. The question for the engineers is then: how large are the currents I_2 and I_3?

The engineers find the answer, because they know Gustav Kirchhoff's rule that the following equation

$$I_1 = I_2 + I_3$$

is valid. One can understand that rule from the analogy to water pipes. The total flow of water in the first pipe must be the same as the sum of the flow in the two pipes that it supplies with water. We just have to solve the above equation to obtain the currents. Because $I_2 = 2I_3$ we have

$$I_1 = 2I_3 + I_3 = 3I_3$$

and therefore $I_3 = 4\,\text{A}$ and $I_2 = 8\,\text{A}$. This is easy! However, imagine an engineer who deals with the power lines of a whole city that are all interconnected. The wonderful thing is that also these complicated problems can be solved by solving linear equations. Now, however, we have many equations with many unknowns, for example, hundreds or thousands of unknowns. Results for these unknowns can be obtained by solving hundreds or thousands of equations, respectively. This can be accomplished by using computers, because there exists a general method that can be used to solve basically any number of linear equations. This method and the use of computers for solving equations are described next.

1.2.1 Linear Equations

We consider first an example of two linear equations with two unknowns x and y. The two equations are

$$4 = 3y + 2x \tag{1.34}$$

and

$$y = 4x - 1. \tag{1.35}$$

To solve these equations we could try a number of possible ways, and, because in this case the solution is simple, we could even guess it. Obviously the equations stay correct if we insert $x = 0.5$ and $y = 1$ as you can see by inserting these values into the equations. The first equation becomes $4 = 4$ and the second $1 = 1$, which is indeed correct. Had the solutions been 8 digit numbers, the probability that we could have guessed them would have been very small. Thus, what we need in general is a method to find a solution of the equations. As it turns out, and as is often the case, there are many methods that lead to solutions of linear equations, and they are usually all mentioned in textbooks. However, there is one method, found by the famous mathematician Carl Friedrich Gauss, that always works, and, because it always works, we need not consider any of the other methods.

There is a famous joke about mathematicians that relates to their tendency to use methods that always work, even if the methods are complicated. The joke starts with the question: how does a mathematician boil cold water? The answer is: the mathematician takes a pot, fills the pot with cold water, puts it on the oven, turns on the heat, and waits until bubbles appear in the water indicating that it boils. Then there is the next question: how does a mathematician boil very hot water that he has already in a pot? The answer is: he pours out the hot water, fills the pot with cold water, and has thus reduced the problem to the known case of the first question. This may sound ridiculous but is in many cases close to an important truth: why worry about new situations if one can easily transform them to a different situation for which the solution is known already.

Back to our example of linear equations, Gauss first rewrote the equations in a form with all unknowns on the left-hand side and only a number on the right-hand side. Thus Eq. (1.34) becomes:

$$2x + 3y = 4, \tag{1.36}$$

because we can simply exchange left and right side of an equation, and we can reorder the terms with the unknowns because of the commutative law of numbers. To bring Eq. (1.35) into the right form takes a little more massaging: we need to use the fact that we can subtract the same quantity from each side of an equation and therefore subtract $4x$ from both the right- and left-hand side to obtain

$$-4x + y = 4x - 4x - 1,$$

which gives

$$-4x + y = -1. \tag{1.37}$$

Rewriting equations this way takes some practice but then is easily done in seconds. In the following, we always will start from the ordered equations that, for the present example, read

$$2x + 3y = 4 \tag{1.38}$$

$$-4x + y = -1. \tag{1.39}$$

Gauss noted now the following: we can multiply the first equation by a number in such a way that the terms containing the unknown x are the negative equal of each other. In our example we multiply Eq. (1.34) by 2 to obtain

$$4x + 6y = 8. \tag{1.40}$$

Note that if we multiply an equation by a number we need to multiply each term on each side of the equation by the same number in order to still obtain a valid equation. Then we have

$$4x + 6y = 8 \tag{1.41}$$

and

$$-4x + y = -1. \tag{1.42}$$

Now we add the two equations, Eq. (1.41) to Eq. (1.42), by adding left to left side and right to right side and obtain

$$4x - 4x + 6y + y = 8 - 1,$$

which is equivalent to

$$7y = 7.$$

Therefore, by dividing left and right side by 7, we obtain $y = 1$, and inserting this result into the original first equation Eq. (1.36), we have

$$2x + 3 = 4,$$

resulting in

$$2x = 1,$$

after subtraction of 3 from each side. Finally, after dividing the equation by 2, we obtain the result $x = 0.5$.

The method of Gauss to solve linear equations seems to take a lot of doing, but the principles behind it are simple: first rewrite the equations with all the unknowns ordered in the same way on the left-hand side. If more than two unknowns are involved, use the ordering $x_1, x_2, x_3, \ldots$. Next multiply the first equation by such a number that the coefficient in front of x (or x_1) equals the negative of the coefficient of the unknown x (or x_1) in the second equation. Then add the equations to obtain a new equation that does not contain x (or x_1) anymore and solve for y (or $x_2...$). Insert the result for y into the original first equation and solve for x.

If we have the equations already given in an ordered fashion, then we do not even need to go through the labor to write down the unknowns. We know that they are there anyway! We just work with the multiplying coefficients of the unknowns on one side and the numbers on the other side. Thus we can write the two Eqs. (1.38) and (1.39) in shorthand as follows:

$$\begin{pmatrix} 2 & 3 \\ -4 & 1 \end{pmatrix} \begin{pmatrix} 4 \\ -1 \end{pmatrix}. \tag{1.43}$$

In the next step, we multiply by a factor in order to eliminate x. This is done in Eqs. (1.41) and (1.42) to give

$$\begin{pmatrix} 4 & 6 \\ -4 & 1 \end{pmatrix} \begin{pmatrix} 8 \\ -1 \end{pmatrix}. \tag{1.44}$$

Leaving the first line unchanged and adding both lines to obtain a new second line (i.e., eliminating x) we have

$$\begin{pmatrix} 4 & 6 \\ 0 & 1 \end{pmatrix} \begin{pmatrix} 8 \\ 1 \end{pmatrix} \tag{1.45}$$

and know now that $y = 1$. Finally, if we insert this value of y into the first of the equations, we obtain $x = \frac{1}{2}$ and have in our shorthand notation

$$\begin{pmatrix} 1 & 0 \\ 0 & 1 \end{pmatrix} \begin{pmatrix} \frac{1}{2} \\ 1 \end{pmatrix}. \tag{1.46}$$

One can easily perform such a procedure by use of a computer. Computers are prepared to deal with objects such as

$$\begin{pmatrix} 4 & 6 \\ 0 & 1 \end{pmatrix} \tag{1.47}$$

that are called matrices and are composed of rows (horizontal) and columns (vertical) of numbers. You can look for matrices in the manual of MATHEMATCA or any other extensive mathematics software tool, and the manual will tell you how to handle matrices and how to solve linear equations. If you need more explanations, the Internet is full of them. Just search for the term "linear equations, Gaussian elimination."

The final matrix, that contains only two ones in the diagonal,

$$\begin{pmatrix} 1 & 0 \\ 0 & 1 \end{pmatrix} \tag{1.48}$$

is called a diagonal unit matrix. Once one has arrived at this unit matrix, one has obtained the solution of the linear equations. In general, solving linear systems of equations is equivalent to "diagonalizing" matrices, i.e., bringing them into a form where all the numbers are 0 except those of the diagonal that are 1.

The method of Gauss is of great importance, because it can be extended to any number of equations with any number of unknowns. Assume, for example, that we have three equations with three unknowns x_1, x_2, and x_3 and assume that the three equations are described by the following matrix:

$$\begin{pmatrix} -1 & 1 & 1 \\ 1 & 2 & 3 \\ 1 & -2 & -2 \end{pmatrix} \begin{pmatrix} 1 \\ 6 \\ -3 \end{pmatrix}. \tag{1.49}$$

This matrix has already a very convenient form because we can eliminate x_1 by adding the first row to the two next rows to obtain

$$\begin{pmatrix} -1 & 1 & 1 \\ 0 & 3 & 4 \\ 0 & -1 & -1 \end{pmatrix} \begin{pmatrix} 1 \\ 7 \\ -2 \end{pmatrix}. \tag{1.50}$$

Now we have performed the trick that we discussed when we joked about how a mathematician boils water, because we have reduced the problem to a problem that we know how to solve: the last two rows represent two equations with two unknowns that we can solve the following way. We start with

$$\begin{pmatrix} 3 & 4 \\ -1 & -1 \end{pmatrix} \begin{pmatrix} 7 \\ -2 \end{pmatrix}; \tag{1.51}$$

then we transform the matrix by multiplying the second row by three and adding it to the first:

$$\begin{pmatrix} 3 & 4 \\ 0 & 1 \end{pmatrix} \begin{pmatrix} 7 \\ 1 \end{pmatrix}. \tag{1.52}$$

This gives us (by inserting $x_3 = 1$ into the first row)

$$\begin{pmatrix} 1 & 0 \\ 0 & 1 \end{pmatrix} \begin{pmatrix} 1 \\ 1 \end{pmatrix}, \tag{1.53}$$

and if we return to the original three equations with three unknowns, we finally have

$$\begin{pmatrix} 1 & 0 & 0 \\ 0 & 1 & 0 \\ 0 & 0 & 1 \end{pmatrix} \begin{pmatrix} 1 \\ 1 \\ 1 \end{pmatrix}, \tag{1.54}$$

which just means that we have diagonalized the matrix and all three unknowns are equal to 1. Naturally equations do not always reduce to such simple numbers, and we just chose this example because it is easy for practice. If you have already had high school classes that deal with linear equations and if it was easy for you to understand this section, then you have been taught well. If this whole section gives you only grief, try again. If it does not work for you at all, STEM may be not your talent, because most STEM fields deal in some form with linear equations of unknowns. This is the way science and engineering questions can be formulated and solved. In fact one can solve hundreds of equations with hundreds of unknowns fairly easily by using commercially available software such as MATHEMATICA. For the above three equations with three unknowns, the solution can be obtained with the MATHEMATICA by just typing the sequence shown in the box:

MATHEMATICA
m = {{−1,1,1},{1,2,3},{1,-2,-2}}
this represents the matrix because the three rows are listed in the way MATHE-MATICA requires it, and
v = {1, 6, -3}
represents the right hand side of the equation, also as MATHEMATICA requires it. In general, such a list as shown for v is called a vector. MATHEMATICA requires then only the command:
LinearSolve[m, v] and pressing shift-enter gives you the solution:
{1, 1, 1}
in fractions of a second.

We have not mentioned yet that some linear equations do not necessarily have solutions. Furthermore, in order to obtain a single valid solution for each unknown, one needs an equal number of equations and unknowns. The interested readers can find more on the Internet by searching for "solution of linear equations." MATHEMATICA also tells you about the solubility of equations. Before finishing this section, we take a brief look at nonlinear equations.

1.2.2 Nonlinear Equations

Nonlinear equations are usually much more difficult to solve than linear ones, and for a general description of methods we refer the interested reader again to the Internet. Software like MATHEMATICA hands you solutions of nonlinear equations on a silver platter. All you need to do is type a command like *NSolve[equation, x]* to solve any nonlinear equation in *x*. If you have two unknowns *x* and *y* in the equations, then you just have to type *NSolve[{equation1, equation2}, {x, y}]*. Now you need two equations for the two unknowns, and you need to use the parenthesis {..} to indicate a list of two unknowns in the language of MATHEMATICA. There is another little trick one needs to know. To indicate that one solves an equation and does not just replace the right-hand side by the left-hand side, as often done in computer programs and as we have explained it in Sect. 1.1.6, MATHEMATICA uses two equal signs. Thus the equation $x^2 - 5x + 6 = 0$ must be written as $x^2 - 5x + 6 == 0$ to indicate that it is indeed an equation. The solution of this equation is shown in the box.

MATHEMATICA
Solve[$x^2 - 5x + 6 == 0, x$] to obtain after hitting shift-enter
x = 2., x = 3.
as solutions. Note that MATHEMATICA can also deal to some extent with algebraic equations e.g. the equation that you obtain by replacing the 5 by a general number *a* in the above equation.

Because MATHEMATICA works in a very general way, it also gives you solutions in terms of complex numbers, i.e., in terms of $i = \sqrt{-1}$ and not only in terms of real numbers. MATHEMATICA can also solve equations with powers of *x* that are higher than 2. MATHEMATICA also solves equations containing complicated functions such as the functions sin and cos that we will discuss in the section on geometry. We just need to remember that MATHEMATICA capitalizes all the functions. A simple example is given in the box.

MATHEMATICA
Solve[Sin[x]==0.5,x] shift-enter gives
x = 0.523599 with x in radians
Or take the more complicated equation
Solve[Sin[x] + Cos[x]==0.5, x] , shift-enter
gives the two solutions x = −0.424031 and 1.99483

MATHEMATICA can also provide solutions in terms of general numbers like *a* and *b* as, for example, by solving the equation $x^2 + ax + 1 == 0$. There is a lot of useful exercising that a student can do here and learn more about nonlinear equations.

If one knows (for whatever reason) that the solution x of a nonlinear equation is much smaller than 1, then there exists a special method to obtain an explicit approximate solution to the nonlinear equation. This method is based on the fact that, for $x \ll 1$, the higher powers of x (such as $x^2, x^3, \ldots$) are still much smaller and can be neglected compared to the lower powers of x. For example, if $x = 0.1$, then $x^2 = 0.01$ and thus is much smaller than x. In such a case, we can, as a first approximation, neglect x^2, and we are left with a linear equation that we know how to solve. A very important trick to remember in this connection is the following: If the unknown x is very small, then we can use the approximation

$$\frac{1}{1 - x} \approx 1 + x. \tag{1.55}$$

Here we have used the symbol $\approx$ to indicate that this equation is only approximately valid. With this in mind, we can solve the following nonlinear equation:

$$\frac{x}{1 - x} = 0.1 \tag{1.56}$$

by using Eq. (1.55) to obtain

$$x + x^2 \approx 0.1 \tag{1.57}$$

and neglecting x^2 we obtain

$$x \approx 0.1 \tag{1.58}$$

as the approximate result. Had we solved the equation with MATHEMATICA by typing $NSolve[x/(1 - x) == 0.1, x]$ then MATHEMATICA would have returned the exact result $x = 0.0909091$ which is not too different from the approximate solution.

There exists another important trick that always works when small numbers are involved. We denote now the small number by ϵ, because that is often done so in mathematics. When you see an ϵ in a math text, you can usually assume that this is a number that is smaller then all other numbers around. For example, if we have a term $(t + \epsilon)$, then you can usually assume that ϵ is much smaller than t. The trick to remember is now the following:

$$(t + \epsilon)^n \approx t^n + n \cdot t^{(n-1)} \epsilon. \tag{1.59}$$

If you are interested in how this approximation is derived, search the web for "binomial coefficient," take the formulas that are given there for $(t + \epsilon)^n$, and then neglect all the (very small) higher powers of ϵ.

With this we finish the discussions of linear and nonlinear equations. There is much more to learn in this area, and nice projects can be found on the Internet. Look, for example, for the stories around Fermat's theorem.

1.3 Geometry

"There is no royal road to geometry."
–Euclid of Alexandria to Ptolemy I, King of Egypt (306 B.C.)

It was more than two thousand years ago when Euclid of Alexandria (Egypt) formulated the basic rules of geometry. Based on the knowledge of his time, he created the mathematical science that helps us to determine the area of a property, the distance of our journey, the height of a mountain, and much more. Even our electronic GPS, which works with satellites and atomic clocks, is based on many of the principles discovered by Euclid. King Ptolemy was frustrated with the difficulties of the rules of geometry, and he demanded that Euclid show him a royal road, an easy way, to understand geometry. Euclid, the great master, stated then that there was no easy way. Nevertheless, there is a road to understand geometry, Euclid's road. This road is as valid as ever, and it is still difficult to understand. Today, of course, we can ask questions on Internet search engines and will obtain any number of explanations. This may make it a little easier to find an understanding than it was for Ptolemy who, by the way, did become a master of geometry.

Euclid started with a few postulates that are often called self-evident true statements, or axioms, a term that we explained in Sect. 1.1. These postulates were taken out of the experience of daily life and give us a hint of the "toolkit" that Euclid had available: a ruler, a piece of string or a rigid stick (both representing a line segment), and something like a pencil to draw. He also assumed that everyone knows (or has at least some concept of) what a point, a straight line, and a plane are. Nowadays, we take a laser pointer and create a laser point with it. If we let the laser light just touch a surface, then we obtain a straight line, and if we turn the laser back and forth, that line will scan over a plane. Euclid could produce a dot with his pencil, a straight line with his ruler, and he could form a plane by using sand. He also could create a circle of practically any size by using the string. Just anchor one endpoint of the string and bind the pencil to the other end of the string. Then, by leading the pencil around, you can form a circle. So much for Euclid's (and our modern) tools of geometry. Euclid's postulates or axioms are:

1. Exactly one segment of a straight line can be drawn through two points.
2. Any segment of a straight line can be extended indefinitely.
3. A circle can be drawn using a segment of a straight line as its radius and fixing one end of the segment as the center of the circle.
4. All right angles are the same.
5. The sum of all angles of a triangle equals the sum of two right angles.

These axioms are illustrated in Fig. 1.7 which shows, from top to bottom, a line segment drawn through two points, an extended line segment with arrows indicating that the extension can go on forever, a circle drawn by use of a line segment, two right angles (the dot · denotes a right angle) that are equivalent or the same, and a triangle with angles α, β, and γ whose sum equals two right angles.

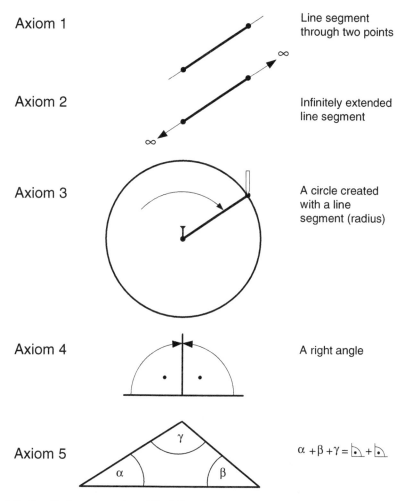

Fig. 1.7 Graphical representation of Euclid's axioms

The simplicity of Euclid's postulates, their direct connection to actual measurements, as well as their power and significance (one can derive a lot of important theorems of geometry from them) is amazing and demonstrates the great mastership of Euclid. It is important for the student to realize that Euclid used only very basic measurement tools, a string, a ruler and some way to measure an angle, and then he formulated his basic propositions with these tools in mind.

Of course, he must have had some deeper insight as the ordinary person would have had. It is still fairly easy to follow his abstractions of infinitely small points and infinitely thin and infinitely long lines. The fifth axiom, however, is a different matter. It took mathematicians thousands of years to figure out the secrets related to the fifth axiom. Nevertheless, Euclid's axioms should not be seen or understood as difficult abstractions that he pulled out of a magic hat. In some formulations these axioms are even difficult to read! For the beginner, it is better to see them as a simple recipe to use a string, a ruler, and some measurement of angles to determine, for example, the area of a property or the height of a tower.

Euclid himself formulated the fifth axiom somewhat differently as stated above but in a mathematically completely equivalent way. This fifth axiom has a long history of work, and for almost two thousand years, it was not clear whether this fifth axiom was really needed or was already contained in the four previous axioms. As it turns out, the fifth axiom is not contained in the previous four and is needed to define all of Euclid's geometry. We will return to this point on some occasions when we talk about geometries that are different from Euclid's, particularly when we talk about Einstein's theory of relativity in Chap. 5.

Below we discuss the geometry of Euclid based on his five postulates starting with simple facts related to area and volume.

1.3.1 Euclid's Axioms and Applications

Euclid's axioms permitted him to derive certain consequences that followed by logical deductions from the axioms. Such true consequences of axioms are often called "theorems." The important point is that these rules or better theorems, together with actual measurements of length and angles, lead then to results that are very much needed in our daily life. Euclid developed a special way of logically proofing things: the proof by contradiction that we have discussed already. We got to know this type of proof when we showed that $\sqrt{2}$ was not a rational number. What we did there was assuming that $\sqrt{2}$ was a rational number, and then we showed that this assumption together with some axioms leads to absurd and false consequences. Therefore $\sqrt{2}$ could not possibly be a rational number. Euclid and many other great mathematicians that followed him used this method in a general way to prove their theorems. We will not provide you with many proofs here. Many famous proofs of geometry can be found on the Internet in a great variety of versions. We just state some of the important results that we use later in the book.

Length, Area, and Volume

We first consider areas of simple geometric figures in a plane (e.g., squares, rectangles, and triangles). Assume that the length of one side of such a geometric figure is a units of length. Euclid proved that then the area A is given by $A = p \cdot a^2$, where p is a given number that is characteristic for the form of the geometric figure.

Let us take first the simple square shown in Fig. 1.8. The area A of this figure is $A = a^2$, and therefore $p = 1$. If we have determined by measurement that $a = 1\,\text{cm}$, then $A = 1\,\text{cm}^2$. This is basically a definition, because expressed in units of centimeters, the unit of area is $\text{cm} \cdot cm = \text{cm}^2$. The same is true for meters,

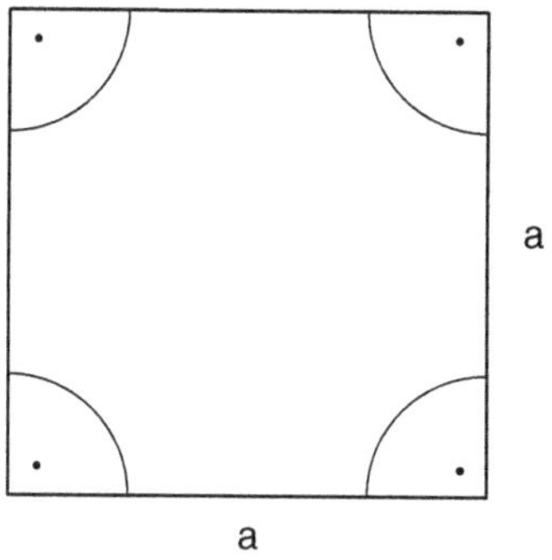

Fig. 1.8 Geometric figure of
a square defined by four equal
sides and four right angles

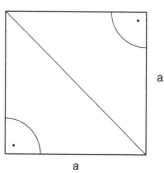

a

a

Fig. 1.9 The square divided
into two triangles, each with
one right angle

a

a

kilometers, and miles. Note that when we talk about length, we do not merely deal
with a number. We must have some measure of the unit of length, and we must
always say in which unit we measure the length. Of course, we could have measured
also in inches or yards.

To have an actual precise length measure available is not a trivial matter, because
the length of real objects often depends on temperature. If we wish to be very
precise, we have to keep the object that defines the length, for example, the object
that represents one "meter," at a constant temperature. This is actually done with the
prototype of one meter (one meter is 100 cm) that is stored in Paris. Whatever unit
of length measure we use, if that length is given, the area of the square figure is by
definition the square of this length, and that is why the figure is called a square.

Next we divide the square into half to obtain two triangles as shown in Fig. 1.9.
Each triangle has now half the area of the square $\frac{a^2}{2}$ which means $p = \frac{1}{2}$. Euclid
found that there is a general principle behind this: The area of a triangle is equal to
the baseline multiplied by its height h and divided by two. In the case of Fig. 1.9,
we have the height $h = a$. In general that is not so. However, the area A_{triangle} of a
triangle with baseline length a and height h above this baseline is always given by

$$A_{\text{triangle}} = \frac{a \cdot h}{2}. \tag{1.60}$$

Fig. 1.10 A *circle* with a *hexagon* inside (inscribed) as well as outside (circumscribed). Archimedes used such constructions to calculate π

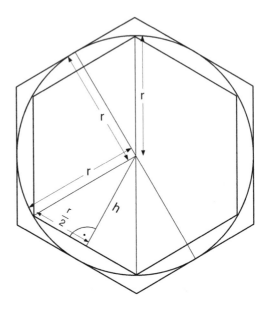

Now we turn to a circle, and we like to know the area of a circle in terms of its radius r which is half of its diameter. We can guess to start with that the area A_{circle} of the circle will be related somehow to r^2 because there needs to be the square of a length in the formula. We could approximately calculate the area of the circle by inscribing many squares in the circle and then counting the number of squares. This was done already in ancient times, and the formula $A_{\text{circle}} \approx 3r^2$ is described in words in the book of Kings of the Bible. The famous Greek mathematician Archimedes provided a different and better approximation. His way of doing it is shown in Fig. 1.10.

Archimedes inscribed a hexagon into the circle and another one outside the circle and then calculated the area of the hexagons by summing the areas of the triangles that form the hexagons. The area of the circle must be larger than that of the inner hexagon and smaller than the area of the outer one. Archimedes also did the calculation with polygons, i.e., geometric figures that have more segments of straight lines than the hexagon. "Poly" is Greek and means many. Archimedes demonstrated his great patience by using a polygon with 96 segments and then found that

$$A_{\text{circle}} = r^2 \pi \tag{1.61}$$

with

$$\frac{223}{71} < \pi < \frac{22}{7}. \tag{1.62}$$

Here the symbol $<$ means that the number left of this symbol is smaller than the number to the right of it. Thus π is larger than $\frac{223}{71}$ and smaller than $\frac{22}{7}$. Using a pocket calculator you can check that this means that π must be close to 3.14, which is very close to the actual value $\pi = 3.1415\ldots$. To really calculate the area of these polygons yourself, however, you would need to know more about triangles, and we will learn this missing information in the remainder of this chapter.

Of course, we do not suggest that you really try to do this calculation with a 96-side polygon. We have already shown in this chapter how to calculate π with the computer and discussed how Ludolph spent all his life to calculate 35 digits by hand. MATHEMATICA gives you any number of digits that you may wish to have in fractions of a second, the triumph of modern technology! The length of the circumference of a circle L_{circle} as related to the radius was also found thousands of years ago, and we quote here just the result and refer the interested reader to the Internet. We have

$$L_{circle} = 2r\pi. \tag{1.63}$$

In words, the length of a circle is twice the radius multiplied by π.

As mentioned, length is measured in meters and area in meters to the square also written as m^2. Of course, you can also measure in inches and square inches if you wish to do so, and the conversion of one into the other is easy to learn although it takes a little practice. Conversions are also easy to find by Internet searches. As you could see from the above discussion, there are all kinds of shapes of objects of which you might like to know the area; you might, for example, like to know the size of your property. There exist even more shapes whose volume you might like to know. A cube with one-meter side length has a volume of one cubic meter or m^3. The volume of a cube is calculated by multiplying the area of one face by the height of the cube. The volume of a cylindric object is calculated in similar fashion. Take the area of the circle at the basis of the cylinder and multiply it by the height of the cylinder. Other shapes are often more difficult to calculate and have been calculated by using the theorems of geometry. The volume of a pyramid is obtained by multiplying the area of the basis by the height and dividing the result by 3. The volume so obtained is still measured in cubic meters. Volumes are important measures for containers. One cubic meter is equivalent to 1,000 l. There are lots of shapes of containers. The containers in the kitchen are smaller and usually measured in liters, at least in many European countries. The British system likes to talk about cups and gallons and so forth. Again, if you have the volume in one type of measure or unit, you can convert it into other units. Think, for example, of a European cooking recipe that measures all volumes in liters. All you have to do is use an Internet search engine and type "liter, cup," and you will find that a liter is a bit more than 4 cups, exactly 4.224 cups. This is easy! However, if you do such things frequently, one cannot avoid asking why we have to use all these different measures. The answer is that the habits developed that way in the past and now we are stuck with them for a while. At some point, the most convenient system, the metric system (working with meters and decimals etc.), will win.

Fig. 1.11 A triangle with one right angle and sides a, b, and c

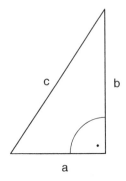

Famous Theorems of Geometry

As mentioned, a theorem states a "truth" that is deducted from axioms. It is important that the axioms are few in number and that they themselves represent some self-evident (obvious) truth. Of course, we do not know about the truth of the axioms until we look at all their possible consequences. Even then we may not be able to show their consistency. The Internet gives you a nice description of the history of Euclid's fifth axiom that is also described in Sect. 5. It took 2,000 years to find out whether or not that fifth axiom was necessary and what it meant for the geometry of our world.

We list three famous theorems of geometry. The proofs can be found on the Internet and are performed there in many different ways. The first is the theorem of Pythagoras. Consider a triangle with one right angle and with sides a, b, and c as shown in Fig. 1.11.

For such a triangle the theorem of Pythagoras states

$$a^2 + b^2 = c^2. \tag{1.64}$$

In words, the sum of the squares of the length of two segments of the triangle equals the square of the length of the segment across the right angle. The theorem of Pythagoras is of great importance because right angles play important roles in daily life and in architectural problems. A tower is built on a plane and forms a right angle with that plane. Any wall of a house forms a right angle with the plane that the house is built upon, and any well-built structure contains usually a lot of corners with right angles.

The second theorem that we list is the theorem of Thales. This theorem is illustrated in Fig. 1.12. The theorem of Thales states: If A, B, and C are the points of a circle and if the segment c connecting the points A and B is equal to the diameter of the circle (two times the radius), then the angle of the triangle opposite to c is a right angle and the theorem of Pythagoras is valid for this triangle.

Using these two theorems, we can solve a large number of problems of geometry. For example, we can calculate the area of the triangles of the hexagons of

Fig. 1.12 Illustration of the theorem of Thales

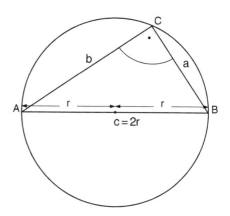

Archimedes shown in Fig. 1.10. The length of the segments that envelope the six inscribed equilateral triangles (only one shown) is just equal to the radius r of the circle which we know (it could be, e.g., one meter or one yard). To calculate the area, however, we need to know the height of these triangles. As is also shown in Fig. 1.10, the segment with the height h cuts the equilateral triangle into two triangles that have one right angle each. From Pythagoras we know that for this triangle, we have

$$\left(\frac{r}{2}\right)^2 + h^2 = r^2, \tag{1.65}$$

and therefore

$$h^2 = r^2 - \frac{r^2}{4} \tag{1.66}$$

giving us h:

$$h^2 = \frac{r\sqrt{3}}{2}, \tag{1.67}$$

and the area of the triangle in question is

$$A_{\text{triangle}} = \frac{r^2\sqrt{3}}{4}. \tag{1.68}$$

The area of the inscribed hexagon is just six times the area of this triangle and equals $\frac{3\sqrt{3}r^2}{2}$. The area of the equilateral triangles that form the circumscribed hexagon is also easy to calculate. We just need to notice that the height of these triangles equals the radius of the circle and again use the theorem of Pythagoras. This results in the area of the outer hexagon to be $\frac{6r^2}{\sqrt{3}}$. The readers are encouraged to do the calculation and to determine from their results the range of numbers that approximate π. The result is $2.6 \leq \pi \leq 3.46$. From this exercise, the reader gets a glimpse of how much more work it must have been for Archimedes to calculate the areas of polygons with 96 sides.

Fig. 1.13 Similar triangles

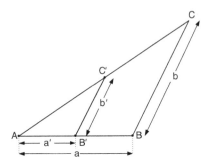

In the following, we also will need a third "truth" or theorem that is related to similar triangles. Again you can find a lot more on the Internet, but all you have to know is the following: Figure 1.13 shows two similar triangles. The triangles defined by the points A, B, C and A, B', C' are called similar if and only if the line segments B', C' and B, C are parallel. Parallel means that if we continue the line segments forever, the two straight lines that we obtain never meet. Then it is a "truth" following from Euclid's axioms that

$$\frac{a'}{a} = \frac{b'}{b}. \tag{1.69}$$

We will use this equation on several occasions.

Angles

We have explored geometrical facts that we can understand by knowing that one angle is a right angle. However, we can measure all angles that we wish with certain instruments that we refer to as "angle meters." What is an angle? Using the precise mathematical definition, we say that an angle is comprised of the union of two rays with a common point as shown in Fig. 1.14.

The value of the angle is denoted by a Greek letter, α in our case. The arrows on the straight line of the figure indicate that rays go on forever. The arrows on the circular lines show in which direction the angle is measured. The counterclockwise direction is defined as our positive direction (i.e., the angles measured in this direction are counted with positive numbers). The dashed clockwise direction (also indicated) is the negative direction; the angles in this direction are counted negative. We also need to make a decision what type of units we use for an angle (as we have used centimeters or inches for a length). This choice can be made as follows. Figure 1.14b shows a standard angle meter that has the shape of one quarter of a circle with a scale on the outer side. This scale may be chosen to be identical to the length of the arc of the circle, starting from 0 and extending to the point in which the second ray of the angle crosses the circle. We know from Eq. (1.63) that the full

Fig. 1.14 (**a**) Illustration of an angle comprised of two rays and (**b**) instrument to measure angles

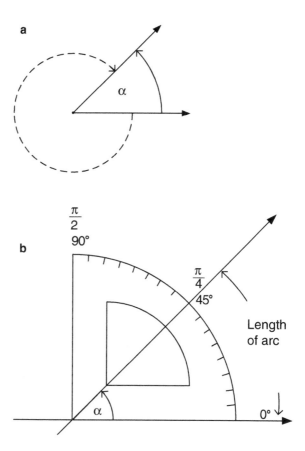

arc length of a circle with radius r is $2r\pi$. If we assume that the circle has a radius of unit length (e.g., 1 cm or 1 in.), then the full length of the arc is 2π (centimeters or inches). The arc length of one quarter of the circle is $\frac{\pi}{2}$ and corresponds to the right angle as indicated in Fig. 1.14. Thus, we can measure the angle α in terms of the arc length, and for the angle α in the figure, we have $\alpha = +\frac{\pi}{4}$, or measured in the negative direction, the angle is $-\frac{7\pi}{4}$. If we measure angles that way, we say that we measure angles in "radians." One can measure angles also in "degrees" by just defining that 2π radians correspond to $360°$. The right angle is then one quarter of $360°$, i.e., it is $90°$.

Angles and the Height of a Tower: The Sine and Cosine

The following is a problem often posed to a surveyor engineer. A surveyor performs measurements that determine the distance and heights of certain objects, such as mountain tops or towers. Assume that we wish to determine the height of a tower and

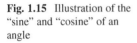

Fig. 1.15 Illustration of the "sine" and "cosine" of an angle

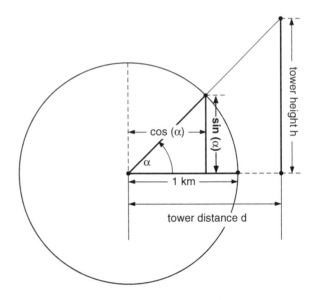

its distance from a certain location. The distance is easy to measure. For example, we might want to know the distance of the tower of Pisa from a certain hotel, because the hotel guests wish to visit the tower and need to know how far away it is. That distance can be measured with a meter (or yard) measure by laying it out repeatedly. For the particular hotel, let us say that the surveyor has measured a distance of 800 m. The hotel guests also wish to know how high the tower of Pisa is. However, just when the surveyor wants to walk up the tower and let a string dangle down to measure its height, he cannot do that because the tower is closed. The tower of Pisa is leaning, and sometimes they close it for restoration and safety reasons. So the surveyor must find some other way to measure the height of the tower, and he can do that by measuring the angle of a straight line from the hotel to the top of the tower. To measure such an angle, the surveyor uses a telescope that contains markings in its field of vision. If you bring these markings to coincide with the top of the tower, then the telescope indicates the angle of the light ray from the position of the surveyor to the top of the tower. One can find photos of such surveyor instruments on the Internet. For the case of the tower of Pisa, the surveyor takes measurements at the hotel and finds $\alpha = 4.0654°$. With a few multiplications and a pocket calculator, the surveyor comes up with the height of the tower $h = 55.86$ m. How did he or she do it?

In Fig. 1.15 we have plotted the angle α of the ray to the tower and a unit circle. The unit can be anything, one meter or one kilometer. The ray intersects the circle at a certain height above the baseline. That height is called the "sine" of the arc, and it is completely determined by the angle α. We denote this height by the mathematical symbol $\sin(\alpha)$. The length of the baseline, from the origin of the circle to the vertical line of the sine, is called the "cosine" of the arc. The cosine is again fully determined by the angle α and, therefore, denoted by $\cos(\alpha)$. From this definition and the application of the theorem of Pythagoras, we obtain the equation

$$\sin^2(\alpha) + \cos^2(\alpha) = 1 \tag{1.70}$$

that is frequently useful when solving problems of geometry.

How can the surveyor find the values of $\sin(\alpha)$ and $\cos(\alpha)$? Well, that is simple enough. The surveyor just draws a unit circle and the angle and then measures the sine and the cosine. Nowadays, of course, we have a better way to do this task. We just ask MATHEMATICA or a pocket calculator, and we obtain the value of the sine and cosine for any angle α. Knowing the sine and cosine, how do we get the height of the tower? Figure 1.15 shows that the triangle to the actual tower is obtained by a continuation of the triangle with the two sides $\sin(\alpha)$ and $\cos(\alpha)$. These triangles are therefore similar, and we know from Eq. (1.69) that for such triangles the ratio of corresponding segments is the same. Therefore we have

$$\frac{\sin(\alpha)}{\cos(\alpha)} = \frac{h}{d} \tag{1.71}$$

and

$$h = d \, \frac{\sin(\alpha)}{\cos(\alpha)}. \tag{1.72}$$

The surveyor has measured the angle α from the hotel to the top of the tower and found $\alpha = 4.0654°$. From the pocket calculator the surveyor knows therefore $\frac{\sin(\alpha)}{\cos(\alpha)} = 0.070955$. Inserting the distance of $d = 800$ m between hotel and tower gives the result of $h = 55.86$ m for the height of the tower of Pisa.

We finish this section with a few remarks on how one can compute values of $\sin(\alpha)$ and $\cos(\alpha)$. Because of the great importance of these "functions" of the angle α, mathematicians have investigated the sine and cosine functions in great detail. They came up with a number of formulae that one can evaluate with a computer. Newton, for example, derived the sine and cosine in terms of infinite sums. If the angle α is given in radians then Newton found that

$$\sin(\alpha) = \alpha - \frac{\alpha^3}{3!} + \frac{\alpha^5}{5!} - \frac{\alpha^7}{7!} + \dots \tag{1.73}$$

and

$$\cos(\alpha) = 1 - \frac{\alpha^2}{2!} + \frac{\alpha^4}{4!} - \frac{\alpha^6}{6!} + \dots . \tag{1.74}$$

We have used here an abbreviation for certain products of numbers, by writing the number with an exclamation mark. For example, 4! is pronounced in words "four factorial" and means $4! = 4 \cdot 3 \cdot 2 = 24$, or 7! (seven factorial) means $7! = 7 \cdot 6 \cdot 5 \cdot 4 \cdot 3 \cdot 2 = 5,040$. The factorial numbers in the denominators of the sum become rapidly large, and the terms of higher power (e.g., α^{20}) are

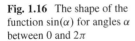 **Fig. 1.16** The shape of the function sin(α) for angles α between 0 and 2π

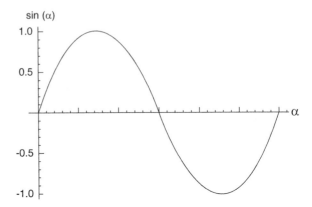

therefore small. One can plot the sine or cosine functions for a range of α by using MATHEMATICA. All you have to do is write the commands shown in the box.

MATHEMATICA
Plot[*Sin*[*x*], {*x*, 0, 2*Pi*}] shift-enter
and you will obtain the graphics shown in Fig. 1.16.

Note that MATHEMATICA capitalizes the Sin[*x*] function and uses rectangular brackets. Instead of any fixed angle α, we have used the variable *x* in the plotting command of MATHEMATICA and specified that *x* varies from 0 to 2π. Then Mathematica plots on a vertical scale (indicated by the vertical line with numbers) the values of sin(*x*) for any value of *x* that appears on the horizontal scale (indicated by the horizontal line with numbers). These vertical and horizontal lines bring us to the concept of coordinate systems in geometry that we describe next.

The procedure to determine the height of a tower that we have discussed here is a special case of the general method of "triangulation." A triangle is fully determined and specified if we know one side and two angles. In our case we measured the distance of the tower and the angle α from the hotel and assumed a right angle as the second angle of our triangle in question. With these three data points we could calculate the height of the tower and any other length or angle of the triangle shown in Fig. 1.15. Naturally one can have variations of this method of triangulation. For example, we might have measured three sides and just wish to have the angles. Triangulation is very important for surveyors and in general was the method to explore the geometry of the world and all the information that you can find on a map, as for example the height of mountains. In order to record all this information, one needs a system of recording, and any such system is closely connected to the concept of "coordinate system." A special case of such coordinate systems will be discussed next.

Functions

Before discussing coordinates, however, we need to add a few remarks about "functions." We have called the sine and the cosine a function of the angle α. The hallmark of the sine and cosine functions is that for any angle α we have exactly one value for $\sin(\alpha)$ and exactly one value for $\cos(\alpha)$. The set of all angles α that we consider is called the domain of the functions, and the set of all values that the functions can assume is called the codomain or also range of the function. This definition applies not only to the sine and cosine functions. Any mapping that associates with any element of a domain exactly one element of a codomain is called a function. For example, if a farmer has 20 chickens and associates with each chicken of the whole "domain" of chicken a number, then the farmer has defined a function. The codomain is simply the set of numbers. These need not necessarily be different. For example, the farmer could find that three of the chickens are so similar that they deserve the same number. The main point of the function is only that we associate with each chicken exactly one number. It turns out that the concept of functions is extremely useful in mathematics and science, and we just got to know this concept here with the geometrical discussion of the sine and cosine functions. In fact, functions are of great importance in questions of everyday life, and we discuss, therefore, the following example of the role of functions for the best choice in presidential elections.

Consider a presidential election and reinterpret Fig. 1.16 in the following way. Instead of just the function $\sin(\alpha)$, the figure represents now the number of jobs created or any indicator that corresponds to a great economy. Consider further that the horizontal axis represents time of service of two presidents. The first president (named P_1) serves during the first half of the figure which shows the positive part of the curve, meaning that the number of jobs is higher than when P_1 took office. The second president, named P_2, serves during the second half of the figure when the curve is all in the negative, meaning the number of jobs and the power of the economy are diminished and all in the red (negative). If we just consider the number of jobs and the power of the economy at any given time (point of the horizontal axis), then we will condemn president P_2 and vote that president out of office. Indeed that is what some would forcefully suggest. However, this would be precisely the incorrect decision. The first president, P_1, has started with the number of jobs and the power of the economy steeply increasing. During the term of P_1, things changed toward the worse: the job and economy improvement slowed down and turned around and went to zero. The second president, P_2, took over when all things went steeply negative and toward the red, which was not that president's doing. P_2 then removed the negative trend and saved the economy. Therefore, P_2 did push things in the right direction. It was P_1 who went the wrong way. What we need to know, therefore, is not just a current value and assessment of economy and jobs. We need to know the whole history of what happened. We need to know the *function* to make the best decision, and we, therefore, need to have some STEM education!

1.3.2 Cartesian Coordinate Systems and Geometry

We have plotted above the sine function as it depends on the angle α given in radians. We can plot all functions in this way. As discussed, a function is a mapping of one set of numbers (or generally some objects), that may be denoted by the variable x, onto another set of numbers (or objects) that are denoted by the variable y. The mapping must be such that for each x we have one and only one y. For example, we may have a list of numbers $(x : 1, 3, 7)$ and that list is mapped onto $y : -1, 2, 8$. This would mean, for example, that we transform $x = 1$ into $y = -1$, $x = 3$ into $y = 2$, and $x = 7$ into $y = 8$. We may also give the mapping by a rule such as $y = 2x + 3$ or $y = x^2$ or, as we did above in Fig. 1.16, with $y = \sin(\alpha)$. If we give a function in this form, we need to specify which values of x or α we are considering. For the graphical representation of such a function, one uses conventionally "Cartesian" coordinates. The name Cartesian derives from the inventor of these coordinates, Rene Des Cartes.

In a plane, Cartesian coordinates are comprised by a horizontal straight line crossed by a perpendicular straight line. The point of intersection is called the origin and denotes the 0 of both the horizontal and vertical scale. Each point of these two perpendicular lines represents a real number. Usually one indicates only the integer numbers on the lines. Of course, one has to choose the unit distance between 0 and 1, and one often (not always) chooses equal unit distances on the horizontal and vertical axis. The horizontal axis is usually called the x-axis, and the vertical axis is the y-axis. Using such a coordinate system we can define every point of the whole plane by drawing perpendicular lines from the point to the x- and y-axes. Assume that these perpendicular lines cross the y-axis at the value $y = y_1$ and the x-axis at the value $x = x_1$. Then we denote the point by the pair of numbers (x_1, y_1) and write this pair of numbers in parenthesis. Thus we have

$$P := (x_1, y_1), \qquad (1.75)$$

where the symbol $:=$ indicates that we have a mathematical definition, in this case of the point P. Figure 1.17 illustrates the very important fact that we can identify and find any point in a plane. In the present case, we have $x_1 = 3.3$ and $y_1 = 3.2$, and the unit distance could be anything you wish or need to choose, for example, one meter or one yard. All of this may sound really abstract. Imagine, however, that you stand somewhere on a large plane. Then, if you have a cell phone or other device with GPS capability, that system knows your coordinates and the coordinates of other objects in the plane, for example, your hotel. Therefore GPS can give you directions how you get from where you are back to your hotel. Once this is understood, the use and importance of a coordinate system becomes clear. In addition, one can of course do a lot more with a coordinate system. For example, we can specify at which coordinates the moon is shining, given a certain date, and at which it is not because of clouds.

What are the reasons to plot functions in a coordinate system? To understand this in more detail, we return to the solution of equations and consider again the

Fig. 1.17 Mathematical and geometric representation of a point in a plane using Cartesian coordinates

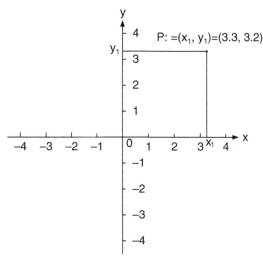

Fig. 1.18 Two *straight lines* defined by the formula of two linear equations with two unknowns (x, y). The intersection of the lines gives a point with the coordinates that are the solution of the two equations

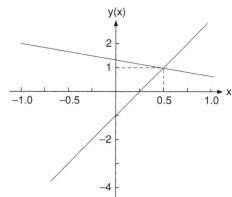

two equations (1.34) and (1.35) that we have solved in the previous chapter. We see that Eq. (1.35) reads $y = 4x - 1$ which is a function that we can plot as described above. If we plot this function in Cartesian coordinates, it forms a straight line. The same is true for Eq. (1.34) which we can rewrite as $y = -\frac{2}{3}x + \frac{4}{3}$. Again, if we plot the function, which we recommend to the reader as an exercise, then we obtain another straight line. If you have MATHEMATICA or similar software available, then you just have to write the sequence shown in the box and you will obtain the graphics of Fig. 1.18.

```
MATHEMATICA
Plot[{4x − 1, −(2/3)x + (4/3)}, {x, −1, 1}] shift-enter
```

In this figure, you can see two straight lines that correspond to the mathematical formula of the two equations. These two lines intersect at the point $P = (0.5, 1)$, which corresponds precisely to the solution of the two linear equations: $x = 0.5$ and $y = 1$. We thus have very neat geometrical interpretations for linear equations with the unknowns x, y: these equations correspond to straight lines when plotted in a Cartesian coordinate system, and their intersection gives the solution of the equations.

We have thus seen that coordinates come in handy when plotting functions. However, this fact alone would probably not have made coordinates that important. It turns out that we need coordinates whenever we wish to talk sensibly about events in nature. For example, we can say: "The moon shines" and then ask the question whether this was true or false. The correct and precise answer to questions is important, if we wish to talk logically to each other. There is a basic logical law that we always use and must use in logical discussions of our daily life. This basic law was pronounced by the Greek philosopher Plato and says: If the symbol A stands for any statement (such as "The moon shines") then A is either true or it is false but there exists no other possibility. We can see this law just like an axiom, this time an axiom of logic. Now let us return to the statement "The moon shines" and let us ask a number of people whether this statement is true or false. We will get a different answer from different people. The New Yorkers will say: "No the moon did not shine because it was so cloudy, the statement is false." Many people in Arizona will say: "Oh yes, the moon shines all the time because we almost never have clouds, the statement is true." People in Tokyo will say the statement is false because we have bright daylight when it is night in the USA. From this we can see that the statement was simply not precise enough. To make it precise we have to add some form of coordinates. For example, if we say "The moon was shining at 9 P.M. in Scottsdale, Arizona", then we have stated a truth to which everyone can agree. Thus coordinates, particularly those that include also time, are necessary to speak logically about the events of nature, which means for us about science and engineering. Of course, these coordinates do not have to be Cartesian coordinates. Other types may suffice, and we will hear about another type of coordinates in the next section.

1.3.3 The Geometry of a Sphere: Describing the Earth

Several hundred years ago, Galileo looked at the moon with his new telescope and saw that the "man on the moon" was not a face but instead was a figure formed by mountains, and he concluded that the moon had approximately the shape of a sphere and had mountains like earth. From all his observations and from the known work of Nicolaus Copernicus and Johannes Kepler, he surmised that earth was also a spherical object, rotating around itself and flying with high speed around the sun. This was, of course, against the then common belief that the earth was the center of all things and was a flat object around which the sun was somehow orbiting. The pope was upset about Galileo's findings and interpretations and put Galileo under house arrest. Galileo's career as a professor was ruined, reminding us of the German poet J. W. Goethe. His famous verse (out of the drama Faust) is in free translation:

Those who exclusively for truth have yearned,
and then discovered it,
have been since ages crucified and burned.

We tend now to see this as a bad mistake of the pope who underestimated the power and wisdom of science and scientists. However, we can be sure that many of the lesser scientists of Galileo's time were on the side of the pope and opposed Galileo also. Today this seems strange. Geometry can be regarded as a most exact science. We measure the length of things with yardsticks, and we can measure angles, and we thus can take precise measurements of all places in the world, and these measurements give the result that the earth is a sphere. We do not need to look at the stars to find that out. Of course, we have a GPS and can use it from our cell phones anywhere on earth, and we also have orbiting satellites that take photographs and show us that the earth rotates around itself and around the sun. There is no doubt that the science of Galileo was correct.

Galileo's fellow scientists could have checked at least a few of Galileo's ideas, such as the spherical shape of the earth and its rotation, without a telescope or any modern equipment. They just needed to walk up a hill at the seaside to see how the visible area expands as one moves to higher elevation. They should have observed the sunset that is later on top of the mountain then down on the beach. These are the hallmarks of the geometry of a rotating sphere.

Figure 1.19 shows a tower of 100 m (or yards) height on top of a spherical object of radius R that represents the earth. Of course, the scale is not realistic because the radius of the earth is, as we know nowadays, about 6,370 km which is 6,370,000 m, much longer than any tower's height. Even a mountain like the Mauna Kea is only about 4,000 m which is much less than the radius of the earth. Because the radius of the earth is so large, the ocean water does actually look flat if we do not look very carefully. However, if we observe what we see and add the observations by a few calculations, then we can easily deduce that the earth is round.

Consider the 100 m high tower from Fig. 1.19 and suppose that we look from the tower to the horizon. The horizon is the distant line that separates the ocean from the air. Assume that we are standing at a height h of the tower, that could be any height between 1 and 100 m. We cannot go easily below 1 m, because of our own height, and not above 100 m because the tower is not any higher. According to Fig. 1.19, the horizon is located at the distance at which a straight line, drawn from the height at which we are standing, touches the water. Such a straight line, touching a sphere or any other curved shape, is called a "tangent." We have plotted in Fig. 1.19 two such tangents, one from a height of 10 m, the line from point H_1, touching the sphere at A, and one from 100 m, the line from H_2, touching the sphere at B. We have also indicated a straight line to the center of the sphere which, of course represents the radius of the earth. These straight lines from A (or B) to the center of the sphere (earth) form a right angle with the tangent, because the tangent just touches the sphere. What we like to calculate is the distance from our point of observation to

Fig. 1.19 The geometry encountered when looking from a tower at the seaside toward the horizon. Two different heights of the tower (H_1 and H_2) are considered. If one knows the radius R of the earth, the distance from the tower to the horizon can be calculated as shown in the text

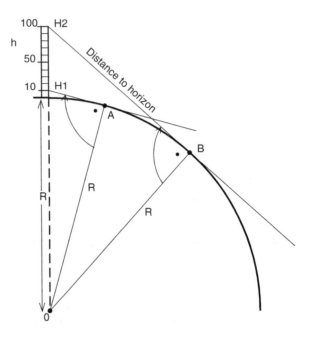

the horizon, i.e., to the points A and B. Before we do this, I would like to remind the reader that geometry problems, such as the one we are considering, are usually solved by triangulation. This simply means that we are using triangles about which we know certain facts, and we use the theorems of geometry to get the other facts that we need. If these other facts cannot be obtained by applying the theorems, then we need to make additional measurements.

The triangles that we consider now are the ones defined by the corner points H_1, A, O as well as H_2, B, O. Of each of these triangles we know the length of two sides:

$$\text{length } O \text{ to } H_1 = R + 1 \text{ meter } \text{and length } O \text{ to } A = R, \qquad (1.76)$$

where R = 6,370 km is the radius of the earth. Furthermore,

$$\text{length } O \text{ to } H_2 = R + 100 \text{ meters } \text{and length } O \text{ to } B = R. \qquad (1.77)$$

We also know that each of these triangles contains one right angle. Therefore, we know three facts about the triangles and can calculate the distance of the horizon. This distance is given by the length of the lines from H_1 to A or, if we are on top of the tower, from H_2 to B, and we obtain it by using the theorem of Pythagoras. To keep the calculation more general, we do not insert the particular heights of observation of 10 and 100 m, but we assume a general height h. The reason is that we wish to have the distance of the horizon d_{hor} for any possible height h of a mountain or a tower or whatever object we consider. The theorem of Pythagoras tells us that

$$d_{hor}^2 + R^2 = (R + h)^2, \tag{1.78}$$

which is a nonlinear equation that can easily be solved. Moving R^2 from the left to the right of the equation, we have

$$d_{hor}^2 = (R + h)^2 - R^2. \tag{1.79}$$

Using $(R + h)^2 = R^2 + 2Rh + h^2$, one obtains

$$d_{hor}^2 = 2Rh + h^2. \tag{1.80}$$

Now we use the fact that the height of the tower, or mountain, is much smaller than the radius of the earth, and therefore we may neglect the very small h^2 to obtain the approximate relation:

$$d_{hor}^2 \approx 2Rh. \tag{1.81}$$

Taking the square root we have

$$d_{hor} \approx \sqrt{2Rh}. \tag{1.82}$$

$2R$ is the diameter of the earth, equal to 12,740,000 m. Therefore, the distance to the horizon equals the square root of the product of the earth diameter and the height of the tower (or mountain). The distance to the horizon increases with the square root of the height of observation above the sea. In other words, the view expands as one moves up a tower or mountain at the seaside. As an example we calculate the distance to the horizon as seen from 1 m height, from 10 m, 30 m, 100 m, and finally from the top of Mauna Kea, about 4,000 m where m stands for meters. From Eq. (1.82) we obtain with a pocket calculator that can calculate the square root: 3.57 km, 11.29 km, 19.55 km, 35.69 km, and 225.74 km, respectively. Here km stands for kilometers, and 1 km is equal to 1,000 meters. You can easily convert these values into miles by looking at the precise conversions at the Internet or by just dividing the kilometer values by about 1.6. Thus, if you stand right at the beach and bend down to look from 1 m height, you can see the horizon in more than 2 miles distance. Walking up a 30 m tower, or from the hotel at Hapuna beach, you see a distance of 19.55 km or about 12 miles, while from the top of Mauna Kea, you can see 225.22 km or about 140 miles. This is all not very unexpected. We are used to seeing a much bigger area of the sea as we climb higher.

However, consider now carefully what one actually should see if the earth was flat instead of round. Why would we discern objects only up to a certain distance and nothing beyond that distance? The only credible explanation for this would be that the light rays cannot propagate over any larger distance through the air. We know that the air above the water contains moisture and the moisture absorbs light, and we therefore may see indeed only a certain maximal distance and not any further. Let us assume this distance to be 10 km on a given day, depending on the temperature

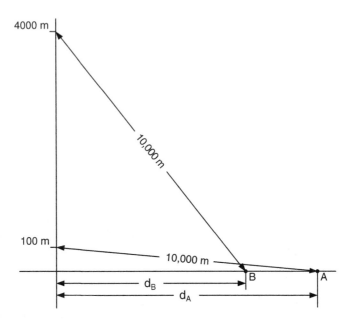

Fig. 1.20 The distances to the horizon that one would perceive on a flat earth. The assumption is made that light rays cannot be transmitted further than 10,000 m because of the absorption of light in the atmosphere (e.g., by humidity or some kind of haze)

and moisture of the air. As we climb up a tower or mountain, we will see in each case no objects beyond the distance of 10 km. The point B will therefore be closer to the bottom of the mountain (tower) than the point A. This is illustrated in Fig. 1.20. The distances d_A, d_B can be calculated again by use of the theorem of Pythagoras:

$$d_A^2 + (100\,\text{m})^2 = (10{,}000\,\text{m})^2, \tag{1.83}$$

which gives us $d_A = 9{,}999.5\,\text{m}$ while for d_B, we have

$$d_B^2 + (4{,}000\,\text{m})^2 = (10{,}000\,\text{m})^2. \tag{1.84}$$

This gives us $d_B = 9{,}165\,\text{m}$ which is far shorter than $d_A = 9{,}999.5\,\text{m}$. From this careful analysis we deduce that our observations (when climbing up a mountain or tower at the seaside) are only compatible with a spherical earth, not with a flat earth.

Thus the scientists of Galileo's times could have checked his theories by climbing up any mountain at the shores of the Mediterranean sea. They could have discovered even more conclusive effects, had they carefully observed a sunset at the foot of a mountain at the seaside. If you are close to sea level and watch the sunset, the sun vanishes completely below the horizon, but the sunlight still shines on top of the mountain! You can see how that works from Fig. 1.21. As the earth turns, an observer at the ocean can see the completed sunset, while the sun still lights up the mountaintop. It takes considerable time until the mountaintop also darkens. As one

Fig. 1.21 Illustration of the different times of sunset at the sea level and on a mountain top close to the beach. Sunset is observed at the beach (point 0) when the tangent line at the point 0 extends directly to the sun. At this time, the tangent line from point M (mountain top) does not point toward the sun. The earth needs to rotate by an angle α for this to happen. Then, after rotation by this angle, the sun set will also be observed on the top of the mountain

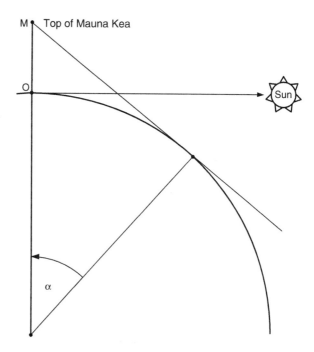

can deduce from Fig. 1.21, sunset occurs on top of the mountain only after the earth turns by the angle α. From the definition of the $\sin(\alpha)$- and $\cos(\alpha)$- functions and from the use of the rules for similar triangles as applied in Eq. (1.71), we know that

$$\cos(\alpha) = \frac{R}{R + h}. \tag{1.85}$$

For the given radius of the earth and a mountain of 4,000 m height, we obtain $\cos(\alpha) = 0.9993725$. The value of α can be calculated by any pocket calculator that features the inverse cosine or, as it is also called, the arccosine, a function that turns the $\cos(\alpha)$ back into α. In MATEMATICA, the arccosine is denoted by ArcCos, and we have ArcCos[$\cos(\alpha)$] $= \alpha$. In this way, we obtain from MATHEMATICA (see box) $\alpha = 0.0354278$ which is the value of the angle given in radians. How long does the earth take to turn such an angle? We can calculate this as follows. The earth turns a full rotation, which is in radians 2π, in 24 h. Therefore the earth turns an angle α in $\frac{24\alpha}{2\pi}$ h. This gives for the angle $\alpha = 0.0354278$ the result of 0.01353 h or about 8 m. This means that the top of Mauna Kea darkens about 8 min after the sun sets down on the beach.

MATHEMATICA
ArcCos[0.9993725] and hit shift-return we
obtain the output:
0.0354278 which is the value of α

The calculation can also be performed explicitly because we know that the angle α is very small. One can then use Newton's formula for the cosine as given by Eq. (1.74) and neglect all powers higher than 2 to obtain

$$\cos(\alpha) \approx 1 - \frac{\alpha^2}{2}. \tag{1.86}$$

We also can approximate the fraction of Eq. (1.85) by

$$\frac{R}{R+h} = \frac{1}{1+\frac{h}{R}} \approx 1 - \frac{h}{R}. \tag{1.87}$$

Comparing Eqs. (1.86) and (1.87) one finds

$$\alpha^2 \approx \frac{2h}{R}, \tag{1.88}$$

and thus,

$$\alpha \approx \sqrt{\frac{2h}{R}}. \tag{1.89}$$

Using $h = 4,000\,\text{m}$ and the radius of the earth, this results in $\alpha \approx 0.0361$ which is very close to the result shown in the MATHEMATICA box. The explicit result tells us also that, for a mountain of only $1,000\,\text{m}$, the time difference of the sunsets on the beach and on the mountain is only about 4 min (half the time difference that we calculated above) because the time difference varies with the square root of the height h. Note that all of this is true only if beach and mountain top are very close. If the mountain top is further inland, then the results are slightly different.

There is no way you can explain a delayed sun set on a mountain top, by assuming that the earth is shaped like an infinite plane instead of a rotating sphere. If you are on Hapuna beach on a clear day, you can see a gorgeous sunset at the beach. At the same time the top of Mauna Kea is still well lit, and the white observatories are visible in the bright orange of the evening sun. It takes about 8 min until darkness approaches the mountaintop starting from the foothills of the mountain. Of course, neither the pope nor Galileo could have observed this effect, because during their lifetime Hawaii was not discovered yet. But the people of Greece could have observed this phenomenon at their highest mountains (e.g., the Olympos) and some actually may have, although their mountains are by far not as high as Mauna Kea and the time difference for sunset is consequently less than 8 min. The Greek mathematician and scientist Aristarch did maintain that the earth was round. Of course only very few, if anybody, listened to him.

We have told the story of Galileo to give a motivation why we should understand geometry as a very first expression of an exact science and how important and interesting the understanding of geometry is for the understanding of our world. The religious leaders certainly have received a lesson that questions of science should be treated with the necessary attention and that one should leave to the scientists

Fig. 1.22 Spherical
coordinate system of the
earth. Geographic
"longitude" is denoted by the
angle θ given in degrees and
measured from the vertical
circle through the village of
Greenwich in England. The
geographic "latitude" is
denoted by the angle ϕ also
given in degrees. Every point
of the earth can be found by
knowing both longitude and
latitude

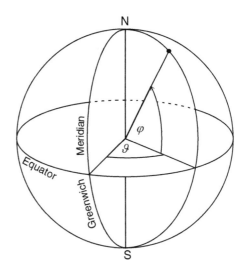

what can be understood by their methods. The scientists also should have learned a
lesson that it behooves them to be humble. Why did Galileo's fellow scientists not
help and stand at his side? Why did most of the scientists of the time side with the
pope, instead of walking up a mountain and observing nature? Of course, the answer
is that it was easier for them to accept authority and the past prejudices, instead of
doing careful and creative analyses. We also have to realize that what can nowadays
be presented at the level of high school homework problems was very very difficult
a few hundred years ago. They did not know the value of the radius of the earth
and could not retrieve it from the Internet. They did not have MATHEMATICA and
even the value of π was not as well known as it is now.

We finish this section with a discussion of a coordinate system that has been used
over hundreds of years by now and that acknowledges the approximately spherical
form of the earth. The earth is not exactly a sphere, for the trivial reason that we
have continents and mountains. Even without mountains, the earth is also slightly
different from a spherical shape, because it rotates and that rotation causes a little
widening of the earth at the equator. For the coordinate system that we discuss now
and that is shown in Fig. 1.22, this does not matter.

The figure shows the earth as a sphere and also displays large circles around
it. Two types of circles are of great importance. One type of circle, also called a
meridian, crosses the north and south poles (denoted by N, S, respectively, in the
figure). Perpendicular to the north-south axis is the equator also shown in the figure
by a circle. Furthermore we have a circle again crossing north and south pole and
also the town of Greenwich that is called the meridian of Greenwich. This meridian
defines the place from which the angle θ that is called the "longitude." is measured
The reason why Greenwich was chosen is the fact that the British navy needed the
earth coordinate system to denote the position of places and to find them with their
ships. They were the pioneers and defined these coordinates. There is a second angle

ϕ which is measured from the equator and is called the geographic "latitude". There is one bold dot shown on the sphere that is defined by the value of the longitude and latitude. Thus we can see that we can find any point on earth if we just know the two angles θ, ϕ. This makes this coordinate system very practical if we use it for navigation on ships.

Of great importance for the characterization of a ships travel was the time of a clock carried with the ship. This clock was set on Greenwich time. Then the ship moved to and arrived at different places at different Greenwich clock times. The local time at these places can be read from the position of the sun. At local time noon the sun is always at its highest point on the sky. Noon in New York is about 6 h earlier than in Greenwich because of the earth's rotation. The clock time was always measured by the very accurate clocks on the ships. There was an important ritual on board a ship at sea in which the captain had to determine when it was noon. By comparing the local noontime with the time on his clock, that shows Greenwich time, the captain can determine the angle θ that the earth must have rotated relative to the position of Greenwich. The captain could also determine the other angle ϕ of the global coordinates from the angle between horizon and sun at noontime. That angle was measured with a special instrument. Then the captain knew the position of the ship in terms of the coordinates θ, ϕ. This is how captains mapped out the geometry of the earth. From their data maps can be created that tell us the shape of the continents and the position of islands. Of course, this works only if you have very precise clocks. If the clock on board did not show the right time, then it was possible that the captain missed the place (e.g., Hawaii) where they wanted to go. Not finding an island could mean that a ship was lost at sea.

A ship's captain also wishes to include a date in their logbook. This is not a simple thing to do if one sails around the world. We have to remember that the earth rotates around itself in 24 h. So if we have the time 12 noon in Greenwich, then on the opposite side of the world, somewhere close to Hawaii, it is midnight, and a new day and date begins there. But how do we count the date? We must fix a longitude, where we start counting the date. The earth rotates toward the east. Therefore, west of this fixed longitude, the date line, we must have the date of the next day as compared to the date east of this line. We can choose this line arbitrarily, and it has been chosen at 180° longitude. However, there is a little problem here. If we would, for example, put the line separating the two dates through the middle of Hawaii's biggest city, Honolulu, then at midnight, some people on one side of this dateline would have one date say December 7 and some on the other side would have December 8. Such a regulation would have bad consequences because, for example, one side of the city would have to submit their tax report but the other side would not yet. What a mess this would create. Therefore, the line were the date changes was arbitrarily chosen in a region of the globe where only few people live. The line also has a little wavy form so that countries along the line keep the same date. It is a nice exercise to look at a globe and the dateline. Then imagine that the New Year starts right toward the west of this dateline in the pacific ocean, while east of it we still have December 31. As time goes on, the New Year then travels to the west toward Japan and China and reaches India, Africa, and Europe. Then it crosses the Atlantic Ocean to arrive at the Americas and continues over the Pacific toward Hawaii and Tahiti. After 24 h, the New Year arrives finally on the east side of the dateline.

We end this discussion of coordinates by noting that time is not measured in Greenwich time at all places of the world. That time would be badly out of whack in Hawaii, where it is dark night when it is noon in Greenwich. The earth is therefore divided into time zones that change one hour for every 15° change of θ. This is only approximately true because of the introduction of daylight savings time (see Internet) at some places. We thus have accomplished to characterize every point on earth by two angles, by the Greenwich time (or local time) and by the date. It is a great exercise to imagine to live at the dateline and start at midnight to imagine all the times and dates at places around the earth.

The knowledge of this coordinate system of the earth and the time and date is sufficient to navigate ships to any given point. It still was, in the past, a considerable task to navigate, because the longitude and latitude could only be determined from the position of sun, moon, and stars and a precise clock time. Nowadays we have it easier because we have a GPS and the Greenwich time and the mechanical clocks have been replace by the atomic clocks of the GPS. This is discussed in more detail below.

1.3.4 Geometry with Computers, from Pixels to the GPS System

The earth can be photographed in its approximately spherical shape using cameras on satellites that are orbiting the earth. The earth can be seen by passengers of our manned space station, and the earth has been seen by a few persons when traveling to the moon. Here on earth, we can look at every place by using the Internet, for example, Google earth, and we also have the GPS to tell us where we are located. The GPS also permits us to find all necessary directions to move from one place to another, no matter what means of transportation we use, ship, car, or airplane. All of this has been made possible by modern technology and particularly by the use of computers. We therefore discuss here a few principles that are used to solve problems of geometry by computers.

Representing Pictures Digitally

Computers were used to solve problems of geometry starting with the date of their invention. We have seen that computers can calculate the sine and cosine functions by performing summations as expressed by Eqs. (1.73) and (1.74). Computers also can calculate π, solve linear and nonlinear equations, and plot line graphs such as shown in Fig. 1.16. The full advantage of computers to deal with geometry became clear when applications started to present and digest photos and movies produced by cameras. The manipulation of images is needed, for example, to show us the

Photo: Michael Aschenbach

Fig. 1.23 Black and white photo, with one pixel of Ian's hair depicted. Also shown is the scale of shades of *grey*, with the actual shade of the chosen pixel corresponding to the number 10. The photo highlights the broad range of STEM education and expectations

positions and directions found with the GPS. The GPS deals with photos of streets, with a car on them that can turn as the real car turns, and images of maps of arbitrary areas. The computer can calculate the pathways of shortest distance and how to proceed from one place to another. We can, of course, not explain in detail how all of this is accomplished and only attempt to highlight some of the ideas that are involved.

Thus, we need to know more about the ways of representing photos and movies electronically and to bring them onto a TV screen, a computer screen, or on the screen of a cell phone. We will give a few simple but typical examples how a computer can handle and process these images. Any picture on the screen of your computer or TV consists of many "pixels." A pixel is a "picture element," a tiny element or dot of the picture. That dot must have the right color and the correct brightness and all the correct attributes that correspond to the point of the scenery from which the photo was taken. The information representing the properties of a single pixel is nowadays stored, transmitted, and received in a digital fashion. In other words, the information is stored by way of digital numbers. This is illustrated in Fig. 1.23 for a black and white photo. The figure shows the photograph of a boy.

One pixel of the boy's hair is chosen and magnified in the inset of the illustration. The scale of 21 grey shades, that are used in our example to create the computer or TV pictures, is also shown. The grey of the chosen hair pixel corresponds to the number 10 of the scale, and this number is stored and transmitted to the corresponding pixel of the computer or TV screen. This transformation of grey shades into numbers is also called quantization, because we denote the grey shade by a number that defines a given quantity of grey. A screen image with great resolution of the details necessitates typically the transmission of more than a million pixels (1 megapixel). To produce the original photograph of such an image, one also needs more than a million sensory elements in the electronic camera and more than one million display elements on the computer screen. For each of these picture elements we need to supply the 21 different shades of grey, or any other number of grey shades that we wish to reproduce. That means that we need more than 21 million pieces of information to take and create the black and white picture and also to reproduce it on the computer screen. Actually we need many more pieces of information to produce photographs and corresponding TV or computer pictures for the following reasons.

For one, we wish to have the picture in color and not just in black and white. If we introduce many different colors with different brightness and other color properties (look up "color hues" on the Internet) then we need to store much more information for each pixel. This means that our electronic instruments (computer, TV) need to deal with hundreds to thousands of millions of bits (hundreds or thousands of megabits) if color pictures are processed. Up to this point we have talked only about one picture or one "picture frame." If we wish to capture motion (e.g., in a video), then we need to deal with about 20 frames in a second. If we study details of very fast motion, such as the wing motion of a humming bird, we might desire to photograph 1,000 frames a second. This means that our camera and computer will need to deal with millions of megabits or terabits. This is just about becoming possible using the most modern camera and computer equipment. It still would not be possible to transmit that kind of information wireless to our home TV. It is currently possible, however, to transmit the 20 frames or so per second and corresponding hundreds of megabit per second for high-definition TV. This capability did necessitate enormous technical progress from the possibilities that existed 40–50 years ago. Remember the executive of the TV company who claimed 40 years ago that all was known about TVs and whom we mentioned in Section "Aim of the book" of preface. At that time one could deal only with about one million information pieces per second for both the cameras and the wireless transmission to the TVs. There was at least a factor of 1,000 missing from today's capability. The competition of the Acme Electronics Corporation was going after this factor and succeeded! Looking at this story from today's vantage point, one can say that the STEM ignorance of the Acme executive was truly unacceptable. The frequently held opinion that any great industry leader can deal with both potato chips and computer chips equally well is just wrong. It is as incorrect as the opinion that every teacher only needs to know how to teach and then can teach mathematics as well as history and art.

As far as pixels are concerned, we have not yet arrived at the ultimate goal of picture representation. Humans like to see in three dimensions, with the highest fidelity of colors and with great dynamical speed to process as many frames as necessary. It will take a long time until technology can give us the very best for our human capability of perception; lots of increased STEM knowledge will be necessary for that.

Computers Manipulating Pictures Mathematically

All of the above applies to TVs, computers, and cameras alike, and we will learn more about the technology of image generation, transmission, and representation in later sections of this book, when we talk about transistors and integrated circuits. Here we add just an example how computers can help us to deal with pictures in a mathematical way. Our example describes how to mathematically rotate a picture by any angle. This is an operation we sometimes wish to do. We might like to rotate a picture so that it better fits into a frame, or our GPS system likes to rotate the picture that it shows, because we take a turn (perform a rotation) with our car. The most general mathematical way to deal with such a rotation is by introducing a coordinate system and then just rotate the coordinates. Let us say we deal with a two-dimensional picture and a two-dimensional Cartesian coordinate system with coordinates (x, y). We then have to assign to every pixel on the screen its appropriate coordinates. To rotate the picture, all we have to do is to find the new rotated coordinates (x', y') and then transfer the pixel information (such as color and brightness) from the coordinates (x, y) to the coordinates (x', y'). That simple! Of course we need to know how we get from the old coordinates to the new coordinates. The mathematicians have figured that out for us. They have shown that for any rotation by an angle ϕ we have

$$x' = x \cdot \cos(\phi) + y \cdot \sin(\phi) \tag{1.90}$$

and

$$y' = -x \cdot \sin(\phi) + y \cdot \cos(\phi). \tag{1.91}$$

It is an easy exercise to use the above equations and plot the rotation of any point of a coordinate system. Particularly easy are rotations by $90°$ or $\pi/2$ rad (which is the same), because for $\phi = \pi/2$ we have $\sin(\phi) = 1, \cos(\phi) = 0$. Therefore the point $(x = 1, y = 0)$ of the x-axis turns according to the above equations into the point $(x' = 0, y' = 1)$ that is a point on the y-axis of our coordinate system, corresponding exactly to a rotation by $90°$, because the two axes of a Cartesian coordinate system form by definition a right angle.

You can imagine what kind of work a computer has to do to rotate the millions of pixels of one high-definition photograph. Naturally there are shortcuts for such processes, and specialized graphics software can do such operations very efficiently. MATHEMATICA also lets you represent objects and rotate them in a simple fashion. Take, for example, the graphic representation of a triangle with the three corner points $(2, 0), (4, 1), (4, -1)$. MATHEMATICA lets you create this triangle by the commands shown in the box. The "polygon" that is mentioned

Fig. 1.24 Rotation of a
triangle by 90° as generated,
for example, by
MATHEMATICA

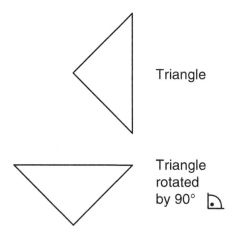

Triangle

Triangle
rotated
by 90°

in the MATHEMATICA line is in this case a triangle, because we specified the
coordinates of three points only. MATHEMATICA could have dealt with a polygon
with more corner points also.

```
MATHEMATICA
Graphics[Yellow, Polygon[2, 0, 4, 1, 4, −1]] shift-enter
```

The result is shown in Fig. 1.24. The graphics figure, the triangle, can be rotated
by the command Rotate$[g, \phi]$ with g just standing for the entire graphics command
and the angle ϕ given in degrees, as shown in the second MATHEMATICA box. The
rotated triangle is also shown in Fig. 1.24. The manipulation of pictures and graphics
objects by computers has been an important field of research and development and
has resulted in software for handling photos with increasing capability.

```
MATHEMATICA
Graphics[Rotate[Yellow, Polygon[2, 0, 4, 1, 4, −1], 90 Degree]] shift-enter
```

We have attempted here only to give a glimpse of what is involved in computer
graphics and how the geometry of the ancients and the coordinates of the Cartesians
are still used to achieve modern objectives such as the rotation of an object on a
computer screen. The great success of thousands of years of geometry, both theory
and measurements, combined with the technological achievements of the past
century can be seen by many an example, one of the most impressive being the GPS.

GPS: The Global Positioning System

The current GPS is US based and uses satellites orbiting the earth as well as stations
on earth to provide positioning and navigation services on a worldwide basis.

Navigation was already important in ancient times and had the goal to guide ships from home to some predesigned place and back. Every captain had some capability of navigating, in more recent times by use of the sun or stars and a clock. It was, therefore, very important for a navigator to have the exact time, and very precise clocks were designed and produced for this purpose over the course of history.

Powerful navies, such as that of Great Britain, had the best clocks. These were wonderful mechanical systems that worked with the precision of seconds over long time periods of time, such as months. These clocks are nowadays in the hands of collectors and museums. Modern clocks of greatest precision capitalize from the precision of oscillations in atoms that we describe when we talk about atomic standing waves in Sect. 2.5. These so-called atomic clocks work with a precision of about nine digits per day which means that they are correct to one millionths of a second over a year.

The GPS system has therefore its own very precise time, the GPS time. It furthermore features currently 24 satellites (plus three in reserve) all over the world. These satellites send signals that can be detected by the GPS instrument that you may have in your car. That actual GPS instrument needs to find 4 satellites out of the 24 total, and it records the signal transmission times t_s of the signals from these 4 satellites. Because we know that these signals are transmitted with the velocity of light c, that we also know with precision, your GPS instrument can calculate the distances to the satellites from the transmission times t_s. The distances are given by $d = c \cdot t_s$, and the GPS instrument can deduce from these distances the actual position of your car. Thus, the satellites have not only replaced the stars for the triangulation of the position, they also are more useful than the stars for the determination of positions, because one can deduce the transmission times and distances from their signal. The actual process of the position determination is called trilateration. Trilateration is more complicated than triangulation as we have learned it and uses the geometry of spheres that have the satellites in their center and have a radius of $d = c \cdot t_s$. All the mathematics that is needed to understand trilateration in detail is similar to what we have learned in Sect. 1.3.3, and teachers are encouraged to assign special projects on this topic to interested students.

We are now used to have a GPS instrument in our cars and can find places within a few meters if we know their address. Such a system also permits us to explore the exact coordinates of its position including the GPS time. The changes of coordinates with time can also be evaluated and used to determine the speed of the car, ship, airplane, or any other object. The 4 global positioning coordinates are given by 3 space coordinates (x, y, z) plus 1 time coordinate t. Using the 4 GPS coordinates (x, y, z, t), you can do a lot more than determining the position of a vehicle. For example, a land surveyor can survey your property and find all the elevations of that property, even if you live on a mountain. From this information, one can determine, for example, which part of the property can be flattened out easily and is therefore best suited for building a house. One calls this the determination of the "topology" of a property. Any geometrical problem that requires the knowledge of the four coordinates (x, y, z, t) can be done in principle with a GPS instrument. This is truly one of the great technical advances of our time. Thus, there may be after all a royal road to geometry.

Chapter 2
Science: The Process of Understanding the Natural World and Its Possibilities

What is science? According to the famous Austrian scientist Ernst Mach science is a description of facts of nature that is as complete as possible and as economical (meaning without unnecessary additions) as possible. This definition does not include any word about creativity and does not mention any of the important tools of science. A very important tool in science arises from the notion of similarities and the forming of analogies. Remember the smaller and smaller grains of sand on the beach and the question whether the waves can grind the grains ever smaller. Democrit of Greece thought that there was a limit to the smallness, a smallest type of grain: the atom. He thus had a "theory" about the matter that surrounds us; he believed that everything was made of atoms. Mach did not mention the words theory or logic nor did he mention mathematics in his definition of science. He even said that theories were unnecessary additions (he compared them to dry leaves), and he did refuse to believe in atoms!

Albert Einstein, on the other hand, thought that Mach's definition of science was a bit "stale," and he asked Mach whether he would not find theories useful, if they would lead to the most economical description of nature. Einstein was also pointing toward the fight of Mach with Boltzmann about the existence of atoms, and he clearly sided with Boltzmann and believed that atoms did indeed exist. Since Boltzmann's times, the concept of atoms turned out to be very important and has guided scientist to very successful findings.

If such great luminaries as Boltzmann, Mach, and Einstein had a discussion of what science really is, then we naturally need to be careful when we wish to define it. Therefore, I describe here what science really is, only as far as I understand it myself. Before I do so, I like to say a word about the phrase "as far as I understand it."

I attended an excellent high school in Baden, Austria, and I had a great chemistry teacher. Her name was Marianne Schwarz. She had received a good education at the University of Vienna, and she continued her education by reading textbooks like "The Nature of the Chemical Bond" by Linus Pauling. When she tried to explain how electrons are forming the bonds between the atoms of a molecule, she showed some nice pictures form Pauling's book and said "As far as I understand it, this means the following." She really impressed me by saying that. Teachers, particularly teachers in Austria at that time, almost never admitted that they did not understand everything fully or could not explain everything to perfection. The reason was and is, of course, that students would use such an admission of ignorance immediately as an excuse that they did not need to know anything about the subject. If a teacher did not really understand it, how and why was a student supposed to know about it? This is a very interesting and important point. All good teachers should, in my opinion, bring the discussion of an important topic toward a satisfactory conclusion. Elementary algebra is, for example, completely understood and can be explained by logical deduction once the axioms are given. Other topics, such as the buzzing of electrons around the nuclei of atoms, or the existence of quarks, cannot be explained with such a degree of certainty and logic, and teachers should admit to the students if a topic cannot yet be completely explained to them. Anything that can not be satisfactorily explained and taught in high school is, of course, not really understood and needs further work by scholars. This is important for the interested students to know, because they are the ones who may find a better explanation later during their best and most productive years. It is also important for any student to know that teachers do not walk on water but try hard to explain what is really known. Students, in turn, need to be impressed if a teacher levels with them and tells them that a subject is not fully understood. They need to be excited that there are always new things to be discovered, and they need to pay particular attention to what the teacher says, instead of flushing things after hearing: "As far as I understand it." So please note that when I say "as far as I understand it," I may not really understand it, and maybe nobody has a totally accepted explanation, but I try my best to explain the current status of understanding.

As far as I understand it, science at its best is what Euclid did. He started with things from everyday life that were useful to measure the size of objects of the surrounding world. If you take a string of a certain length, then you can measure the distance from one place to another. You do this by repeatedly using the string and then you find, for example, that the distance to the neighboring house is hundred times the length of the string. If the length of the string is one "meter," then the distance to the neighboring house is 100 m. Of course, you need first to agree on what one meter is. This is, in principle, a definition. As mentioned, the actual meter measure is made out of platinum and stored in Paris at constant temperature so that it does not expand or contract when the temperature changes. Once one has a measure of distance, one can measure a lot of things. As we know we can then also determine the area of the property that you own. Furthermore, as we know, Euclid worked also with circles, with the length of an arc, and with angles. This initial use of strings and straight objects such as rulers, as well as circles and arcs, was followed by ideas. Euclid used the abstract idea of a point, of a straight line, of an infinitely extended line, of an ideal circle, and so forth. Using these ideas he wrote down simple rules that apply to these ideas, his axioms. Then these simple rules were used to derive theorems, a logical truth, such as the Pythagorean theorem, and the rules together with the theorems permitted us to derive and *predict* a lot of important consequences. We could even calculate the distance from a tower at the beach to the horizon of the ocean. It is important to note that Euclid used only few (five) axioms and few logical–mathematical rules to derive the laws of geometry. Thus he used a very "restricted" form of language with very careful definitions and much higher precision than we are used to have in ordinary language. This precision is very important for science, and therefore the mathematical–logical language is very important for science.

Of course, the most important point of any scientific approach is to check whether the theoretical results agree with all the observations, with all the actual measurements. Consider our example of calculating the distance to the horizon as seen from a tower (Sect. 1.3.3). To check this calculation one needs to place markers at certain distances out in the ocean and then see whether these markers become visible at the horizon when moving up to certain heights of a tower at the seashore. Of course, this is not an easy experiment. An example of a more straightforward experiment or measurement related to geometrical science would be to measure the area of a triangle by inserting as many little squares into the triangle as possible and by counting the squares and thus measuring the total area. Then we can check if that area is also obtained by the law of Euclid's geometry: multiply the length of the baseline by the height of the triangle and divide by 2. Euclid's geometry was checked in this way over and over in millions and millions of experiments in the thousands of years after Euclid. All these checks came out correct. But no science is ever totally correct, no matter how self-explanatory, no matter how beautiful, logical, and mathematically justified it may be.

More than 2,000 years after Euclid, mathematicians and scientists were still puzzling over Euclid's 5th axiom which says that the sum of the three angles of a triangle equals two right angles. They did not understand whether this should be called an axiom or whether it actually followed from the four other axioms and what it would mean if it were not an axiom or if it were not true? As it turned out, and as we know now from the work of Einstein, the geometry of the universe is not Euclidean and, if very large distances are involved such as those between stars, the sum of the three angles of a triangle does not have to be equal to two right angles. For us on earth, however, it is true to many digits and can be measured to many digits by using laser light to represent the straight lines that form the sides of the triangle. However, we can today also measure, and have measured, how light bends when going around the sun and how then the definition of a straight line becomes more difficult, and how then the 5th axiom can be violated. Details are given in Sect. 5.

To summarize, Euclid's science involved a process of using elements of the world that surrounds us, such as strings and sticks, and then forming limiting abstractions of these elements such as a straight line. Then, using these abstractions, Euclid defined rules relating them to each other. These rules or axioms form the basis of a theoretical framework that can be dealt with using logic and mathematics (e.g., arithmetic and algebra). The results are then carefully checked by measurements with instruments that correspond to the abstractions; for example, a straight line can be simulated by a laser beam. If a problem is found, and the results of the theory do not agree with experiment, then the theory is corrected and a new improved theory emerges such as Einstein's non-Euclidean theory of the universe.

Euclid's work is often seen as pure mathematics and not really as science. This would be only true if one just takes Euclid's axioms *without their connection to real things* and deduces logical consequences. Indeed, there exists an enormous body of work that has been performed that way and is therefore regarded as pure mathematics. The reason why Euclid's work is so special is that it can be seen as both: as great science connected to all that surrounds us and, on the other hand, as pure logic and mathematics. One can look at Newton's work from a similar point of view: Newton developed the mechanics of planet motion and the corresponding laws of physical science. In the course of this work he discovered and developed the "calculus" which is pure and beautiful mathematics. Science and mathematics are intertwined and gain from each other. Newton's work, however, brings out one more important point of science, a point that was not quite as "visible" in the work of Euclid. This point is that science is *predictive*. The laws that Newton found let us predict the orbit of the planets and where they will be visible on the sky. They also let us predict the path of comets, even comets that we have not seen previously. This predictive quality is the main feature of science, is the feature that makes science useful to mankind.

Many scientific approaches to nature use logic plus some framework of symbols and rules (a theory) and differ quite significantly from the work of Euclid and Newton because they do not use mathematics. The use of mathematics, and a logical framework based on a few axioms, is the signature of scientific maturity and guaranties that precise predictions can be made. The characteristic feature of science is always that the results of the theory, its predictions, can be and are compared to measurements. These measurements connect the theory to nature and prove or disprove the truth content of the theory. If a discrepancy is found then the theory is abandoned or extended until the theory agrees again with the known data that are obtained by observing nature. The observations can be done with our eyes and ears or with elaborate equipment, such as a microscope or telescope, that extends the capabilities of our senses. Thus a theory of science is dealing with abstractions, but clearly connects to nature because its results and predictions have been (and can be) tested over and over by experiments. After many confirming tests, we usually believe that the theory is correct, and then we do not doubt the theory. Indeed, we extend doubt to all that are against the theory. Often this way of thinking is justified, and it is silly to doubt the theorem of Pythagoras when calculating the height of a tower. However, we always need to remember the story of the 5th axiom of Euclid. Einstein found a problem with Euclidean geometry after it had been checked out for more than two thousand years. We also know that Galileo was right, when he refused to believe the then "known fact" that the sun was orbiting around the earth and the earth was standing still; and all his colleagues and even the Pope who opposed him were wrong.

This brings me to the difficult problem of the relation of science and religion to each other. As we can see from Galileo's case, this question can only be approached with great caution from all sides. We have to give science what belongs to science and to religion what belongs to religion. Under most circumstances this is easy, because science and religion can be clearly separated. Science deals with occurrences in nature that can be experimentally explored and repeated, and science permits us to *predict outcomes* of experiments with a large measure of certainty through the knowledge of some basic laws. If we drop a stone, for example, we can predict with great certainty how fast the stone will fall, and it would be illogical to assume that god will have to govern the falling of the stone by his direct intervention and actions, whenever we choose to drop it. Similarly, if we mix two chemicals and obtain a third one and we can repeat this experiment over and over, then we are exploring natural law and not the directly induced action of an all powerful being. These laws of nature that we explore with scientific methods can be very beautiful and certainly humble us, because our understanding of them is always "anthropomorphic," meaning limited by the human ways of thinking. Boltzmann, the great theoretical physicist, looked at the laws of electromagnetism that were discovered by Faraday and Maxwell and cited the famous verse of Goethe's drama Faust: "Was it a god who wrote these signs?" We can, of course, not prove or disprove the existence of god with our scientific methods. Inversely, the religions of the world cannot deny the existence of a natural law that is accessible to the methods of science and lets us freely experiment and *predict* experimental outcomes.

There are some scientific areas that border on the realm of religion. Scientists have proposed that the universe was created by a "Big Bang" out of an extremely small nucleus and thus virtually out of nothing. Indeed many of the astronomical observations of the universe are consistent with such a theory. There exists, for example, a microwave background radiation in space that could be the remnant of this explosive expansion of the universe. Naturally, religions may assume that the "Big Bang" was the act of creation by god and may discuss this in their instructions. Such themes may also be topics in philosophy, and we admire the wisdom and modesty of Socrates when he said "I know that I know nothing." Science education, however, must exclusively deal with repeatable observations of nature, with deductions of natural laws, and with their justification by further experiments and observations. Naturally, it is important to emphasize that the Big Bang theory is much less convincingly proven than, for example, the laws of falling stones. Scientists cannot now, and never will be able to, perform the crucial experiment, the recreation of the Big Bang. All our evidence can only be indirect, and therefore we are at a borderline between what can be theorized and what can be proven by science.

Such considerations also apply to the often discussed topic of evolution. The theory of evolution was formulated by Charles Darwin and maintains that biological life-forms like birds and humans, and also smaller biological entities, such as biomolecules or cells, have naturally evolved from more humble beginnings to the currently existing forms. Darwin started his book *"On the origin of species"* by explaining how plants were domesticated and made useful by human selection. The natural law, basic to the theory of evolution, is assumed to be the selection of nature and the corresponding survival of the "fittest" or best-suited forms of life. This selection of nature works similarly to the selection process that humans used to domesticate plants and animals and results in live forms that have a larger complexity and a winning edge.

We know now that the information that is necessary for the formation of living beings, including humans, is stored in a giant molecule denoted by the acronym DNA, and we know that the DNA of parent life-forms determines the DNA of offspring life-forms. We also know that there are great similarities and connections between the DNA of humans and animals. Yet, the development of human life from more humble beginnings through DNA changes is frequently debated, and some teach the direct intervention of god to create humans and other life-forms. Of course, if we look at the complexities of human life and the DNA, we may well exclaim, as Boltzmann did, "Was it a god who wrote these signs?" However, in 2010, DNA has been artificially created in a biochemistry laboratory of the Venter Institute. The scientists implanted this completely artificial DNA (of a bacterium) into a host cell without DNA. The host cell was awakened to life by this implantation and began to grow and reproduce. We have therefore arrived, at least with some forms of DNA, at the point at which we can experiment at will and predict the outcomes of these experiments with great certainty. We have a scientific understanding of DNA!

Will that end discussions about evolution? Probably not! It is unlikely that we will be able to experimentally reproduce, in the laboratory and within a short time period, the DNA changes that may have occurred over millions of years and that have led to the current DNA forms. Therefore, some experiments that may be crucial to decide such a debate of creation versus evolution of life may not be possible. Nevertheless, science education must be confined to the methods of science, its deductions, and its predictions. It is the predictions of scientific method that make all the difference, and only theories that predict facts that can be experimentally checked should be taught in science classes. Teachings of creation by an all powerful being, as valid as they may be, explain everything but predict nothing. This was already pointed out by the mathematician Pierre–Simon Laplace to Napoleon in the early eighteen hundreds.

This author believes firmly in the theory of evolution. I am also convinced, however, that STEM teachers must be careful not to overstate their case. They should not claim that the evolution of DNA, millions of years ago, is as well understood and established as the science of falling stones.

In spite of the always present limitations of our knowledge, however, it is clear that science provides an extremely useful tool to pursue many great goals of mankind. Science provides also the foundation for engineering methods and enables us to develop technologies that provide us with energy, housing, medication, and recreation, all based on scientific *prediction*. Science thus helps us in our pursuit of happiness. This is what STEM is all about; at least this is what it should be about.

2.1   Physics: Force, Velocity, and Energy

The name physics is closely associated with concepts such as force, velocity, energy, particles, and waves. We will explain these concepts, and what they do for us, in an approximately historic fashion and start with the concept of force.

2.1.1   Force and Energy

The concept of force is probably as old as mankind. Force is necessary to protect yourself from nature, from dangerous animals, from storms by building some kind of roof, to fight enemies, and even to just lift a stone. One certainly thinks in all these connections of muscular force, such as the strength of our arms to lift objects or of our legs to run. The early cultures surprise us often by great buildings such as the pyramids and structures such as Stone-henge, because it is difficult to imagine that such structures were built with the forces that we have normally at our disposal. They show that it must be possible to exert much larger forces than usually attributed to a biceps. To explore these possibilities, we need to understand the nature of what we call force in greater detail.

Some of the features of the concept of force are revealed in Fig. 2.1. This figure shows a seesaw with different board length and weights on each side. The long board with length L_2 needs fewer persons to balance the persons on the short board with length L_1. The persons sitting on the boards exert a force because gravity pulls them down. The important rule for the balance of the forces on each side is

$$F_1 L_1 = F_2 L_2. \qquad (2.1)$$

We note that the product of force and length has to be equal on each side to balance the seesaw. We will see below that this fact can be derived from one of the most basic laws of physics, the law of conservation of energy. For now, however, we are

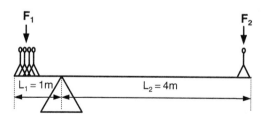

Fig. 2.1 A seesaw having different board length on its two sides. One side has a board length L_2 that is four times as long as the other that has length L_1: $L_2 = 4 L_1$. One finds for this case that one person on the L_2 side can balance four equally heavy persons on the L_1 side. Because of gravity, the persons sitting on the board exert forces F_1 and F_2, respectively, as indicated, and the seesaw is balanced if $F_2 = 4 F_1$

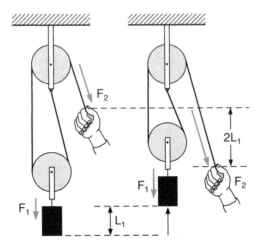

Fig. 2.2 Pulley hoist consisting of two wheels that can rotate around their axis. The axis of one wheel is fixed, for example, mounted to the ceiling, and the second wheel is connected to the first by a rope that winds around both wheels (if n discs are involved around n discs). The weight is pulled down by gravity and thus exerts the force F_1. It can be pulled up by the force $F_2 = \frac{F_1}{2}$ or for n wheels by a force $F_2 = \frac{F_1}{n}$. The ratio $\frac{1}{2}$ of the distance L_1 to the pulling distance $2L_1$ becomes intuitively clear if F_2 points vertically downwards. It stays the same, however, if the force F_2 points in any direction

only interested in the forces that we can exert and how we can increase these forces. We can change the length of the boards so that $\frac{L_2}{L_1} = 10$. Then we need to exert 10 times the force on the short side compared to the force on the long side. This principle tells us why one can move huge stones with a crowbar. The same principle explains why we can exert large forces with wrenches. In fact, many tools of any household are just based on this principle.

Another example is presented by a pulley hoist as shown in Fig. 2.2. This is a well-known tool that can be used to pull up heavy loads. To pull up a load that exerts a force F_1 over a distance L_1, one needs to pull the rope with a force F_2 over a distance L_2. One finds from the experiments that

$$F_2 = \frac{F_1}{2} \tag{2.2}$$

and

$$L_2 = 2\,L_1. \tag{2.3}$$

Taking the product of the left- and right-hand sides of the two equations gives again Eq. (2.1). If we pull up a load with just a fraction of the force necessary to pull it directly, then we have to pull the rope a longer distance corresponding to this fraction. This is universally so for all experimental arrangements; we can think of hoists with more than two wheels that can pull heavier loads. It took a long time until it was understood, that the product of force (in the direction of the movement) and the distance (that one actually moves) corresponds to the energy that one invests in the process. Energy was up to then, for a variety of reasons, not clearly defined. For example, if it is very hot when persons are pulling loads, it may appear to them that they need a lot more energy in order to accomplish the same task. Physics does not deal with feelings, and based on the rule of Eq. (2.1), as well as other experimental facts, energy is defined by

$$E\,(\text{Energy}) = F\,(\text{Force}) \cdot L\,(\text{Distance}). \qquad (2.4)$$

The distance is the length over which the force is applied. If the force changes during the pulling or moving, then one must take the sum of all the distances multiplied by the different forces in the direction of the movement. In the limit of changes over very small distances, this sum needs to be performed according to the rules of Sect. 5.2.2.

Energy is a very important quantity and represents the single most important concept of physics. No machinery, no matter how ingenious, can create energy out of nothing or destroy energy into nothing. This is a simple formulation of the most basic law of physics that we call the law of energy conservation. This law extends beyond mechanical machines and is of general validity. The interested reader should consult the Internet. For mechanical machines energy conservation means the following: we cannot construct any machine, with seesaws, hoists, or whatever, that creates energy. All we can increase (or decrease) is the force. The search for a mechanical machine, that gains energy out of nothing, has been the object of the lifework of many people who wished to create a "perpetuum mobile," a machine that perpetually turns wheels and produces energy. Such a machine cannot be built, because it would violate physic's most basic law, the law of energy conservation. You might ask: how do we know that for sure? Euclid's fifth axiom was not correct, and maybe energy conservation is not correct either! To such an objection one can only say that the law of energy conservation has been tested like no other law of nature, and it has always been found correct.

Forces are not just numbers that characterize a magnitude. The direction of a force is also of great importance. Quantities that are characterized by both magnitude and direction are mathematically represented by "vectors." Like numbers, the mathematical abstraction of a vector follows certain axioms. We will not discuss these axioms here, but rather highlight the main properties of vectors and their application by the following example. Figure 2.3 shows a person pushing a cart on a street or plane that is inclined by an angle α. The force that the person needs to push the cart up the street points in the uphill direction parallel to the street. This force depends, therefore, on the angle α, i.e., on how steeply uphill the cart needs to be pushed. If $\alpha = 0$ then we need practically no force at all to push the cart. As α increases one needs a larger and larger force to push the cart forward. If we wish to push a large weight up a hill and use only little force, then we need to have a small angle α. The architects of the pyramids knew this. They built streets toward the pyramids that had indeed a small angle α of inclination and thus permitted them

Fig. 2.3 Pushing a cart up an inclined plane illustrates how the total force F_t (that acts on the cart in the direction of the inclined plane) can be seen as being composed of a horizontal force F_1 and a vertical force F_2

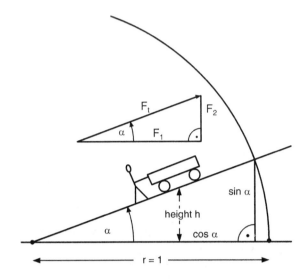

to push the big stones upwards to the top of the pyramids. This also illustrates that the force is a vector because its direction is important, and, in our case, the direction also determines how much force we need, i.e., the magnitude of the force.

All of these facts can be understood from the law of energy conservation. Because a horizontal movement does not need any energy (if we forget about the friction of the cart on the ground), all the energy is needed for pushing the cart to a greater height against the forces of gravity. This upward pushing is accomplished by the magnitude of the upwards pointing force F_2 shown in Fig. 2.3. The energy E that one needs to push the cart to a height h is given by

$$E = F_2 \cdot h, \tag{2.5}$$

which explains why we can do the work with a small force F_2, if we have a small inclination α of the street. The exact geometrical and mathematical relationships of the forces that are involved in the example of Fig. 2.3 can be derived as follows. We have drawn in the figure a circle with a radius of unit length (e.g., one meter) and have also indicated a triangle. This triangle includes one right angle and the angle α, and its longest side is equal to the radius of the circle. The lengths of the other sides are equal to $\sin(\alpha)$ (vertical) and $\cos(\alpha)$ (horizontal), respectively. This triangle is similar (in the mathematical sense) to the triangle that shows the forces. From the rules given in Sect. 1.3.1 for similar triangles, we obtain then the following relation for the magnitudes of the forces:

$$F_2 = F_t \sin(\alpha) \tag{2.6}$$

and

$$F_1 = F_t \cos(\alpha). \tag{2.7}$$

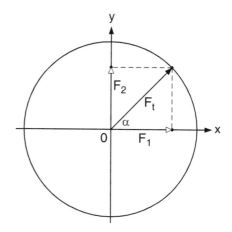

Fig. 2.4 Radius vectors are defined as the *directed line* going from the origin of the coordinate system to any point. The fact that these vectors indicate a direction is expressed by the *tip of an arrow* at the endpoint of the line

This decomposition of the total force F_t into a horizontal component F_1 and a vertical component F_2 is useful because it helps us to calculate the force that is necessary to do the work. The work equals the energy that is needed. This way of dealing with forces is most efficiently done with the mathematical concept of a vector. It is convenient to give the explanation of vectors by using a Cartesian coordinate system and defining a "radius vector" following the illustrations of Figs. 2.3 and 2.4.

Radius vectors are the directed lines that start from the origin of the coordinate system and extend to any point (x, y) of the plane of the coordinate system. Radius vectors are usually denoted by bold-faced letters like **r** and are characterized and represented by the coordinates (x, y) of the point; thus we have $\mathbf{r} = (x, y)$. The vector of Fig. 2.3 that points from the center of the circle in the direction of the inclined plane toward the point that intersects the circle has the coordinates $(\cos(\alpha), \sin(\alpha))$ and, therefore, $\mathbf{r} = (\cos(\alpha), \sin(\alpha))$. Such a vector provides us with the length and the direction of a line. The physical sciences deal with many different vectors. As we just have learned, the force is a vector. Therefore electric and magnetic fields are also vectors because they are forces that have a direction and a magnitude. How do we deal with such general radius vectors? Exactly the same way as we deal with vectors that represent a length and a direction! There is just a little trick necessary. We plot the coordinate system with a line pointing from the 0 to the endpoint of the vector. However, now, we do not deal anymore with distances but with forces. We therefore just replace the unit distances on the x, y axis by the unit forces. Figure 2.3 shows the result. The vector from the 0 point of the coordinate system to the circle has now the magnitude F_t, and we call it the vector $\mathbf{F}_t$. Thus we only need to multiply everything by the length of the vector which in the above example is the magnitude of the force. The vector $\mathbf{F}_1$ has the coordinates $F_t(\cos(\alpha), 0)$ and the vector $\mathbf{F}_2$ has coordinates $(0, F_t \sin(\alpha))$. We can therefore write

$$\mathbf{F}_t = (F_t \cos(\alpha), F_t \sin(\alpha)), \ \mathbf{F}_1 = (F_t \cos(\alpha), 0), \ \mathbf{F}_2 = (0, F_t \sin(\alpha)). \quad (2.8)$$

We define then the addition of vectors in the following way. Geometrically speaking, we add vectors $\mathbf{F}_1$ and $\mathbf{F}_2$ by putting the lower end of vector $\mathbf{F}_2$ at the right end of the vector $\mathbf{F}_1$, and the vector $\mathbf{F}_t$ is obtained exactly that way as shown in Fig. 2.4. Algebraically this means vectors are added by just adding the x- and y-components separately. Thus the vector addition

$$\mathbf{F}_t = \mathbf{F}_1 + \mathbf{F}_2 \qquad (2.9)$$

is for the radius vectors equivalent to

$$(F_t \cos(\alpha), F_t \sin(\alpha)) = (F_t \cos(\alpha), 0) + (0, F_t \sin(\alpha)), \qquad (2.10)$$

which represents the algebraic way to add vectors. Note that F_t is just the magnitude or length of the vector $\mathbf{F}_t$. It is a good exercise to add vectors both in a geometrical fashion as shown above with the cart and algebraically as just explained.

In this way we can add forces that point in arbitrary directions and obtain the total resulting force. This rather involved procedure of adding vectors is of great importance in physics. As mentioned, many physical quantities are vectors, because they are involving not only a magnitude but also a direction. The velocity of physical objects is also a vector and so is the acceleration that we will define later. For someone who likes physics, it is therefore very important to get familiar with the concept of vectors. However, in most of what follows, we have avoided this rather sophisticated type of calculation, by assuming to start with that only one direction, for example, the x-direction, is relevant.

2.1.2   Momentum: The Mechanics of Billiards

We discussed in Sect. 2.1 the very important law of energy conservation. There is a second law for the mechanics of objects that is related to energy conservation and also of great importance. This is the law of conservation of momentum. Momentum is, in contrast to energy, a vector, i.e., a quantity for which direction is of importance and is denoted by a bold letter, for example, by $\mathbf{p}$. Thus we have a symbol for the word momentum, but what is it and what is conserved? Momentum is related to both the velocity of an object (also a vector) and to the energy of the object as illustrated in Fig. 2.5. We therefore discuss first the definition of velocity.

The concept of velocity is well known from our daily life. The velocity of an object is obtained by dividing the total distance that the object travels by the time that it takes for the travel. Consider an object that travels in the x-direction of a coordinate system. Then, if the object starts at the point x_1 at time t_1 and arrives at point x_2 at time t_2, we calculate the velocity v from

$$v = \frac{x_2 - x_1}{t_2 - t_1}. \qquad (2.11)$$

Fig. 2.5 A physical object with total energy content E moves with velocity **v**. The velocity is a vector and therefore denoted by the bold-faced **v**. The "momentum" is also a vector and is by definition proportional to the velocity. The constant of proportionality is described in the text

It is customary to denote the differences $x_2 - x_1$ by Δx and $t_2 - t_1$ by Δt, respectively. Thus we have

$$v = \frac{\Delta x}{\Delta t}. \tag{2.12}$$

For the case of very small differences Δ (see Sect. 5.2.1 for a detailed explanation), we write

$$v = \frac{dx}{dt}. \tag{2.13}$$

Generally the velocity is a vector **v** in two or three dimensions, i.e., in a plane or in space, respectively.

Even if we consider only the x-direction, the velocity is not a simple number. The velocity of a car is, for example, given in kilometers (or miles) per hour. Other units can also be used. If we wish to have Δx in meters and Δt in seconds, then we obtain the velocity in meters per second. Units like this are important for science and engineering problems. The use of such units usually presents some problems to students, because there is a new concept involved here. When we talked about adding and subtracting numbers, we stated that we can substitute all kinds of things for the numbers, for example, apples and oranges. We only needed to make sure that we add and subtract only the same kind, i.e., either apples or oranges. Clearly it makes no sense to subtract five apples from ten oranges. Physical quantities such as the velocity are composed of both space and time measurements, and the velocity is obtained by dividing distances by times. Therefore, the unit for the velocity is composed of two different separate units, for example, of the units kilometer and hour. We can still add and subtract such quantities. However, we need to make sure that all these quantities are given in the same units. It does not make any sense to add or subtract kilometers per hour from meters per second.

We return now to momentum and define the momentum as a vector **p** that is proportional to the vector of the velocity **v** and also proportional to the total energy content E of the object. The object is shown in Fig. 2.5 as a "blob" with arbitrary shape symbolizing that it may consist of a single particle, such as an electron or of a complicated collection of many bodies that interact with each other by the forces of nature such as a molecule. It may also be a big object like a billiard ball or even a planet.

Thus we have the definition

$$\mathbf{p} := KE\mathbf{v}. \tag{2.14}$$

Proportional means equal except for a constant that we have named K in this case. The symbol $:=$ means that this equality is valid by definition and does not represent by itself any law or rule. This particular definition of momentum encompasses the work and results of many famous physicist over many centuries. It is therefore not obvious why we have defined "momentum" that way and what benefits this definition gives us. It was Einstein who realized the full meaning of the concept of momentum. Galileo and Newton had already realized the following. If that "blob" of energy that we have shown in Fig. 2.5 moves somewhere out in space and no force acts on it, then it will move on forever with the same velocity. Thus, the momentum $\mathbf{p}$ is a quantity that is conserved and does not change if we do not apply any forces. Einstein also found the constant K from his theory of relativity that is explained in Sect. 5.1. He found that

$$K = \frac{1}{c^2}, \tag{2.15}$$

where $c = 300{,}000 \, \text{km/s}$ is the velocity of light. As we have explained also in Sect. 5.1, Einstein derived the very famous relation

$$M = \frac{E}{c^2}, \tag{2.16}$$

where M is the mass of the blob that we could measure if we would try to accelerate or weigh it. Therefore we have

$$\mathbf{p} := M\mathbf{v}. \tag{2.17}$$

This equation was the definition of Newton. Galileo and Newton were the first to realize the importance of the following law of physics: if an object moves and there are no forces acting on it, then the object will move on with the same momentum forever. There is a great deal of abstraction here, because objects usually are influenced by forces. Gravity acts on all objects on earth and most objects are subject to a force of friction that tends to stop the motion by creating heat energy (see Sect. 2.3). Even far out in space there is usually some gravitational force, for example, that of the sun, acting on objects. So Newton's abstraction was a big step and not obvious at all. But why is this law so important? This is discussed in the following by using the examples of the movement of billiard balls, rocket engines, jet engines, and other mechanical phenomena.

Billiard balls are made out of a fine material that has virtually no friction, neither with the billiard table nor with other billiard balls. Gravity acts on them. However, because the billiard balls are on a very horizontal table, the gravitational forces are canceled out by the table that pushes against the balls. Therefore on a billiard table we can observe Newton's famous law by moving a billiard ball: it moves on with the same speed as long as no forces act on it, meaning as long as it does not hit something. What if a billiard ball hits centrally another one that stands still?

You might think that somehow both will then move on. This is not so! The ball that hits stops entirely and the other one moves on with the same speed that the first ball had. The reason is that both momentum and energy of both billiard balls need to be conserved, and this is only possible that way.

If we push the billiard ball over a distance, that means we exert a force on that ball over that distance. As we know, force times distance equals the energy that we supply. All that the billiard ball does, as a consequence of this energy supply, is that it moves with a certain velocity. One therefore says that one has changed the energy supplied by the push into moving energy or "kinetic energy." We describe now a few properties of this kinetic energy and then give the equation from which one can calculate it easily. How can one measure kinetic energy? There are, of course, many elegant ways to do so. We mention here a very inelegant but very important way. Consider a car that moves with a certain velocity v. If the car smashes into a wall, then the kinetic energy goes to zero because the car is stopped. Energy cannot be destroyed, so where is it? If you look at the remnants of the car then you see that steel has been distorted, windows have been smashed, and other damage has occurred. Clearly you need energy (forces over a distance) for all of this. Also, and this is very important, the parts of the car that have been severely distorted have heated up. Thus a lot of energy has been transformed into heat (see Sect. 2.3). If the car would not have driven into a wall, but stopped by use of the brakes, then all the energy would have been transformed into heat at the brakes. The brakes are therefore heating up when used. If you use the brakes of a car too frequently, for example, when going downhill, then the brakes will get very hot and may start burning. The same problem happens, if you forget that the brakes are on while driving. Thus one can measure the kinetic energy by just measuring the heat that is produced when braking or the damage plus heat that is generated when smashing the car. Modern cars, so-called hybrids, can turn the braking energy into electrical energy that can be used again. The kinetic energy is thus turned into electrical energy that also can be measured. This would be a more elegant way of measuring kinetic energy.

We turn now to the details of the rules for the kinetic energy. We know that there is more damage to the smashed car that drives into a wall with higher velocity v. Careful measurements of the damage and heat generation for a given velocity show the following. If we double the velocity, the damage and heat is not just twice as large but four times as large. If we triple the velocity the damage and heat is nine times as large. You can guess that for four times the velocity we have sixteen times the damage and heat. Thus the kinetic energy that we denote by E_{kin} increases with the square of the velocity. One also finds that the kinetic energy increases with the mass of the car. A truck can cause much more damage than a small car. Here one finds that if one doubles the mass the damage doubles, if one triples it, the damage triples, and so forth. Thus the kinetic energy is proportional to the mass. As we will see later, for the ordinary velocities that we deal with on earth, the factor of proportionality is just $\frac{1}{2}$ and we have

$$E_{\text{kin}} = \frac{1}{2} M v^2. \tag{2.18}$$

Consider now the motion of two billiard balls when they collide. We start with one ball moving with velocity v_1 that hits a second ball centrally. The second ball is standing still and has therefore velocity $v_2 = 0$. After the collision we denote the velocities by v_1' and v_2', respectively. Both billiard balls have the same mass M. Because momentum is conserved, meaning that the total momentum before and after the collision is the same, we have

$$Mv_1 = Mv_1' + Mv_2'. \tag{2.19}$$

Canceling M on both sides gives

$$v_1 = v_1' + v_2', \tag{2.20}$$

and squaring left and right side of this equation we obtain

$$v_1^2 = v_1'^2 + 2v_1'v_2' + v_2'^2. \tag{2.21}$$

From Eq. (2.18), the law of conservation of energy, we get, after again canceling out the equal mass

$$v_1^2 = v_1'^2 + v_2'^2. \tag{2.22}$$

Subtracting now Eq. (2.22) from Eq. (2.21) we have

$$0 = 2v_1'v_2' \tag{2.23}$$

which means that either v_1' or v_2' or both need to be 0. Because both cannot be zero (that would mean the energy has vanished), the only solution that makes sense is that $v_1' = 0$. Then we obtain from Eq. (2.20) the result $v_2' = v_1$. This means that the first ball stands still and the second moves away with exactly the same velocity that the first had. This is a well-known result and is often shown in experiments. You can try it also yourself on a billiard table. The experiment becomes even more astonishing if one does the experiment with three balls as shown in Fig. 2.6. There is a common toy found in select gift shops. Several stainless steel spheres are hanging on a string next to each other. If one takes the first one and lets it swing so that it hits the others, only the last one flies off. When this last one returns and hits, the first one swings back and so forth.

2.1.3 Acceleration: The Mechanics of Falling Stones and Planets

The title of this section may strike the reader as strange. "Falling stones and planets" sounds like planets are falling exactly as stones do. This is no mistake! The laws of falling stones and planets orbiting the sun are really very similar, almost identical, and it was Newton who had the great idea that the orbiting of the planets can be

power of the orbiting radius), that the gravitational force of the earth (or sun) must be weaker the farther we go away from the earth (or sun), and that it decreases with the square of the distance. This is the major finding that we must remember to understand Newton's calculation of the moons orbit. Thus, he knew that he could not use the value $g = 9.81 \frac{m}{s^2}$ for the falling acceleration of the moon, but he had to reduce g with the square of the distance from the earth.

Newton actually deduced from the astronomical data the full law that tells us the magnitude of the gravitational attraction between two objects, and we state this law here for completeness. The gravitational force F by which two objects such as the earth with mass M_{ea} and the moon with mass M_{mo} attract each other is

$$F = k_N \frac{M_{ea} M_{mo}}{d^2}, \qquad (2.37)$$

where k_N is called the gravitational constant. The gravitational constant can be expressed in units of $m^3 \, kg^{-1} \, s^{-2}$. In words, this reads meters to the third per kilogram and per square seconds. The value in these units is $k_N = 6.673 \cdot 10^{-11} \, m^3 \, kg^{-1} \, s^{-2}$. Note the decrease of this force with the square of the distance d. We need now a connection between force and acceleration to obtain the precise relation between acceleration and distance of the objects in question. Newton found this law that is also needed to calculate the Moon's orbit. It is Newton's most famous physics law and gives the connection between any force F and the acceleration a of an object with mass M that is subjected to this force:

$$F = Ma. \qquad (2.38)$$

Knowing that the acceleration of the moon by the gravity of the earth decreases with the square of the distance of the moon from the earth, Newton calculated the orbit of the moon as follows. The distance to the moon is 60 times the radius of the earth, and Newton knew this from Kepler's third law. Therefore we can calculate the acceleration a with which the moon "falls" by dividing g by $60^2 = 3{,}600$. We know this from Eqs. (2.37) and (2.38). Then we obtain from Eq. (2.34) the velocity of the moon by replacing R' by $60R$ where $R = 6.37 \cdot 10^6$ m. This gives for the velocity of the moon $v = 1{,}020 \frac{m}{s}$. The length of the moon orbit is $2\pi R' = 2\pi 60 R = 2.40 \cdot 10^9$ m. The time the moon needs then to go full circle is obtained by dividing this orbital length by the moons velocity, which gives $2.35 \cdot 10^6$ s. This is about 27.25 days, which is very close to the moons orbiting time.

The precise calculation for the moon is more difficult and was a problem even for Leonhard Euler, one of the greatest mathematicians of all time. The reason is simply that not only the earth attracts the moon but also the sun. Therefore one deals with three objects, sun, moon, and earth. This is called a three-body problem. If one wishes to be more precise one even has to include the other planets and ends up with a many-body problem. These problems have no exact solution. However, one can solve them by solving Newton's equations by using a computer. This can be done basically to any desired accuracy and has been done with greatest precision to send astronauts to the moon. Of course, the moon mission that resulted in the first moon landing on July 20, 1969, required more than this calculation. It did require the technology of the rocket engines that propelled the giant Apollo rocket and the moon lander, and it required the technology to sustain the astronauts in space and to return them safely to earth. When Neil Armstrong made the first step on the moon and looked up to see the earth he said these famous words: "That's one small step for a man, one giant leap for mankind." Indeed it was. Here we can see the symbiosis of science, engineering, and technology in the best light, and we can see the enormous developments that are necessary to bring the idea for the laws of motion in space to the first step of an astronaut on the moon.

Of course, the orbits of the planets around the sun can be understood from Newton's laws as well. They are more complicated to find than the circular satellite orbits that we calculated above, because the shape of these orbits is ellipses and only almost circles. However, all the calculations for these planetary ellipses follow the same principles that we have just discussed. Details of all of these orbit calculations for planets can be found on the Internet and provide nice student projects, particularly for those who have mastered Sects. 5.2.1 and 5.2.2.

In summary, we have learned in this section, that we can calculate the orbit of a satellite just by applying Galileo's laws for falling stones. The calculation of the orbit of the moon requires also the knowledge of how gravity diminishes with distance. Newton provided these laws that give us the understanding of massive objects orbiting around each other.

2.1.4 Spinning Objects: From Gyros to Electrons

Spinning discs with an axle, also called gyros, never have ceased to amaze me. They fascinated me as a child, as a student, and they fascinate me today, while I write this book. The really astonishing effects can be seen only when the disc is rather heavy and spins very very fast. Then, the fast rotating disc appears to be a totally different object as compared to a nonrotating disc. For example, if the rotating disc has an axle with a thin tip and if we put the tip down on a smooth surface (e.g., a table), the gyro stays upright for a long time, while a nonrotating disc immediately tips over. This is illustrated in Fig. 2.8.

Even more astonishing is the fact that a disc that is specially mounted as shown in Fig. 2.9 does not drop downwards, but circles slowly keeping the vertical distance from the ground almost constant. If the disc does not spin, it drops down immediately.

This looks almost magical, and you have to see real experiments to believe it. Such gyros are, of course, available commercially and hopefully ready for experimentation in every science classroom. Newton concluded from such experiments that the spinning disc behaves different because it spins in space, and this space is at rest compared to the disc. Mach did not believe that and suggested that it was the spinning compared to the other masses in the universe that made the difference.

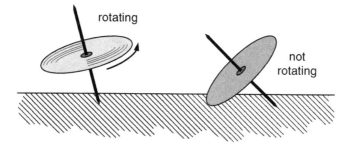

Fig. 2.8 A rotating disc, with an axle that touches a surface with its tip, will not tilt over toward the surface, at least not for a while. A disc that does not rotate will quickly drop down and touch the surface

Fig. 2.9 Spinning disc mounted on an axis that can freely rotate both vertically and horizontally. If the disc spins very fast, it will not fall down but rotate around the central stand

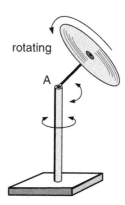

Einstein was at first impressed by Mach's view, but later changed his mind. I think it is safe to say that we really do not know the root cause why spinning discs are behaving so differently to the resting ones. As far as I understand it, this must be linked to what space and vacuum really are, and you can find more about it in Sect. 5.4.

It is a fact, however, that the behavior of spinning objects derives directly from the inertia of moving bodies: a body with mass M that is not subjected to any forces will move on forever with the same velocity because momentum is conserved. This connection of momentum conservation and behavior of spinning discs can be seen from the following two famous experiments.

The first experiment (experiment 1) is performed with a gyroscope. A gyroscope is a spinning disc whose axle is mounted such that it is free to take any orientation. The way this mounting is done is shown in Fig. 2.10. If the disc spins very fast, it does not matter how the outer frame is rotated or moved. The axis of the spinning disc will always point in the same direction. This is why the gyroscope can be used as a compass and tell you the direction you are going, and it is actually often much more reliable than a magnetic compass needle. The direction of the Hubble telescope that is orbiting in space, for example, is controlled by a gyroscope. In

Fig. 2.10 A gyroscope consists in essence of a spinning disc with an axis that is mounted in such a way that it can freely turn in any direction

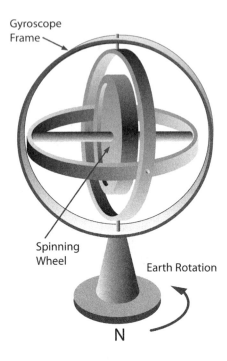

Gyroscope
Frame

Spinning
Wheel

Earth Rotation

N

experiment 1 we put the gyroscope of Fig. 2.10 directly at the north pole, with its stand pointing in the direction of the axis of the earth. One of us has to go to the north pole also and watch the rotating disc of the gyroscope. What that person will see is the axle of the disc performing a horizontal rotation. If the axle points in a certain direction at the start of the experiment, it will do so again after 24 h. If the disc does not spin then the axle does not rotate. What is happening here? Of course the earth has rotated around itself in 24 h and the observing person has rotated with the earth. The spinning disc, however, kept a constant direction and was not influenced by the rotation of the earth. The non-spinning disc, on the other hand, turned just as the observer turned with the earth and therefore appeared to the observer as standing still. What has that to do with inertia and momentum conservation? This is illustrated by the second experiment (experiment 2) that is also performed at the north pole but with a pendulum instead of the gyroscope.

A pendulum is consisting of a heavy object hanging on a rope. For our case it is important to use a long rope (at least a few meters long) and a very heavy object (at least a few kg or pounds). The rope compensates for the action of gravity and holds the heavy object against the gravitational force. We can therefore compare the motion of this object in some respect to the motion of an object free of forces. Such an object will follow the law of momentum conservation and continue to move without changing directions. For the pendulum this is not quite true, because as it swings away from its lowest position, the heavy object goes against gravity and loses kinetic energy until it stands still and then reverses and moves in the

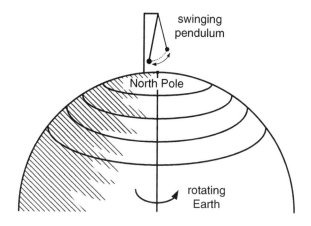

Fig. 2.11 A pendulum swinging at the north pole will swing in a plane. For an observer at the north pole, this plane appears to rotate a full turn in 24 h. It is, of course, the observer and the earth that rotate, while the pendulum swings in the same plane

opposite direction. However, the actual circle along which the heavy object moves stays in the same plane because gravity accelerates only toward the center of the earth. Therefore if we have a pendulum at the north pole, it will swing in one plane and that plane will seem to rotate a full turn in 24 h exactly as the gyroscope. This experiment is illustrated in Fig. 2.11. Of course, it is again the earth that rotates while the pendulum stays in one plane. This similarity between pendulum and gyroscope suggests that the "strange" effects that gyroscopes show are in essence due to the inertia of massive bodies and the law of momentum conservation. To do these two experiments, you actually *need not* go to the north pole. You can see the rotation of the earth clearly in your own home as long as you are not living too close to the equator. The problem is only to have a pendulum with long enough rope and heavy objects so that the pendulum oscillates long enough to see the earth rotation, that is at least for several hours.

The special properties of spinning objects are very important in many subareas of physics. Even electrons, protons, and atoms do have a so-called "spin" and the properties of this spin are reminiscent of those spinning objects. There are, however, important differences between spinning electrons and classical gyroscopes that are discussed in quantum mechanics courses at the university, which provide a complete theory for the electron, proton, and atom spin. This is an advanced topic that was pioneered by P.A.M. Dirac and is discussed in Sect. 5.3.

2.1.5 General Properties of Waves: From Sound to Tsunamis

We know waves from watching water in the wind. The water moves up and down at any given spot showing valleys or minima and mountains or maxima. At the same time the minima and maxima move into some direction with a certain speed also called the velocity of the wave. At any beach you can see that the velocity of the waves is about a few meters per second. This is the velocity with which a surfer rides with the wave. Ordinary water waves are thus rather slow. Electromagnetic waves have the highest possible velocity, the speed of light, as we will learn in Sect. 5. The speed of waves in water, in solids, or in air depends on how the liquid, solid, or gaseous material that carries the wave actually moves. This movement can occur in different ways even if we have the same material. For example, for ordinary surface water waves, the particles of water move up and down, and the motion is rather slow, as mentioned, a few meters per second.

There are also other possibilities of creating waves in water. A strong earthquake can lead to an enormous compression of water because of a rapid movement of the bottom of the sea. For example, in the earthquake of Japan in March 2011, a large area of the bottom of the sea moved upwards within a very short time. The water cannot react immediately to this rapid movement and is therefore compressed by enormous forces. We know that compression and subsequent dilatation of air gives rise to the ordinary sound waves. These travel with a speed of 343 m/s (referred to as Mach 1). Pressure-dilatation waves also propagate in water. The water-pressure wave after an earthquake moves mostly below the water surface. It is much faster than the ordinary water waves at the surface and can be several hundreds of kilometers per hour. The actual speed, which matters in catastrophes called Tsunamis that often follow earthquakes at the bottom of the sea, can be determined very precisely by computer simulations. When the fast-moving pressure-dilatation waves approach the shorelines, they convert their enormous energy into surface water waves of ordinary speed (around 20 km/h). These ordinary waves may then be of vicious duration and power, depending on the magnitude of the earthquake. The destructive Tsunami of the 2011 Japan quake destroyed Japanese shorelines and cities and still caused powerful destruction, when the pressure wave arrived about 8 h later, thousands of miles away in Hawaii. In addition, and this is an important part of the destructive action of Tsunamis, they have very long wavelength. Therefore the first sign of a Tsunami is often that the water leaves the coastlines as if there would be very low tide. Then the water returns, and because of the long wavelength returns, and returns, and returns.

Mathematical Description of Waves

There are three important values that are characteristic for waves: the velocity v, the frequency ν, and the wavelength λ of a wave. We have discussed already some aspects of the velocity. The frequency is defined as the inverse time period that a wave needs to complete one full cycle. Consider, for example, a water wave that has a maximum height at a certain location and time. Then the water moves downwards and upwards again until it reaches again the maximum. If this process takes 5 s then the frequency is given by $\nu = \frac{1}{5\,s} = \frac{1}{5}\,s^{-1}$. The frequency of waves can vary in wide limits and may be extremely high for some electromagnetic waves, as we will see below. The wavelength is the distance from one wave maximum to the next maximum, or from one wave minimum to the next, and is usually denoted by the Greek letter λ.

Mathematically a wave is typically represented by the sine (or cosine) function. The box shows a MATHEMATICA plot of the function $\sin(x)$ for $0 \leq x \leq 2\pi$. The resulting graph has already been shown in Fig. 1.15.

> MATHEMATICA
> $Plot[Sin[x], \{x, 0, 2 Pi\}]$ shift-enter
> as we have done in the chapter on geometry.

If we wish to plot the wave in such a way that we have exactly one full wavelength over a given unit distance, then we need to plot the function $\sin(\frac{2\pi x}{\lambda})$, because then we have a wave rising from 0 at $x = 0$, encompassing a maximum at $x = \frac{1}{4}$ followed by a minimum at $x = \frac{3}{4}$ and returning back to 0 again at $x = 1$. Thus $\sin(\frac{2\pi x}{\lambda})$ represents the mathematical formula for a wave as a function of distance. However, the wavy motion is also a function of time. At any given point x the wave moves up and down with the frequency v. Therefore, in order to capture this time dependence, we need to insert the term $2\pi v t$ into the sine function to arrive at

$$\text{Wave} = A \sin\left(\frac{2\pi x}{\lambda} + 2\pi v t\right). \qquad (2.39)$$

Here we have added a so-called amplitude factor A. This is because the sine function is at most equal to 1. If we wish the water wave to be 5 m high at the crest and 5 m deep in the valley, as it sometimes occurs in Hawaii and at other places, then we need to use $A = 5$ m to describe that wave. The amplitude A may also slowly vary with time and distance. For example, the amplitude of the Tsunami pressure wave may become smaller after a very long travel, simply because the energy of the wave is distributed over a larger area. We assume for most of our discussions that A is constant.

To understand how well Eq. (2.39) describes a wave, it is useful to perform calculations for fixed times and variable space coordinate x and conversely for fixed space coordinate and variable time coordinate t. This can be done nicely with any software that plots mathematical functions or with a little more effort even with a pocket calculator. Because one does not wish to repeatedly write the 2π factors when dealing with this type of equation, one defines the so-called angular frequency $\omega = 2\pi v$ and wave number $k = \frac{2\pi}{\lambda}$ to obtain

$$\text{Wave} = A \sin(kx + \omega t). \qquad (2.40)$$

An important relation that is valid for all waves follows from the fact that the velocity v of the wave must be equal to the product of the wavelength and the frequency, simply because the wave makes v full oscillations per unit time and each oscillation corresponds exactly to one wave length λ. Thus we have

$$v = \lambda v. \qquad (2.41)$$

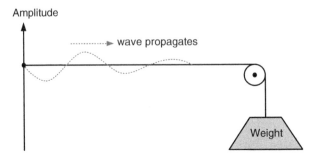

Fig. 2.12 A string that is fixed on one side and pulled by a weight on the other side can perform wavelike motions when excited. This happens, for example, to the string of a guitar. The wave that propagates from the point of excitation on the left to the other side of the string is described mathematically by Eq. (2.39). The amplitude of the wave may stay constant. Often, however, the amplitude decreases with traveling distance

This relation is also valid for light waves. Because we denote the velocity of light by c we have

$$c = \lambda \nu. \tag{2.42}$$

Strings and Standing Waves

Equation (2.39) describes a one-dimensional wave because it considers only the x-direction. One can experimentally create such a wave by using a long string that is held fixed at one end and pulled by a weight on the other end. Then one can hit the string at some point, exactly as one plays the string of a guitar, and a wave will start propagating away from that point. This effect is shown in Fig. 2.12.

Of course, this is exactly what happens when a guitar is played. However, guitar strings are very short, and the wave is immediately reflected from at the ends of each string. A reflected wave propagates in the negative x-direction if the original wave has propagated in the positive x-direction. The equation for the reflected wave is therefore

$$\text{Reflected wave} = -A \, \sin\left(-\frac{2\pi x}{\lambda} + 2\pi \nu t\right), \tag{2.43}$$

which is identical to the original wave except for the negative sign of the term that contains the space coordinate x. We also have changed the sign in front of the equation. This sign change comes from the fact that the reflection at the end of the string turns the amplitude of the wave around. The result that one obtains if one adds both waves together is the vibration pattern that the string of a guitar shows. Adding wave and reflected wave gives

$$\text{Standing wave} = \text{Wave} + \text{Reflected wave} = 2A \, \sin\left(\frac{2\pi x}{\lambda}\right) \cos(2\pi \nu t). \tag{2.44}$$

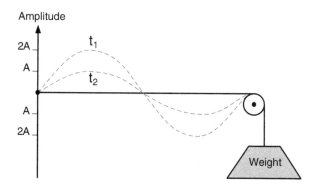

Fig. 2.13 A "standing" wave plotted for two different times t_1 and t_2 that are chosen such that $\cos(2\pi \nu t_1) = 1$ and $\cos(2\pi \nu t_2) = \frac{1}{2}$, respectively

This result can be obtained by using the so-called theorem of subtraction of the sine function, which tells us that $\sin(\alpha) - \sin(\beta) = 2\sin(\frac{\alpha-\beta}{2})\cos(\frac{\alpha+\beta}{2})$. The resulting wave is called a standing wave. The reason for this name is that the maxima (and also minima) of this wave do not move anymore. They stay always at the same place, meaning that the velocity of this resulting wave is 0. One can see this easily by plotting Eq. (2.44) for different times as done in Fig. 2.13.

The figure is plotted for the case that the wavelength of the exited wave equals the length of the string. It is also possible that the string is just half a wavelength long and the standing wave has just one maximum or minimum exactly in the middle of the string. This is actually the usual form of excitation when a guitar is excited by plucking it right in the middle of the string.

The frequency of the vibrating string is determined by the string tension (the weight stretching the string), the mass of the string, and its length L. Thus the musical tone that a string excites in air through its vibration also depends on these factors, because it is determined by the frequency of vibrations. The possible wavelength of a vibrating string is, of course, also determined by the length of the string, and we typically have $L = \frac{\lambda}{2}, \lambda, \frac{3\lambda}{2}, 2\lambda, \ldots$ It is known since Pythagoras that the tone of a string with length L is higher than that of a string with length $2L$; musically speaking it is an octave above. There are many interesting projects with vibrating strings, and the Internet is full of information about it. More complicated objects than strings, such as metal plates, drums, the body of a violin or a piano, vibrate in standing waves also. These standing waves are not just waves in one dimension but, in general, in three dimensions, and they have very complicated shapes. The complicated shapes are, in turn, the reason why such a world of sounds can be created by instruments like the guitar, the violin, a piano, or an organ made of pipes.

Standing waves play also a major role in modern physics and particularly in the area of quantum mechanics. One of the theories that is a contender to explain all phenomena that we know is the so-called string theory. Its name derives from the fact that elementary particles are described by using concepts of vibrating strings as we just discussed them. The interested reader is encouraged to surf the Internet and learn more. We also have added material in Sects. 5.3 and 5.4.

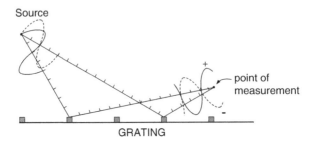

Fig. 2.14 A wave emitted from a source is interacting with a grating. A point of measurement of the wave intensity is also indicated. The *waves* are shown close to the source and also close to the point of measurement by *dashed lines*. The *full lines* with distance markers show the wave direction. The height of the two waves that are reflected from the grating is equal and opposite at the point of measurement and therefore the waves cancel each other resulting in zero total wave intensity at this point. If the wave would be a light wave, the point of measurement would be *dark*. This effect is called destructive interference

Diffraction of Waves

All the phenomena described above are important for waves. The one single phenomenon that is usually considered the hallmark of waves is diffraction. The word diffraction, or diffraction of waves, simply means that the wave is influenced in a specific and important way by some structure of a material that interacts with the wave. This structure could consist of wooden posts that influence water waves, thin wires that influence electromagnetic waves, or it could be just the atom layers of a diamond that influence light and make it glitter in beautiful colors.

A very important diffraction structure is the so-called diffraction grating. A diffraction grating is simply a number of lines engraved into some material. This can be straight lines such as metal lines painted on top of glass, or lines created as scratching the glass, or scratches in a thin metal layer on plastic, and so on. The lines may also be drawn perpendicular to each other, thus forming a two-dimensional crosswise pattern or even a three-dimensional pattern, such as a crystal lattice. The crystal lattice may either be artificially made, as described in our section on nanostructures (Sect. 3.4), or be just a natural crystal lattice, such as diamond. In other words a diffraction grating is just a geometrical structure with regular arrangements of lines of different atoms or even of missing atoms typically in two or three dimensions. Photons that hit the reflection grating interact then with a (large) number of atoms of these lines and, as a consequence of this interaction, are redirected and reflected. This redirection and reflection does not have any arbitrary direction. It occurs just in certain directions that depend on the spacing of the lines and on the wavelength of the waves that are incident on the grating. The reason for the different redirection or reflection into the various directions is that the waves that are redirected from many points of the grating may amplify each other in certain directions, while they may weaken or destroy each other in other directions. This latter effect is shown in Fig. 2.14.

In this figure we have included a source of waves and also two paths the wave can take toward a point of measurement. These waves could be, for example, microwaves, i.e., electromagnetic waves as they are used in microwave ovens. One path of the waves is 23 cm long while the other is only 21 cm long. The wavelength of the wave shown is chosen to be 4 cm. We can see from Fig. 2.14 that the two waves would end up with their electric fields pointing in opposite directions at the point of measurement. Therefore the two fields would cancel each other at that point, and no wave would be measured at this point. One calls this destructive interference of the waves. One actually does not need microwaves to see such an effect. One can also perform such an experiment with water waves. You can create waves with two fingers while sitting in the bath tub, and you will observe spots on the water surface that do not show any wave motion, because the reflections from the various walls of the tub cancel each other at these spots.

The calculation of interference effects proceeds as follows. We use Eq. (2.39) to describe the wave. Now we have two waves:

$$\text{Wave}1 = A \, \sin\left(\frac{2\pi \, 23}{4} + 2\pi \nu t\right) \tag{2.45}$$

and

$$\text{Wave}2 = A \, \sin\left(\frac{2\pi \, 21}{4} + 2\pi \nu t\right), \tag{2.46}$$

where A is just giving the highest amplitude of the wave that is obtained when the sine function equals one. We have in this equation also time t and frequency ν of the wave. However, the term containing time is for both waves the same and therefore not important for our discussion. We therefore put at first $t = 0$ and forget about this term. We can then calculate the sum of the two waves at the point of measurement and find

$$\text{Wave}1 + \text{Wave}2 = A \, \sin(11.5\pi) + A \, \sin(10.5\pi) = 0. \tag{2.47}$$

This result can be found by use of a pocket calculator or by MATHEMATICA as shown in the box.

```
MATHEMATICA
Sin[11.5 Pi] + Sin[10.5 Pi] shift return
Output = 0
```

You can also add the terms with any given time in the sine functions, and the result will still be 0. Actually calculators and also MATHEMATICA may give you some very small number very close to (but not quite) zero. This is because computers can often only calculate approximate numbers. At any rate, the two waves annihilate each other at the point of measurement.

Had we used a wave with different wavelength, say 0.5 cm instead of 1 cm, we would have obtained a different result. Then we would have had

$$\text{Wave1} + \text{Wave2} = A \sin(23\pi + 2\pi\nu t) + A \sin(21\pi + 2\pi\nu t) = 2A \sin(\pi + 2\pi\nu t).$$

$$(2.48)$$

Now our result does depend on time as you can easily check by use of MATHE-MATICA or a pocket calculator that features the sine function. The maximum result that you can obtain equals $2A$ and the minimum equals $-2A$. In other words, the amplitude of the wave has doubled because the two waves have added and helped each other. One calls this case constructive interference.

Note that the wavelength of wireless phones happens to be also around 1 cm. That means that interference effects with close-by walls may influence your cell phone receiving strength. The wavelength of visible light is much smaller around 600 nm. However, you can still perform the same experiment with light. Of course, then the spacing of the lines of the grating needs also be around 600 nm. The point of observation that we have chosen in the above example would give no light for the longer wavelength and brighter light for the shorter. Sunlight is actually a mixture of all light colors. Red light corresponds to the longer wavelength and blue light to the shorter. The point of measurement would therefore appear blue, and points of observation more to the right in Fig. 2.14 would be red. How can we make such an experiment? There is indeed such a fine grating in every household, because DVDs that are used to play movies contain the information for the pictures in the form of very closely spaced dots that are positioned on very closely spaced circles. Indeed if you hold a DVD toward sunlight you see all colors of the rainbow that are created because light with different color is reflected by the grating in different directions. If you have a laser pointer, then you can do an additional experiment: shine the laser pointer in a grazing angle onto the DVD and watch the reflections on a wall. If you had a mirror instead of the DVD, you would see exactly one point as the reflection of the laser pointer. However, now with the DVD, you can see several points often up to five or more and all in the same color because the photons of the laser have only this one color. You can calculate that there exist several points at which a grating causes the light to constructively interfere; this is done by just calculating the sum of two waves with one given wavelength λ for a sequence of observation points. You will see then constructive and destructive interference effects and therefore minima and maxima of light as you proceed more to the right in Fig. 2.14.

The Doppler Effect

We end this discussion of waves with a few remarks on another important effect that was discovered and investigated by the Austrian physicist Christian Doppler. If a person moves with the wave, for example, exactly as fast as one of the highest points of the wave, then to this person, the wave will appear to stand still. If the person moves in the same direction as the wave does, but a little slower, then the frequency of the wave that person measures appears to be lower. The reason is simple: if you walk in the direction of the wave and count the highest or lowest points of the wave that pass you during a given time period, then you will count a smaller number as compared to the case when you stand still. The frequency of the wave that you measure is thus decreased because it is given by the number of highest or lowest points that you counted divided by the period of time during which you counted. If you walk fast enough in the direction that the wave propagates, you can even make it happen that no highest or lowest point of the wave passes you. Then, from your view, there does not seem to exist an up and down motion, and the frequency of the wave appears to be zero. Inversely, if you walk against the wave, then you count a larger number of wave maxima or minima while walking than when standing still. Thus the frequency of maxima or minima that you count is increased. That frequency difference that occurs in situations where we encounter waves and moving objects is called the Doppler effect.

The Doppler effect can be observed and used in a large variety of situations. For example, one can send electromagnetic waves toward a moving object and can conclude from the Doppler effect, in this case the change in the frequency of the reflected waves, how fast the object is moving. Such a measurement system, that sends out electromagnetic waves and measures reflected waves (including their frequency), is called radar. The frequency range of radar is typically in the GHz range. Doppler radar is used to determine the speed of approaching storms and the speed of winds in a hurricane. Doppler radar is also used by police to determine the speed of cars.

2.2 Electromagnetic Phenomena

The ancients knew already that one could rub a little block of amber, looking like a yellow stone, against a fur, against a towel, or against one's coat and then draw electrical sparks out of it. That little piece of material also tends to attract one's hair or even raise one's hair in all directions when it is dry. Also known to the ancients was that small metallic looking pieces of certain materials would attract each other. We now call these pieces magnets, and I remember how fascinated I was as a boy when I saw magnets for the first time. The excitement was even greater when I built my first battery-driven electromagnets. I put paper between the magnets and put my finger between them, and they still attracted each other while my fingers felt nothing. This was like magic to me. Astounding electrical phenomena were actually always known to mankind. The lightening of thunderstorms was appreciated as very special and, for example, attributed to the Greek god Zeus.

We know now that electricity and magnets are not magic, and we understand electricity and use it in our daily life. In fact the basic electrical particle that underlies the cause of all these phenomena, the electron, is all around us. It derives its name from the Greek word for amber which is "electron." Electrons are not only part of amber but of all materials that we know. Everything we touch, everything we see, everything we smell, is somehow connected to electrons. How did that fact stay hidden from us for such a long time? How did mankind tame electricity for its purposes? These and other questions are the topics of this section.

Electricity was more difficult to figure out than gravity, the force that rules the motion of the planets, and has been explored in scientific detail by James Clerk Maxwell and Michael Faraday 200 years after Newton (also in England). A major reason for the difficulties to understand electricity is the magnitude of the force related to it. This force is actually much much larger than the gravitational forces between two massive bodies such as earth and sun that is described by Newton's equation (Eq. (2.37)). In contrast to the forces of gravity, however, there exist two opposite types of forces between electrical particles that are the sources of these forces. Depending on which kind of electrical particles one deals with, one obtains attractive or repulsive forces. One distinguishes electrical particles that are "positively charged," meaning that they are the sources of one type of electrical force, and particles that are "negatively charged" and are the sources of the opposite type of force. Opposite means here that the forces point in opposite directions. Positive and negative charges attract each other, while positive charges repel other positive charges and negative charges repel other negative charges. It now so happens that the most basic atom, the hydrogen atom, is composed of one particle with negative charge, that is the electron that we mentioned above, and one particle with positive charge, that is called the proton. This means that the sum of charges that are contained in the hydrogen atom is zero, and this fact is also true for all other atoms. Atoms are extremely small and cannot easily be decomposed into their parts, because the positive and negative charges attract each other and stick closely together. This is the reason why electrical charges and their presence in all atoms were not recognized earlier. In fact, Maxwell himself still thought that no electrical charges would be present in metals! This is a very incorrect notion. Metal wires carry the electrical currents (flowing charge) that supply whole cities with power.

Why is the hydrogen atom of such basic importance for the understanding of electricity? Hydrogen atoms are the most abundant type of atoms in the universe. There exists a silly joke that only stupidity is more abundant in the universe than hydrogen. Our sun and billions and billions of similar stars in each of the trillion of galaxies are mostly composed of hydrogen. We will learn in Chap. 5, that deals with Einstein's theory of relativity, that a very special mechanism involving hydrogen makes the sun and the stars shine. Most of the other atoms that we know have probably been generated from hydrogen during the aging of the stars and during explosions of stars (look up "supernova" on the Internet). Because hydrogen is composed exactly of one positive and one negative electric charge, we can therefore deduce that the number of positive and negative charges in the universe is about the same. Charge is, in addition, conserved in all physical processes, meaning that the number of positive and negative charges stays the same. This is the law of charge conservation, which is held almost at the same level of importance as energy conservation is.

Positive and negative charges attract each other with a very significant force. This is why the electron and the proton of hydrogen are very close to each other, and one needs a considerable energy to separate them over larger distances. This is also the reason why atoms were thought to be indivisible for a long time. Typically the separation of protons and electrons in hydrogen atoms is less than 10^{-8} cm or equivalently 10^{-10} m. Just as an aside, because one speaks a lot about nanoscience and nanometers, 1 nm is 10^{-9} m, and therefore the separation of electrons and protons in the hydrogen atom is less than 0.1 nm. We will learn more about atoms and their constituent charged particles in following sections. The electron and the proton are the origin of opposite electrical forces and, because they are opposite, these forces originating from close-by electrons and protons (as they are in atoms) cancel each other at distances far away from the atoms. It is for this reason that under normal circumstances we do not notice electrical forces, in spite of the fact that all matter that we know and that surrounds us on earth contains electrons and protons.

Only when we separate electrons and protons can we see the enormity of the electrical force. This separation takes energy. We will learn later how to calculate this energy. For now we note only typical values. It takes the energy of about $2 \cdot 10^{-18}$ J to separate the electron and proton of the hydrogen atom. A small glass of water contains about 10^{25} hydrogen atoms. To separate then all the protons from the electrons and put them into two different containers will take the energy of about $2 \cdot 10^{7}$ J. That energy would heat a kitchen stove for more than an hour. It is therefore understandable that (a) we normally do not see or feel the electrical forces, because the positive and negative charges are not being separated and (b) if the charges are separated, then they may be useful to heat our kitchen's stove. In fact, electrical forces are now all pervasive in our daily life and do for us a lot more than just heating a stove. The question for such applications is therefore, how do we separate positive and negative charges and how do we generate electrical forces?

2.2.1 Galvanic Elements: The Lithium Battery

Historically, an important generator of electricity was the so-called galvanic element, named after its discoverer Luigi Galvani from Italy. Batteries that we use in our daily life for cars or laptops, are based on the principle of galvanic elements: the generation of electrical power and currents by chemical processes. A major difficulty in the understanding of galvanic elements arises from the fact that one needs to understand some chemistry to start with. One needs to know a bit about the nature of atoms and the nature of combinations of two or more atoms to form so-called "molecules" and "chemical compounds." However, an important part of chemistry itself needs to be explained by describing experiments that involve galvanic elements. The methods of explanation of all of these facts involve, therefore, a "circle" that one needs to master by first accepting some of the chemical terms without detailed explanation in order to understand the basic electric phenomena, and then reading up on the chemistry later.

We know that the most important atom, the hydrogen atom, consists of one negatively charged electron and a positively charged proton. We have mentioned this above, without emphasizing the surprising fact that an atom is *not* indivisible, as the ancient scientist Democrit thought, but can be decomposed into electrons and protons. It turned out that this fact was a most important discovery. All the materials that we know are composed of atoms, and all of these atoms contain electrons and protons. The major chemical differences of atoms arise simply from the fact that they are composed of different numbers of electrons and protons. The helium atom has two electrons and two protons, and the next atom is lithium with three electrons and three protons. For reasons that we will discuss in Sect. 2.4 on chemistry and quantum mechanics, one of the three lithium electrons is rather easy to remove from the atom. It takes only $\frac{1}{3}$ of the energy to separate one lithium electron from the lithium atom as compared to the energy that is necessary to remove the electron from hydrogen. The separation energy for a helium electron, on the other hand, is about twice that of hydrogen. Lithium is therefore special and is at room temperature a solid metal while both hydrogen and helium are gases. The small atom size and the metallic properties make lithium useful to produce galvanic elements that we nowadays just call batteries. A lithium battery most probably powers your laptop and, in the future, may possibly power your car.

Originally, batteries consisted of metals embedded in a liquid in which a salt of the metal was dissolved. Salts are, of course, very well known to everyone. The salt used for cooking is sodium chloride, consisting of a sodium (Na) and a chlorine (Cl) atom, and is denoted in chemistry by the symbol NaCl. Sodium is a metal just as lithium is. In liquids, the sodium and the chlorine atoms dissolve and separate. However, the atoms in the solution are now charged, because one sodium electron can easily be removed (just as one electron of the lithium can be). The sodium becomes then positively charged, and the chlorine takes that electron and becomes negatively charged. This preference of the electron to move from the sodium to the chlorine has some deep chemical reason that we will discuss in detail later. Thus, if you drop salt into water, the water will then contain positively charged sodium atoms denoted as Na^+ and negative chlorine atoms denoted by Cl^-. One calls such a liquid (water plus salt) an electrolyte, because it contains elements of electricity and conducts electrical currents. The dissolved and charged atoms Na^+ and Cl^- are called ions. Ions are simply atoms carrying charge. The same effects happen if you dissolve a lithium salt instead of NaCl, or any other salt.

The principle of the workings of a lithium battery (and a large number of batteries based on different materials) is shown in Fig. 2.15. Lithium atoms that are contained in some chemical compound (Ch1) on the left side of the battery dissolve into the electrolyte leaving an electron behind at that left contact side that is labeled in car batteries as "black." Black indicates that a negative charge consisting of electrons is accumulating at that contact. The dissolved lithium atoms have lost these electrons and become, therefore, positively charged ions. The electrons propagate via a wire to the lightbulb (car, laptop, or whatever is powered by the battery), and the positively charged ions propagate within the battery toward the right contact. At the right contact, that is labeled in car batteries as "red," the electrons that have moved

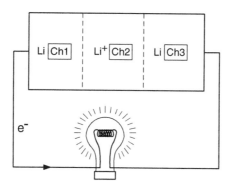

Fig. 2.15 Principle of a lithium battery: The *rectangle* indicates the battery that has two electrical contacts. The *left* contact supplies negative electrons indicated by e^- and is often labeled by the color *black*. The other contact (to the *right* in our figure) is positive and usually labeled by the color *red*. At the *black* side, lithium in the form of a chemical compound indicated by (Ch1) loses one electron and proceeds as positively charged lithium Li^+ through another chemical (Ch2). (Ch2) permits Li^+ to pass but rejects electrons. Li^+ arrives at the red (the positive) side. The negative electron, on the other hand, proceeds via the outer circuit and the lightbulb. It finally arrives at the right side and neutralizes the positively charged Li^+ that together with (Ch3) forms the compound Li(Ch3) as indicated. The flow of many electrons that is generated that way can be used to power a lightbulb or a personal computer and the like

through the lightbulb (or other devices) and the positive ions that moved inside the battery reunite to form again complete lithium atoms with help of chemical (Ch3). The chemical processes that are involved in all of this electron and ion generation create thus an electrical current in the wire that supplies the necessary power to drive our chosen equipment. In this way chemical energy is turned into electrical energy, and the chemicals need to be carefully chosen to achieve this.

For safety reasons one can not use pure lithium-metal contacts in batteries. Lithium metal would burn in air explosively. This is another important reason why battery producers use chemical compounds of lithium, meaning they use lithium atoms connected to a number of other atoms. In our Fig. 2.15 the chemical compounds are just denoted by (Ch1), (Ch2), and (Ch3), because we are here mostly interested in the electrical current generation and not in the chemistry of these compounds. The compounds, which in principle can be solids, liquids, or even gases, determine in the final analysis the performance of the battery, how much power the battery can deliver, and how often it can be discharged and recharged. The design of battery materials is a very important area of STEM, and many chemical engineers attempt currently to produce the most competitive batteries.

The amount of electrical current that can be drawn from a battery depends on the number of lithium atoms that can be dissolved. Typically the design is such that a fraction of the total available lithium dissolves per hour, so that the battery can work for several hours. To give an example, we assume that the total available number of lithium atoms at the left contact is given by Avogadro's number of $6.022 \cdot 10^{23}$ atoms which corresponds to a few grams of lithium (see Sect. 2.3.2). Assume further that a small percentage of these lithium ions dissolve into the electrolyte say 4 % per hour. Then about $2.4 \cdot 10^{22}$ positive lithium ions propagate per hour toward the red (positive) side of the battery. At the same time the same number of electrons flows through the lightbulb or our laptop. This corresponds to about $6.69 \cdot 10^{18}$ electrons and ions per second. These are all very reasonable numbers, and this is what actually happens when you power a lightbulb or your laptop with a lithium battery. However, you are usually given the electrical current in units of amperes, not by the numbers of ions and electrons per second. We will return to this important unit of ampere below.

The chemical equation that describes the discharge of a lithium battery at the left (black, negative) side is

$$\text{Li(Ch1)} \leftrightarrow (1 - x)\text{Li(Ch1)} + x\text{Li}^{+}(\text{Ch2}) + xe^{-}. \tag{2.49}$$

Here, the double-sided arrow means that we can read the equation in both directions. If we use the battery to power lightbulb or laptop and discharge it, you must read the equation in the direction toward the right $\rightarrow$. Lithium dissolves in the form of x % of positively charged ions into the electrolyte (represented by the term $x\text{Li}^{+}(\text{Ch2})$) and leaves xe^{-} electrons at the contact that supplies wires toward the lightbulb or laptop. The symbol e^{-} indicates the negative charge of one electron. The number of the electrons that are generated is, of course, determined by the percentage x that dissolves. For example, if 4 % are dissolved, then $1 - x$ represents 96 % of lithium that is left over and is represented by the term $(1 - x)\text{Li(Ch1)}$.

At the right side of the battery (red, positive), the following equation describes what happens:

$$(1 - x)\text{Li(Ch3)} + x\text{Li}^{+} + xe^{-} \leftrightarrow \text{Li(Ch3)}. \tag{2.50}$$

The positive lithium ions, that originate from the left side while the battery is discharged, are neutralized by the electrons that have been propagating through the lightbulb or laptop and are now returned at the right side. There, the resulting lithium atoms form the compound Li(Ch3).

Up to now we have only considered the process of discharging the battery that produces electrical power. The inverse process is also possible. We use electrical power from a so-called recharging equipment to restore the battery to its original charged state. The chemical equations for the recharging are also given by Eqs. (2.49) and (2.50), but now the equations need to be read from right to left $\leftarrow$. On the left side of the battery (with Eq. (2.49) being the relevant equation), we can supply negative electricity in the form of electrons from the recharging equipment. This negative electricity attracts and subsequently neutralizes positively charged lithium ions that thus are returned as lithium atoms and restore the original Li(CH1). On the right side of the battery, electrons are extracted by the positive recharging equipment, resulting in positively charged lithium that is dissolved and propagates now to the left while the extracted electrons are transferred by the recharging equipment toward the left contact.

Two important consequences follow from the above description of the lithium battery that should be remembered by everyone even if the details of the process are forgotten: (a) Batteries have a positive contact labeled as "red" contact or by a "+" and a negative contact labeled as "black" or by a minus "−." The black is connected in a car to all of the metal of the car and often called the "ground" contact. (b) The recharging equipment also usually has a red or positive and a black or negative contact. When recharging the battery you must connect red to red (positive to positive) and black to black (negative to negative)!

What we have learned up to now is that one can construct a chemical machinery that results in a flow of charged particles. There is energy involved in that flow. In one direction of the flow we can operate, for example, a lightbulb and use the energy. In the other direction of flow, we need to supply energy to the system by using the recharging equipment that we usually plug into a power outlet of our home. What are the energies that are involved in this flow of charge? The chemical reactions in the galvanic element give rise to an energy difference between the left and the right contacts. Each electron that flows from the left to the right gains that chemically generated energy and can, in principle, transfer it to the equipment that we have attached to the galvanic element. That chemical energy depends on the chemicals that one uses for the galvanic element and is typically 2.4–$4.8 \cdot 10^{-19}$ J for each electron. This is a very small decimal number, and one can understand that the average person will not know what to do with such a number or what it means. For this and historical reasons, one uses different units for the energy that is supplied to electrons. Instead of Joules, one uses electron volts or eVs. The typical energy that each electron gains in a galvanic element is 1.5–3 eV, and one eV corresponds to $1.602 \cdot 10^{-19}$ J. Electrical engineers are usually not talking about energies of electrons but just about the "voltage." A battery supplies 1.5 V (V for "volt") means that each electron that propagates from the negative to the positive side gains 1.5 eV. This energy gain gives us some idea that energy and power can be supplied by batteries. We know that it is not dangerous to touch a battery that supplies 1.5 V. It just stings a little if we put two wires from a battery to our tongue. However, a power outlet that supplies typically 110 V is not to be touched without great danger. The thousands of volts of a high power line are even more dangerous, and accidents related to these power lines have killed people.

As you could see, the movement of ions and electrons in batteries is not so easy to understand. The chemistry of batteries is an even more complicated subject, because it involves a lot of chemicals and a lot of possibilities of interactions between them. What one wishes to have is a battery that can store the largest possible energy and does not degrade with frequent recharging. For example, a battery that powers a car would need to supply enough power to drive about 100 km or even miles. It should at least be possible to charge it $2,000$ times because then the car could be driven $200,000$ km (or better $200,000$ miles) before the battery needs to be replaced, which is very expensive. Ideally, the battery should be, of course, as inexpensive as possible and mass producible. Because all these requirements are difficult to meet, more research and development by STEM experts is needed in this area. Recent developments of new types of batteries that involve so-called nanostructures do appear very promising.

2.2.2   Basic Concepts Related to Electricity

The number of electrons that flow is also inconveniently large and is therefore not used in the language of engineers and in daily life. One uses the unit of A for "Amperes" to indicate how much charge is flowing in one second from one electrical contact to the other. The unit of charge therefore becomes As standing for "Ampere seconds", simply because we have to multiply the charge flow during one second by the number of seconds, in order to obtain the total charge that is transferred from one contact to the other. Typically we have a flow of 1–10 A through a lightbulb or a laptop, when we power them with a battery. The charge of a single electron is $-1.602 \cdot 10^{-19}$ A s (Ampere seconds). (One often denotes the positive value of this charge by the letter e and speaks of the elementary charge.) To obtain the typical 1–10 A, we need then, as we know, a large number of electrons. To obtain 1 Ampere second of charge, or as one says a "current" flow of 1 Ampere, one needs to multiply the charge of one electron by the number $6.24 \cdot 10^{18}$ which is then equal to the number of electrons that actually flow. As with all units, it takes a while to get used to them. Units are (or at least should be) chosen to be convenient. However, the "convenience" depends on what we are actually doing with the units, and with electrical energy we are doing many different things. The operation of lightbulbs was very important in the history of electricity. Nowadays we have an enormous number of applications that are powered by electricity. This fact becomes very noticeable when we have a power failure.

What is the definition of electrical power? It is simply the energy that can be and is supplied per second. The electrical energy E_{elec} that is supplied according to our discussion above is given by the energy that is supplied per charged particle (electron) multiplied by the number charges; thus

$$E_{elec} = V A s. \tag{2.51}$$

In words, the electrical energy that is generated in every second is given by the number of volts that the galvanic element or power-outlet supplies, multiplied by the charge that flows which is measured in ampere seconds A s. The power to do things depends, of course, on the energy that we have available per second that means the energy that we have obtained divided by the number of seconds during which it was delivered. Thus the electrical power is given by:

$$Power = V A = W. \tag{2.52}$$

In words, the power is measured in volt amperes (volts times amperes) which is also called "watts." All these names are derived from the scientists and engineers that developed the knowledge and use of electricity. If you have a given voltage, as, for example, the voltage of 110 V of a standard US power outlet, and if you have a given power that some appliance takes, say 1,100 W, then you can calculate the electrical current (charge flow) to be 10 A ($= \frac{1,100\,W}{110\,V}$). Sometimes one needs to know all these values to buy the fitting appliance and the wires with the appropriate thickness.

Fig. 2.16 Electric circuit
with lightbulb (or other light
emitter) driven by a galvanic
element (battery) that is
represented by the symbol on
top of the figure. The + sign
indicates the positive
electrode of the battery and
the − sign the negative. By
convention, the *longer
vertical line* | always
indicates the positive side.
The voltage of the battery is
also shown on top and is, in
this case, 9 V

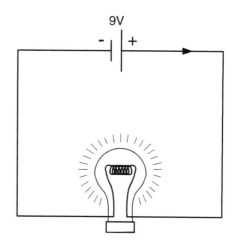

Lets look, therefore, at some typical values of the power that we need to operate
frequently used devices or instruments. The lightbulb of a typical lamp on our desk
uses the power of about 20–40 W or equivalently 20–40 V A (volt ampere). A laptop
uses about the same power and corresponding energy. The energy is obtained by
multiplying watts by the number of seconds used. This also gives you the energy
in Joules. If we wish to light up a whole room, we need more power and buy a
100–200 W lightbulb. Actually, there are different types of lightbulbs. The ordinary
lightbulbs that have a glowing wire need a lot of energy, because much of the total
energy consumed is turned into heat, and only a small fraction of the energy is
turned into visible light. New types of lights, as, for example, light-emitting diodes
(LEDs), do not produce much heat and need only much lower power. The author
anticipates that light involving heated wires will fade out of history. It is simply too
wasteful by generating more heat than visible light. To heat the home takes much
energy and if we are doing it electrically we need about 10,000 W per home, again
depending on the heating equipment. A TV will use around 300 W and a typical
personal computer (PC) 100 W. Knowledge of the power consumption is important
for decisions that we have to make in daily life, and it is good to remember volts,
amperes, and watts and their meaning.

We usually do not need to know the details of what happens with atoms and
chemistry if we deal with electrical equipment. As we could see from the above
discussion, we need just the volts that the battery or any other power source supplies
and some way to calculate the current, the amperes, in order to determine what we
need to know about the electric energy and power. The complicated description
of a battery in Fig. 2.15, that also describes some of the chemistry of the galvanic
element, can thus be replaced by use of simpler symbols that are sufficient for the
purposes of electrical engineering. The "electrical" version of Fig. 2.15 is shown in
Fig. 2.16. To really use such electrical circuits just as the electrical engineers do, we
need to solidify some of the concepts introduced above and also to introduce new
concepts.

Fig. 2.17 Electric currents
flowing toward and away
from a node. A node is a point
at which a number of current
carrying wires are connected.
Currents flowing toward the
node are defined as being
positive, while currents
flowing away from a given
node are counted as negative

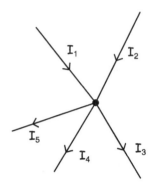

We first discuss the concept of the electrical current. We know from Fig. 2.15 that we have essentially two differently charged particles that are the root cause of the electrical current. Within the battery we have positively charged lithium ions that are moving, and in the metal wire the negatively charged electrons move. If one investigates the chemical details of the galvanic elements further, one notices that there are also negative ions in the chemical Ch2. This is absolutely necessary because charges exert such a great force on each other that one cannot separate positive and negative charges, and big objects such as a battery need to contain therefore just about an equal amount of both charge types. This is also true for metal wires. Wires contain also positive charges, in the nuclei of their atoms (protons), which keep the wire overall neutral. All that matters for the electrical currents, power, and energy is the movement of negative and positive charges relative to each other. This is important to remember. Therefore, if we talk about electrical circuits, we usually do not need to know the actual movement of all the charges. All we need to know is how they are moving relative to each other. It is the historical convention (originating from times when nobody knew that electrons and protons existed) to assume that electrical currents are carried by the positive charges, while the negative charges are just there to keep the objects electrically neutral. Figure 2.16 shows, therefore, the arrow that represents the direction of the electrical current pointing from plus ($+$) to minus ($-$).

It is important to know that electrical currents behave in several ways completely analogous to streams of liquids. We have already discussed this analogy in Chap. 1.2 in connection with Fig. 1.6. There we have stated that the total current I_1 of water of a pipe, that branches out into two pipes (with currents I_2 and I_3, respectively), is conserved, meaning that $I_1 = I_2 + I_3$. This is a very plausible rule, that means simply that no water is lost. It also applies for electric currents and means then that no charge is lost either. Gustav Kirchhoff formulated this law in a more general way that is very valuable for electrical circuits. Figure 2.17 shows a number of wires that all are connected in one point, a "node." We define now all currents that are flowing toward the node as positive currents and all of those that are flowing away from the node as negative independent of what charges are moving. Kirchhoff's rule states that, as in the case of liquids, the total current flowing toward the node equals the total current flowing away from the node. With the sign convention for the currents that we just discussed, this means that the sum over all currents at a node is zero. If we have N such currents then Kirchhoff's rule is:

$$\sum_{n=1}^{N} I_n = 0. \tag{2.53}$$

Resistors

The concept of electrical resistance of wires has also its clear analogies to the transport of fluids in pipes. A very thin wire has typically a higher resistance than a very thick wire made of the same material, just as a very thin pipe or tube cannot carry as much water per second than a very thick pipe can. Of course, the quantity of water that is actually flowing through a pipe depends also on the pressure difference between the two ends of the pipe. For electrical phenomena, the amount of current flow through a wire depends on the voltage difference between the two ends of the wire. Usually the stream of water that flows is directly proportional to the pressure difference, meaning that with twice the pressure difference we have twice the amount of water flowing. The same is usually true for electricity. With twice the voltage difference, the electrical current in a wire doubles. We say here "usually" because there exist many exceptions. For example, if the pressure becomes too large, the pipe explodes. Similarly, wires can burn out if the voltage becomes too large. However, even before that happens, a wire normally heats up when large currents flow and the resistance of the wire changes with the heat. Nevertheless, if we do not need to worry about such exceptions, the electrical current is proportional to the voltage. One then can define the resistance R of a piece of wire or any equipment by

$$R = \frac{V}{I} \tag{2.54}$$

which can also be written to express the proportionality of current to voltage:

$$I = \frac{V}{R}. \tag{2.55}$$

This equation is commonly known as Ohm's law, because it was Georg Ohm who noted in 1827 the proportionality of current I to voltage V. The unit of resistance is therefore the Ohm, denoted by the Greek letter Ω, and we have $1\Omega = \frac{1V}{1A}$.

If one is interested in electrical phenomena, it is important to remember a few typical values of resistance. A 100 W lightbulb has a typical resistance of $100\,\Omega$. One meter of thin wire has about the same resistance, while a thicker wire may just have about $1\,\Omega$ or less. The resistance will also depend on the material that the wire is made of. Wires are made out of metals that are usually great conductors of electricity, meaning they are having a low resistance. Copper is a very good conductor and therefore often used for the electrical wiring in the household.

Fig. 2.18 Circuit with one galvanic element (battery) with voltage V and one resistor with resistance R. The current I flowing through the resistor is given by $I = \frac{V}{R}$

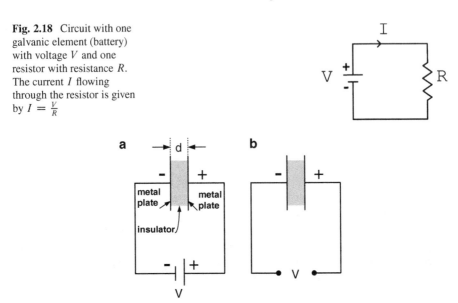

Fig. 2.19 (a) Two metallic plates, mounted parallel to each other and separated by an insulating (air or other insulator) region of thickness d, are called a capacitor. The capacitor is "charged" by the battery that supplies electrons to the left side and takes electrons away from the right until the voltage on the capacitor equals that of the battery. (b) The battery from (a) is removed. The voltage V can now be measured on the plates of the capacitor

A simple electrical circuit with one resistor is shown in Fig. 2.18. This circuit points toward another very useful rule which says, that in any closed loop such as given in Fig. 2.18, the voltage of the battery is equal to the product IR of the current and the resistance through which the current flows. This rule can be extended to more than one galvanic element and more than one resistor as discussed in Sect. 5.5.1.

Capacitors

A capacitor consists, in principle, out of two metal plates that are mounted parallel to each other at a distance d, as shown in Fig. 2.19. Also shown in Fig. 2.19 part (a) is a battery that forces the flow of charges onto the capacitor, until the voltage caused by the charges on the capacitor plates equals that of the battery. In part (b) of the figure, the battery is removed, leaving the capacitor charged and exhibiting the voltage V and a positive charge $+Q$ on one plate as well as an equal negative charge $-Q$ on the other plate. Both charges are, of course, measured in Ampere seconds A s. The capacitance C, which measures the power of the capacitor to store charge, is defined as:

$$C = \frac{Q}{V}. \qquad (2.56)$$

The unit of capacitance is the Farad, named after Michael Faraday. One Farad is the capacitance that shows a voltage of 1 V on the plates, whenever the charge Q on the plates equals one Ampere second.

Capacitors have many applications in electrical circuits. For one, they can be used instead of batteries when it is important to recharge quickly. Capacitors can be charged almost instantly, because there is no chemical process connected with their charging. In the past, the standard capacitors could not store much charge. Recently, however, with the use of methods from nanoscience, it has been possible to create structures that have a very large capacitance. These structures are called super-capacitors. The secret is to be able to produce structures with very large area in a very small volume. Twice the area can store twice the charge. Small volumes are needed to reduce the distance between the plates which increases capacitance. Small volumes also make the capacitor useful for smaller equipment, to power toy airplanes and the like. Capacitors are needed otherwise in virtually any circuit application, ranging from the storage of charge in computer memories to the creation of electrical oscillations in wireless communications.

Currents are flowing to and from a capacitor only during charging or discharging. If one applies a constant voltage, then the capacitor charges up and the current stops flowing. If, on the other hand, one applies an alternating voltage, i.e., a voltage that varies with time from positive to negative, then a current that charges and discharges the capacitor is flowing continuously. Capacitors are, therefore, useful in electronic applications that let alternating currents (ac currents) flow and block constant or so-called direct currents (dc currents).

Several other applications (e.g., electronic memories that use capacitors) will be discussed in later sections. Here we finish with a neat recent discovery about capacitors. One can produce now very tiny structures, nanostructures, and therefore also very tiny capacitors. Remember that one nanometer is 10^{-9} m. We will discuss the production of such small structures in Sects. 3.2.2 and 3.4. It is possible to produce capacitors with a capacitance of Atto-Farad meaning 10^{-18} F. If we transfer exactly one charge (one electron) from one plate of the capacitor to the other, then the voltage that is necessary to do that is according to Eq. (2.56) given by

$$V = \frac{Q}{C} = \frac{1.602\,10^{-19}\,\mathrm{A\,s}}{10^{-18}\,\frac{\mathrm{A\,s}}{\mathrm{V}}} = 0.1602\,\mathrm{V}. \qquad (2.57)$$

This means that if we apply less than 0.1602 V, no electron can be transferred from one capacitor plate to the other, and therefore no lasting currents can flow, because electrons *cannot* be divided into smaller entities with smaller charges. However, above that voltage, one electron can be transferred. Using higher voltages one can transfer one electron after the other like putting pearls on a chain. The neat experiments connected to this effect do not only tell us much about the existence of single electrons but also let us measure the effects of single electrons in molecules. This may lead to new applications in electronics and chemistry.

Fig. 2.20 Two electromagnets that can be created by wrapping insulated copper wires around a pencil or other round object. The electromagnets will attract each other when connected to batteries as shown

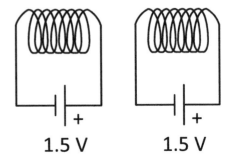

1.5 V   1.5 V

Electromagnets, Inductors

The following experiment is easy to do. Take some thin (about 0.01 cm) insulated copper wire and wind it (always in the same direction) around a pencil in about 50 layers with 10 windings each. This is schematically shown (one layer only) in Fig. 2.20.

Take away half a centimeter of the insulation at each end of the wire. All you need now is a battery, such as AA in the USA that provides about 1.5 V. Connect the ends of the wires to the ends of the battery, and you have produced an electromagnet that can pick up little pieces of iron or steel, such as paper clips. One also can influence the needle of a magnetic compass with this electromagnet, and, if you build a second identical magnet, you can do even more interesting experiments. For example, these two electromagnets will attract each other if you put the tail end of one to the front end of the other as also shown in Fig. 2.20.

I still remember the day that I made my first electromagnets. As soon as I noticed the force of attraction between the magnets, I was astounded. There was nothing in between them, yet there was a considerable force of attraction. Even more amazing, when I exchanged the battery connections of one electromagnet, then they repelled each other. Nothing on the outside of the wire-winded magnets gave away that something so special would happen. I put paper in between the magnets and nothing changed, the force went right through it. It appeared that one could exert some influence with one magnet over the other in spite of putting obstacles in between them. Did the magnets influence each other over a distance with nothing going on in between them? Or was something going on in between the magnets that was just not influenced by the paper? The following experiment provides a clue. Put the bottom of a steel pan in between the magnets. Then, if we turn off the power to the left magnet (disconnect the battery) the right magnet is still attracted to the steel pan but about the same way with and without the left magnet (be it switched on or off). From this it is clear that it does matter what is in between the magnets. Faraday was one of the great inventors and discoverers of electromagnetic phenomena, and he thought about it very deeply. A famous insight of Faraday is shown in Fig. 2.21.

Faraday surrounded the coil representing the electromagnet by iron dust. The little iron pieces start to form lines indicating the magnetic force everywhere.

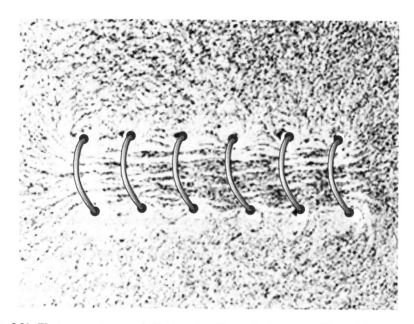

Fig. 2.21 Electromagnet surrounded by iron dust. Notice the formation of lines of dust corns that indicate the presence of the magnetic force

Faraday deduced from this that an electromagnetic field H is surrounding the coil at every point in space. He interpreted the interaction of one coil with the other as a consequence of the immediate field and not as an influence at a distance.

The question whether or not influences at a distance exist, or whether all influences are due to force fields that act only locally, is a very interesting one. Historically, gravity was conceived by Newton as influence at a distance. Faraday and James Clerk Maxwell, however, showed that influences at a distance were not necessary for their theory of electromagnetism, and Einstein was convinced that both gravity and electromagnetism should be understood without action at a distance. The discussion of all of these possibilities are still ongoing, and today's physics is divided on the point of influences at a distance. A number of scientists working in the area of quantum mechanics believe that influences at a distance are a possibility or even a necessity. This author believes that Einstein was correct, and nature can indeed be understood without any influences at a distance. I developed this conviction in spite of the fact that the action of magnets on each other appeared to me almost spooky in my childhood.

The experiments of Faraday (and many different experiments in the subsequent centuries) showed that the magnetic field H is created by the coils as a consequence of the electric currents that are flowing in the wires. A further interesting experiment can be performed as follows. We can superimpose the two coils on top of each other by taking two equally long wires and wrapping them parallel to each other around a pencil. Then, if we apply the batteries to both coils, we have a magnet that is twice as strong as that of any single coil. However, if we apply one of the batteries in the

Fig. 2.22 Two coils that are very close to each other or even wrapped in parallel. One coil is connected to a battery, the other to an instrument that can measure current I or voltage V

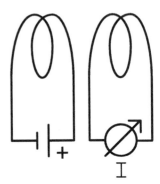

opposite way as compared to the other, then we obtain no magnetic field. This is understandable because then the electric currents in the two coils flow in opposite direction and cancel each other out; they interfere with each other with the result of zero magnetic field.

The situation becomes even more interesting if we perform experiments that are time dependent or in other words, if we have a magnetic field $H = H(t)$ that depends on time t. This experiment requires two coils close to each other and is illustrated in Fig. 2.22. One coil can be connected to a battery, the other is connected to an instrument that can measure a current I. Such instruments are available in hardware stores and can usually measure not only currents but also voltages (as well as the resistance of a wire). After the battery in the circuit of the first coil is switched on, a current starts to flow and a magnetic field starts to build up. This buildup of the magnetic field takes energy and therefore does not occur instantaneously. It is as if the coil resists the magnetic field buildup. This "resisting" depends on how many loops such a coil has and on other circumstance is called the "inductance" L of the coil. The inductance is an important circuit element. Thus an inductive resistance is only present when there is a change in current and magnetic field.

During the time when the magnetic field changes in the first coil, a very important effect occurs also in the second coil. The instrument of the second coil indicates a current! One says that the current is "induced" in the second coil by the changing magnetic field of the first coil. Another way of expressing this fact is to say that the current is generated in the second coil by "induction." The inductance of coils in electrical circuits comes thus into play only when the electrical currents and corresponding magnetic fields change, i.e., when we deal with alternating voltages and currents. This is similar to the situation with capacitors that we discussed above. Note that there is no connection between the wires of the two coils, yet changes of the electrical currents of one coil cause electrical currents to flow in the other. The electrical current flows only if a force is present that accelerates negative (electrons) or positive (ions, protons) charges. The force per charge is called the electric field F. Thus a change of the magnetic field generated in the first coil "induces" an electric field and consequently an electric current in the second coil. The coils can even be spatially separated and the experiment still works, as long as the magnetic field that is shared by the coils is large enough.

It is important to remember the following fact. If one measures the voltage of the second coil instead of the current, then this voltage depends on the number of windings that each coil has. If both coils have an equal number of windings as shown in the figure, and if the coils are very close, then the voltage measured at the second coil equals that on the first during the time of changing magnetic field. If, however, the number of windings of the second coil is doubled, then the voltage is also doubled, if it is tripled, the voltage is tripled. The reason for this effect is similar to the situation with the seesaw. The longer end of the seesaw can exert a stronger force than the shorter. The electric field that is induced in the windings of the second coil represents the force and is the same throughout the second coil. If the second coil is longer (more windings) then the electric field adds up to a higher voltage at the ends of the coil. Of course, we cannot change the available energy, but we can change the available forces exactly as we could do mechanically by use of a seesaw or a hoist. Thus we can "transform" voltages, change them from larger to smaller, and vice versa by use of coils. This is done in cars to obtain the high voltages necessary for the spark plugs. The function of "transformers," that one often finds close to one's house, is also based on the principle of induction. They transform the currents and voltages from the high voltage of the power lines to the standard voltage in the household (110 V in the USA). The high voltages of the power lines must be lowered for household applications for safety and other reasons. The lowering can be achieved by using two coils with very different windings. Lets say that the coil that is connected to the high voltage state power line has 10,000 windings and the second coil that is connected to the household application has only 100 windings. Then the voltage for the household is lowered exactly by the ratio of the number of windings of each coil that is exactly by a factor of 100 in this example. The ignition coils in cars work the other way around. To create a spark one likes to increase the battery voltage by large factors.

Electric Motors

Another interesting application of coils and magnetism is the electric motor. Electric motors convert electrical into mechanical energy, for example, the rotation of some disc that drives washing machines, refrigerator pumps, cars, or even trains. There are two types of electric motors. One type uses alternating electric currents, that is, currents that change their direction. The power outlets in our households provide alternating currents (ac) that change their directions 50–60 times per second. Batteries provide only direct currents (dc) that flow in one direction only. One can, however, change a dc current into an ac current with relative ease. This is being done to create electric motors that can run on batteries. The principle of such motors is shown in Fig. 2.23.

A coil is mounted between two fixed magnets in such a way that it can rotate. One wire of the coil is electrically connected to one metallic half of a disc and the other wire to the other half of that disc. The two halves are separated by an insulating region, a region that does not conduct electricity. Two metallic springs slide around

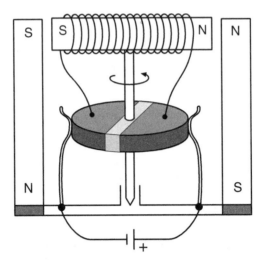

Fig. 2.23 A coil magnet is mounted between two other magnets (e.g., magnetized iron bars or additional coil magnets). The contact wires of the rotating coil are connected to the separated metallic surfaces of a disc that rotates with the magnet coil. The separation of the two metallic disc surfaces is achieved by a stripe of insulating material (*light grey*) between them. Metallic springs are sliding along the metallic disc boundary. As the disc turns, the current in the coil changes its direction and so does the magnetic field of the coil

the boundary of the disc and are connected to the battery. Assume now, that we start out with the position shown in Fig. 2.23. The coils north pole is close to the north pole of the fixed magnet (and the same is true for the south pole on the other side). Before the motion starts, the coil is turned just slightly as indicated by the arrow in the figure. The pairs of north and south poles repel each other, and the mobile coil starts moving away from the fixed poles, turning now more rapidly in the indicated direction. After half a turn, the north pole of the coil comes close to the south pole of the (left) fixed magnet and the south pole of the coil close to the (right) fixed north pole. Now the pairs of poles attract each other and the rotation continues. The rotation would then stop because of the attractions of north and south poles. However, as soon as a full rotation is complete, the springs that slide on the disc change to the other half of the disc. Then the direction of the current changes which in turn changes the magnetic field of the coil; the poles repel each other again and the motion continues.

From the description of this principle of electric motors, we can see that we need alternating currents through a coil to create rotating motion. If you use the ac power from the power outlet, the alternating current is available to start with, and the electric motor does not need any disc with sliding contacts. If you use dc electricity from a battery, then you need to use an electric motor that creates the alternating current from its own motion as described above. The springs that slide around the disc are the key to create that alternating current mechanically. There are nowadays also more modern ways, not involving mechanical parts but only electrical devices, to create alternating currents if needed.

2.2.3 Maxwell's Laws of Electromagnetism

Many of the initial experiments that are now common knowledge in the area of electromagnetic phenomena have been performed and explained by Faraday. Faraday was a self-made man with great genius and intuition. He and several other important contributors found the rules of electromagnetism and experimental ways to test these rules. Maxwell transformed these rules into mathematical equations called partial differential equations. As an example, we will discuss Maxwell's wave equation, the equation that describes all electromagnetic waves, in Sect. 5.3. Here we list the rules that are the basis for, and equivalent to, Maxwell's equations:

1. Electrical charges Q are the source of electric fields F that surround these charges. The electric field at a point in space is defined as the force per unit charge.
2. There exist no magnetic charges in analogy to the electric charges. The sources of magnetic fields are electric currents as illustrated in Fig. 2.21.
3. A magnetic field H encircles any electrical current.
4. An electric field F encircles any magnetic field $H(t)$ that changes with time t.

These are the laws that govern electric and magnetic fields and currents. Maxwell brought these laws into a strict mathematical form that was later improved by Hendrik Lorentz and Albert Einstein. Maxwell definitely had a type of fluid in mind, when he derived his equations, and we have outlined above how similar fluids and electricity behave. As a consequence, Maxwell also thought that the electrical phenomena take place in a fluid-like "ether" that fills all of space. Einstein's final version of electromagnetic phenomena is completely abstract and does not refer to any substance such as the ether. Einstein only talks about electric and magnetic fields, as well as their connections to each other and to electric charges and their motion.

2.2.4 Electromagnetic Waves, Wireless Communication

Wireless Communications

One of the most important consequences of Maxwell's rules is the possibility of wireless communications by use of electromagnetic fields (waves). Wireless communications are of great importance for technology and our daily life.

The first step, to understand how this particular application of electromagnetic waves works, is to understand how the waves can be created by use of the circuit shown in Fig. 2.24, a so-called resonant circuit. The circuit contains a capacitor which consists in essence of two parallel metal plates, an inductor, consisting of a wire coil, and a switch which is just a little piece of metal that can open and close the circuit. Assume that the capacitor is charged, which can be done by connecting it briefly to a battery. One is then left with positive charges on one plate and

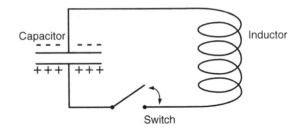

Fig. 2.24 Electrical resonant circuit with capacitor (parallel metal plates), inductor (coil), and switch (a wire connection that can be opened and closed)

negative charges on the other as indicated in Fig. 2.24. Up to this point, the switch is open. Now we close the switch. As soon as that happens, the electrons that have accumulated on the negative side of the capacitor start to move via the wires toward the positive end of the capacitor. This, however, means that an electrical current must flow through the coil, which in turn means that a magnetic field H is generated in and around the coil until the capacitor has lost all charge. Then the current stops, and the magnetic field starts to decrease, meaning that we have a time-dependent magnetic field $H(t)$ in and around the coil. According to Maxwell's rule (4) from above, the changing magnetic field causes an electric field and a current flow in the wire. This current flow occurs in the opposite direction (because the magnetic field is now collapsing) and has the consequence that the capacitor is now also being charged in the opposite direction. Then the capacitor starts to discharge, and the process continues until we arrive at the original situation, with one plate negatively and the other positively charged. If no energy is lost anywhere, this electromagnetic oscillation repeats itself over and over.

Heinrich Hertz understood from his detailed knowledge of the work of Maxwell, that an oscillating electrical circuit would lead to electromagnetic fields and waves in its neighborhood. To demonstrate the existence of electromagnetic waves in free space, Hertz constructed two identical resonant circuits of the type shown in Fig. 2.24 and placed them in two different rooms of his laboratory. He then showed that if he charged the capacitor in one room and started the oscillations, the oscillator in the next room would start oscillating by itself, even if the capacitor of this second circuit was not charged up by anyone. It was therefore clear that the electromagnetic oscillations were transmitted from one room to the next, just like oscillations of water waves or sound waves can be generated, transmitted, and detected. Wireless communication was born!

The formation and propagation of the electromagnetic waves created by Hertz is a consequence of Maxwell's rules (3) and (4). A graphical representation of the idea is shown in Fig. 2.25 that illustrates graphically how these rules lead to electromagnetic waves. Electric fields cause electric currents in Maxwell's ether (so-called displacement currents), and these currents are encircled by time-dependent magnetic fields. Time-dependent magnetic fields are, in turn, encircled by electric fields. In that way, an electromagnetic wave is created that propagates through space, as we understand it from the analogy to water waves and other types of waves. The velocity of electromagnetic waves, however, is extremely high, much higher than that of sound or water waves, and it took some time until it could be

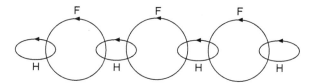

Fig. 2.25 Electrical (F) and magnetic (H) field lines encircling each other as demanded by the Maxwell–Faraday rules. Electromagnetic waves propagate through space in this way

reliably measured. The value of the velocity of electromagnetic waves in vacuum is generally denoted by c, and the measured value is $c = 3 \cdot 10^8$ m/s. This is an enormous speed. Such a wave travels from the earth to the moon in about one and a quarter seconds. All electromagnetic waves in vacuum travel with the velocity of light. The enormous speed is also an important factor for the use of wireless communications. It enables us to communicate almost instantly over large distances.

What is the frequency of the oscillations of a resonant circuit? The number of oscillations per second can be calculated from the following formula that we give without proof:

$$\nu = \frac{1}{2\pi\sqrt{LC}}, \tag{2.58}$$

C is the capacitance in Farads and L is the inductance in units of Henry (named after the American scientist Joseph Henry). The frequency is usually denoted by the Greek letter ν, and we have therefore also used that letter. The units of the frequency are s^{-1} (seconds to the minus one) because the frequency counts the oscillations per second. We have not discussed the unit Henry because this unit is not as easy to describe as, for example, the unit Farad for capacitance that was defined in Eq. (2.56). The interested reader is referred to the Internet with its detailed explanations. For very large capacitances and inductances one obtains (from Eq. (2.58)) a few oscillations per second. One oscillation per second is also called one Hertz. For very small values of inductance and capacitance, the oscillation can be in the Giga Hertz range, which means we have 10^9 oscillations per second. Kilohertz (kHz) stands for 1,000 Hz, megahertz (MHz) for one million hertz, and one terahertz (THz) is 1,000 GHz. Modern wireless communications work mostly in the GHz range.

Why did it take more than hundred years after the discovery of Hertz to develop radio communications, and why do we only now have cell phones that can be used wirelessly almost anywhere? There are multiple answers to these questions. For one, the resonant circuits will not oscillate for very long because energy is lost to the propagating electromagnetic waves and also because of heat generation by the electrical currents in the resonant circuit (see Sect. 5.5.2). To maintain the oscillation, one needs some mechanism that feeds energy to the resonant circuit. This mechanism of feeding energy is called amplification, and the devices that feed the energy are called amplifiers. With help of such amplifiers, the oscillation can go on forever. The Internet features many descriptions of amplifier circuits. The basic device that makes such amplifiers tick is the transistor of which we will hear more in Sect. 3.2. A second big hurdle to overcome was the need for very high frequencies of oscillation. One needs high frequencies for the following reason. If you think of digital information, information transmitted by zeros and ones, then, in order to create a one (a measurable signal), you need at least half a wavelength. You can think about this by observing water waves. The one could be the wave maximum and the zero the minimum. Thus, if you wish to transmit a gigabit of information, you need at least a frequency of one gigahertz. Electromagnetic waves with gigahertz frequency are called microwaves. Our cell phones work indeed in the gigahertz range, and it took the development of very special transistors that could amplify with such high speed. Other engineering advances were also necessary to make wireless communications possible. Antennas had to be developed that transmit (send) the waves from the basic oscillator circuit to the surrounding areas and the world, and other antennas that receive the transmitted signals in an optimal way and excite the receiving oscillator circuit.

Modern communication tools such as the Internet do use wireless communications via satellites and in Wi-Fi home networks, all working in the gigahertz frequencies. Some internet communications require much larger information transfer, measuring in the terabit (tera = 1,000 giga) range. To accomplish this, one needs communication by light waves. Light has a very high frequency and is, for this reason, capable of communicating a lot of information. It has the disadvantage that it is impeded and absorbed by obstacles that are in its way. This problem has been solved by using optical fibers. Optical fibers are made out of very transparent glasslike material, and they look like thin wires. There exist optical fiber connections between continents; the fiber cables are placed, for example, under water on the ocean floor between the USA and Europe. Of course, the fiber connection cannot be regarded as wireless anymore. The wires are just replaced by glass-type fibers. The communication is still achieved by the use of electromagnetic waves.

Range of Electromagnetic Waves: Radio Waves to γ-Rays

All electromagnetic waves are of similar nature. However, depending on their frequency, they appear very different to us. The generation of the electromagnetic waves is also very different for the different frequency ranges, and the resonant circuit described above is only useful for the "lower" frequencies, that is, for frequencies up to about 100 GHz. A given range of frequencies, such as the frequencies of the light of a rainbow, is called the "spectrum" of the electromagnetic waves. We will see in Sect. 2.5 that the frequency of the electromagnetic waves is closely related to their energy, and increasing frequency means increasing energy.

We have described already the waves of relatively low energy that are useful for radio and TV communications as well as cell phones. The resonant circuit of Fig. 2.24 can oscillate with a frequency as low as you like, if only capacitance and inductance are chosen large enough. For a 1 F capacitance and a 1 Henry inductance the circuit oscillates precisely once every second and the frequency is then 1 H: $\nu = 1\,\text{Hz}$. The electrical power that you can draw from a power outlet in your home oscillates at 60 Hz in the USA and at 50 Hz in Europe. Radio waves oscillate still much faster, typically in the range up to a few hundred megahertz (MHz). 100 MHz correspond to 10^8 oscillations per second. The cell phone interacts with its receiver station by sending out waves in the gigahertz (GHz) range. 1 GHz is $10^9\,\text{Hz}$. In this range we can use electromagnetic waves to generate heat in microwave ovens. Waves in this frequency range are called microwaves.

Electromagnetic waves at still higher frequencies, such as terahertz (THz) and above, are generated, for example, by heat and are emitted by a hot stove. One calls this type of radiation also infrared radiation, because its frequency is below that of visible red light that starts with frequencies of about $4 \cdot 10^{14}\,\text{Hz}$. The light that our eyes can see ranges from red to violet which has a frequency of about $8 \cdot 10^{14}\,\text{Hz}$. Beyond that frequency we have the so-called ultraviolet radiation and—still higher up in frequency—the spectral range of X-rays. X-rays are important for medical diagnostics and are used to perform CAT scans as described in Sect. 4.3. As one goes further up in frequency, one approaches the so-called γ-rays starting about at $10^{19}\,\text{Hz}$. These latter rays of the highest frequency arise from events that involve the nucleus of atoms and radioactive processes. Often these processes are connected with the creation and also destruction of stars, and the γ-rays from stars are also called cosmic rays or cosmic radiation. They also are generated on earth, during the decay of radioactive atoms such as uranium or plutonium.

2.3 About Heat, Temperature, and Atoms

By now, we have learned about the laws of mechanics including the motion of planets, the laws of electricity including the workings of batteries, and about waves of all kinds including the electromagnetic spectrum. These are mostly phenomena that we can observe with our eyes and generally with our senses. We needed, however, some knowledge about the invisible atoms, electrons, and protons to explain, for example, the lithium battery. The existence of atoms is today generally accepted and agreed upon. Modern microscopes can produce computer images of atoms. Not too long ago, however, two famous scientists, Mach and Boltzmann, were fighting about the existence of atoms and Mach, and most people, did not believe in atoms. No powerful microscope was then available, and all information about atoms was very indirect. Boltzmann derived his ideas about atoms from phenomena connected to heat and how these phenomena could be explained by invoking atoms. Heat-related machines are very important in our daily life, ranging from car motors to jet engines and from refrigerators to air conditioning systems. This is what this section is about: how can we explain heat-related phenomena from the basic principles of mechanics and electricity, how can we use heat-related phenomena to engineer powerful machines, and how can we at the same time, find out more about atoms.

2.3.1 Heat and Temperature

It took scientists a long time to figure out what heat really is, and we are not going to discuss how they actually did find out. We present here only the final result: heat is nothing else but the kinetic energy of the atoms and molecules of the substance that exhibits heat. Consider a gas such as air as the substance. If we feel that a given volume of air is hot and another volume of air is cold, this means that the atoms and molecules of the hot air move with higher velocity than those of the cold air.

To be more exact, we need to have a method and an instrument to measure "hot" or "cold" or any "temperature." This method must provide us with a precise number that is the measure for hot or cold. Historically the temperature has been linked to the freezing and boiling of water. There are three temperature scales in common use. The temperature scale formulated by Celsius, and named after him, fixes its 0 point at the freezing temperature of water, while the water's boiling point is defined to be 100°C (in words: one hundred degrees Celsius). It is implied in this definition that the water is subject to a normal air pressure of precisely one atmosphere (see below). The Celsius scale is used predominantly in central Europe. The temperature scale that is used worldwide in science is derived from the fact that there exists an absolute zero of temperature, meaning that it is not possible to reach a lower temperature. The fact that the absolute zero cannot be reached by any procedure is known as the third law of thermodynamics (the first and second law are discussed below). This absolute zero is denoted by 0 K, where K is the abbreviation for Kelvin or degrees Kelvin. The absolute zero of the Kelvin scale equals -273.16°C which is far below the freezing point of water and even far below the temperatures of the north or south poles in coldest winter.

If you are living in the USA or in Great Britain, you may complain that the Fahrenheit scale was not mentioned yet, although this is the only well-known scale in your country. This is the third temperature scale of importance. You can easily find the conversion of all scales (Celsius, Kelvin, or Fahrenheit), as well as more about temperature scales and methods of measuring temperature, on the Internet. Here we need to be brief because we need to cover more modern topics in the limited space of this book. We also like to be scientific and therefore just use the Kelvin scale in the following. Except for the zero point, the Kelvin and Celsius scale are the same. You can obtain the Kelvin temperature from the Celsius temperature by just adding 273.16°.

2.3.2 Heat, Energy, and Avogadro's Number, Ideal Gases

The connection of heat and energy is illustrated in Fig. 2.26 that shows a movable piston within a cylinder that also contains a gas.

The principle is simple. If we increase the temperature of the gas in some way, it will expand and push the piston upwards. This push upwards can be used to perform

Fig. 2.26 A metal cylinder containing a gas and topped by a piston represents an important element of machines that convert some portion of heat into mechanical power

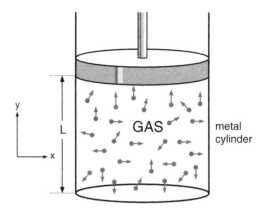

some work, for example, lifting a weight just as we discussed in Sect. 2.1.1. Because the performed work corresponds to energy and because energy is conserved, it follows that the energy comes from the heat and that heat is related to energy. We have already stated that heat corresponds to the kinetic energy of the atoms of the gas. This is also illustrated in Fig. 2.26 where we have included dots with arrows symbolizing atoms having a velocity in the direction of the arrow. We now derive from this basic knowledge the force that is exerted on the piston by the impact of the atoms. To do this, we need to remember Eq. (2.38) which says that the force F equals mass M times acceleration a; that is, $F = aM$. The acceleration that an atom encounters when hitting the wall is calculated in the following way.

Assume that the atom moves with a velocity v_y in the y-direction of the coordinate system which is the direction perpendicular to the piston as indicated in the figure. As soon as the atom hits the piston, it will be reflected and moves then in the opposite direction having the velocity $-v_y$. Thus the change in velocity is $2v_y$. To calculate the change in velocity per time period Δt, we need to know how often the atom hits the piston in that time period. This time period is given by

$$\Delta t = \frac{2L}{v_y}, \tag{2.59}$$

because any given particle must move up and down the full length L (from the bottom of the cylinder to the piston), in order to hit the piston just once for sure. From this relation we can calculate the acceleration a, which equals to the velocity change per time period:

$$a = \frac{2v_y}{\Delta t} = \frac{v_y^2}{L}. \tag{2.60}$$

The force exerted by the atom on the piston can then be calculated from Newton's law for the acceleration:

$$F = Ma = M\frac{v_y^2}{L}. \tag{2.61}$$

This is the time-averaged force exerted by a single atom roaming around in the cylinder and hitting the piston. The number of atoms in a volume of a few thousand cubic centimeters is incredibly large. Therefore, huge numbers of atoms hit the piston at any instant of time. Their velocity and energy depends on the temperature T (that we measure in Kelvin). The force exerted on the piston is, therefore, a statistical average and sum of a large number of atomic forces. This average force can be calculated if we know the relationship of $M v_y^2$ to the temperature. It is intuitively obvious that the average kinetic energy in the y-direction, which equals $\frac{M v_y^2}{2}$, must be proportional to the temperature. If we denote the constant of proportionality by k_B, we have

$$M < v_y^2 >= k_B T. \tag{2.62}$$

Here the symbol $< \ldots >$ indicates that a statistical average over many events has been taken. The atoms move actually all with different velocities, but if we average over all these velocities, then this is the result we get for the average force that one atom exerts on the piston:

$$< F >= \frac{k_B T}{L}, \tag{2.63}$$

while N atoms exert then an average force of

$$< F >= N \frac{k_B T}{L}. \tag{2.64}$$

The constant k_B is now called Boltzmann's constant, a number that you can look up, together with its history, on the Internet. It was actually named by Max Planck, who wished to honor Boltzmann's pioneering work. Boltzmann himself was too absorbed with his fight to validate atomic models for gases such as the one we have just described. He never pursued any measurement of the average kinetic energy of atoms or molecules himself.

The pressure p exerted on the piston is defined as average force divided by the area A of the piston. Thus we have

$$p = \frac{< F >}{A} = N \frac{k_B T}{L A} = N \frac{k_B T}{V_{ol}}, \tag{2.65}$$

where $V_{ol} = L A$ is the volume of the gas in the cylinder. We can reformulate the last equation to obtain:

$$p V_{ol} = N k_B T. \tag{2.66}$$

This is the basic equation for an ideal gas. Ideal means here that we have assumed that the atoms interact with each other exactly the same way billiard balls do in their elastic collisions. We have described these collisions by the equations for energy and momentum conservation in Sect. 2.1.2, and we have used the essence of these equations in the derivation of Eq. (2.66). If the atoms did interact in some other way, the resulting equation for pressure and volume of the gas would be different. For example, the atoms could attract each other over larger distances by van der Waal forces (see below). That would make things more complicated. Indeed, as we will discuss below, real gases do have such atomic interactions, and the above equation for ideal gases is only approximately valid. However, the equation agrees incredibly well with the experiments as long as the temperature is not too low. This agreement represents a great success of the methods of science. Just remember that the laws that we used were derived from the way stones are falling and planets are moving. Now we have used these laws to calculate the pressure on a piston; a pressure that is caused by an enormous number of atoms swarming around in an otherwise empty space. This statistical bombardment of the piston by atoms is what causes the pressure. Equation (2.66) is one of the first equations of physics that involves statistics and probability, because it involves averages over large numbers of particles. It tells us that the pressure that we measure is just some average value caused by atomic bombardment. On a fine scale there must therefore be fluctuations, because of the fluctuations of the bombardment of the piston with different numbers of atoms.

Pressure is usually measured in "atmospheres." One atmosphere is the force per unit area (per square centimeter) that is exerted by the air above as measured at sea level. So, consider that you have a container of gas with a pressure of 1 atmosphere closed off by a piston on top and you are standing at the sea side. The piston then will not be moving, because the air from above bombards the piston on the other side and the pressure from above and below balance each other exactly. If you evacuate the container, then the piston will be pressed down by the significant force of the air above it. The pressure of the air at sea level corresponds almost exactly to the pressure that 10 m of water above the piston would exert on the piston. Therefore, if you dive 10 m below sea level, the pressure is about 2 atmospheres, one arising from the air, the other from the water above you. For a more detailed definition and other units of pressure, consult the Internet.

The number N of Eq. (2.66) of atoms in a gas can be determined by using our knowledge of Galvanic elements. Consider Eq. (2.49). This equation tells us how many lithium atoms are freed at one side of the Galvanic element for each electron that we supply in the form of electrical current when we recharge the battery. Therefore, we can count the number of charges that we supply through the electrical current during a given time period, and that number equals the number of atoms that we deposit at that side. Lithium is actually difficult to obtain in gas form. However, we can do the same experiment by using hydrogen. Then we can generate, for example, about one gram of hydrogen and then count the number of electrons that generate the hydrogen by measuring the electric current. This number is equal to the number N_A of hydrogen atoms in about one gram of hydrogen and is called Avogadro's number, which is

$$N_A = 6.0221415 \cdot 10^{23}. \tag{2.67}$$

You may have noted a little inconsistency here. We said "about one gram of hydrogen" and then gave Avogadro's number to seven decimals. The reason behind this is that nature does not supply us with hydrogen (or any other element for that matter) in its purest form, with every atom being the same and consisting of precisely one proton and one electron. There are usually also heavier or lighter atoms present in any given material with the same number of electrons and protons per atom. The difference in weight arises from an additional particle that is called a neutron. This particle does not change the electrical properties of the atom because it is neutral. It does change, however, the weight. In the case of hydrogen the weight almost doubles when a neutron is added, and one calls the atom then deuterium. Deuterium is in our drinking water and everywhere, but only in small quantities (about 1 Deuterium atom for every 6,000 hydrogen atoms). Deuterium is called an isotope of hydrogen. The topic of isotopes is an interesting one for advanced projects and Internet searches. Avogadro's number was actually determined before neutrons and isotopes were known. We proceed in the following by ignoring the weight corrections due to isotopes and just use the word "about" to indicate that the actual weights of gas atoms that are measured in nature may be slightly different because of isotope effects.

If we concatenate Avogadro's number with atoms other than hydrogen, we obtain different weights measured in grams. For example, we obtain about 4 g of helium, 6 g of lithium, 12 g of carbon, and so forth. The reason why the weight is close to a whole number is the following. The atoms consist of electrons that weigh very little and of neutrons and protons that have almost equal weight. Hydrogen has one proton and Avogadro's number of hydrogen atoms weigh about one gram. Therefore we can conclude that the weight for the other atoms indicates the sum of protons and neutrons. The number of protons and neutrons is equal for the above-mentioned atoms. We can deduct from this that helium contains two protons and two neutrons, lithium contains three, and carbon six protons and six neutrons. The atoms also contain as many electrons as they contain protons because the total charge must be zero as we know from our discussions of Galvanic elements. Thus, once we have a hypothesis that atoms exist and that they consist of protons neutrons and electrons, we can deduce the number of all these constituents from rather simple measurements. Of course, the inverse process, the actual deduction of the number of atoms, electrons, protons, and neutrons from measurements, has been much more complicated and has required great ingenuity from the scientists that went that arduous path. To describe how they actually found all of these facts would lead us too far away from our major objectives.

Avogadro's number is an incredibly large number. 10^{23} is a number with 23 zeros. I would like to illustrate this by the following stunning example. Think of the great roman leader Julius Caesar after whom the month of July is named. In a few days of breathing, Caesar was inhaling oxygen and exhaling, by easy estimates, more than a few grams of carbon dioxide molecules and therefore at least Avogadro's number of them. Lets assume now that these atoms of Caesar's breath have distributed themselves in the last 2,000 years in equal amounts all over the world's atmosphere. Now we calculate the volume of the atmosphere of the world. The radius R of the world is $6.37 \cdot 10^6$ m. The surface area of a sphere with that radius is $4R^2\pi \approx 5.1 \cdot 10^{14}$ square meters (m²). Because the atmosphere is approximately

10,000 m high, that gives us an air volume V_{air} of about $V_{air} =\approx 5.1 \cdot 10^{18}$ cubic meters (m^3). In one cubic meter of air we have therefore about $\frac{6.02 \cdot 10^{23}}{5.1 \cdot 10^{18}} = 1.2 \cdot 10^5$ carbon dioxide molecules that Caesar had in his lungs. Because we breathe a cubic meter of air within an hour or so, we breath every hour more than 100,000 atoms that also Caesar was breathing. The story gets a little funnier, if you realize that the gas exhaustion of Caesar did also include gases that Caesar did not just exhale from his lungs, and, in addition, we would also breath gases from his dog, if he had a dog.

There is a great lesson to be learned here about dealing with atoms or molecules. Because there are so many atoms in every reasonably large volume of our surroundings, our body will encounter numbers of atoms that appear to be large, as the 100,000 atoms of Caesar, but are, in fact, very small compared to Avogadro's number. It follows therefore that our body can usually deal with such numbers without problems. For example, if we inhale air through a piece of garden hose made out of rubber, then we will breath in thousands and probably millions of sulfur atoms because rubber contains sulfur. This will not influence us, however, in any major way. If it did, a gardener would live shorter than other people because of just handling the garden hoses. Current science permits us to measure incredibly small quantities of atoms. In some cases, measurements are sensitive to a few atoms. Only in rare cases will 100,000 or fewer atoms hurt our body. There are, of course, poisons that are so potent that they may do harm even in small quantities. One needs therefore to apply STEM reasoning if one deals with TV—or newspaper—reports that traces of a certain substance such as sulfur have been found above normal levels in living tissue. Maybe the tissue has just touched a garden hose. Then, there is no need to get excited about small traces of ordinary chemicals. Of course, with potent poisons, one cannot be careful enough.

We can see that the pressure that a gas excerts on a simple piston, together with the idea of atoms, permits us to discover a lot about the atoms. It so happens that pistons driven by a gas are the building blocks of very important mechanical motors including the steam engine and the engines of cars. Both types of motors are based on the fact that heat is a form of energy.

2.3.3 Steam and Combustion Engines

An engine is defined as a machine or instrument that turns energy into mechanical motion. We just discuss the principles of a few such machines that are extremely important for our daily life and power our cars and airplanes.

Steam Engines

The steam engine is of historical importance and is still in use for some applications. For the steam engine, the gas of operation is, of course, steam. Steam is nothing else but water heated so much that it changes from its liquid form to become a gas.

Fig. 2.27 Principle of the steam engine. A container with water is heated so that pressurized steam is generated. The steam is transferred to a cylinder with piston. The pressure pushes the piston upward and the piston is connected to a disc (wheel) that turns. When the piston passes the opening at the left of the cylinder, steam is released while the disc keeps turning because of its inertia and pushes the piston below the release opening. Then the steam pressure increases again and pushes the piston upwards. The process continues and the disc keeps turning

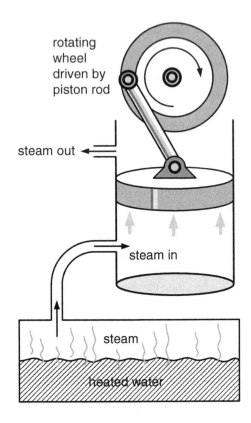

The water molecules are moving so fast that they cannot stick together as they do in the liquid but move farther apart from each other. Water molecules are composed of hydrogen for which we use the chemical symbol H and of oxygen for which we use the chemical symbol O. Each water molecule consist of two hydrogen atoms and one oxygen atom. It is therefore represented by the symbol H_2O. In the liquid, these molecules are attracted to each other and stick very closely together for reasons that we will discuss below. The principle of the steam engine is shown in Fig. 2.27.

In a cylinder, a piston is pressured upward by the hot steam coming from a boiler that is heated so that water evaporates. The piston is connected by a metal rod to a disc (wheel) that can rotate around an axis. The connections of the metal rod to both the disc and the piston are flexible. As the piston pushes upward the disc starts to turn and continues turning until the steam is released through an opening in the upper portion of the cylinder, and the rotating disc moves the piston back to the original position. Then the process repeats itself and the disc keeps turning. The power of the turning disc can be used to drive a train or other useful machinery. Steam engines have revolutionized mankind. Since their invention, the power of the muscle could be replaced by the greater power of machines. Factories using steam engines where built all over the world, and railroads have transported people and loads around the country much faster than horses could. Steam engines are

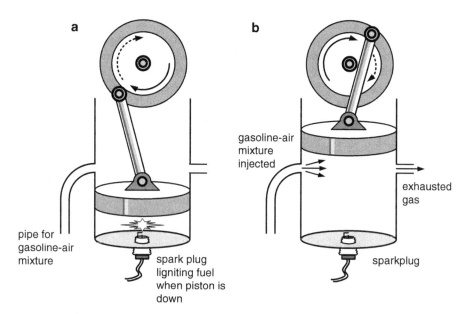

Fig. 2.28 Principle of the two-stroke internal combustion engine, an engine that still drives lawn mowers and other vehicles. (**a**) In stroke one, a spark plug ignites the mixture of gasoline and air that has been injected into the metal cylinder and compressed. The electrical spark is generated at the right moment when the piston is close to its lowest point. The ignited hot gas pressures the piston upwards. (**b**) In stroke two, the hot gas is exhausted through the upper opening on the *right* side of the cylinder, while a new air–gasoline mixture is injected through an opening on the *left*. As in the case of the steam engine, the piston is connected to a rotating disc that continues to rotate through its inertia and pushes the piston with the gasoline–air mixture down. Close to the lowest point of the piston, the spark plug is ignited again and the process is repeated keeping the disc spinning

used in modern nuclear power plants and supply a significant percentage of the worlds energy. The importance of steam engines for mankind would certainly justify a detailed description of the more sophisticated modern implementations, and the reader is encouraged to surf the Internet to find these descriptions.

Combustion Engines

The next revolution of this type was caused by the invention of the so-called "internal combustion engine." The principle of this engine is similar to that of the steam engine. However, instead of steam, a combustible gas, usually a mixture of gasoline (or alcohol) and oxygen from the air, is injected into the cylinder and then electrically ignited at the right moment by a spark plug. The exploding gas pushes then the piston upward. This is illustrated in Fig. 2.28.

This machine is called also a two-stroke engine, because there are essentially two positions that are important for the machine to run. First, when the piston is compressing a mixture of gasoline (or alcohol) and air, a spark is generated electrically at the bottom of the cylinder by a spark plug. The gas mixture combusts then and the hot gas generates a high pressure. This fact can be seen from Eq. (2.66) that tells us that the pressure rises when the temperature rises while the volume stays at that moment the same. Thus, second, the piston is pushed up with great force until it gets to the opening in the cylinder where it releases the combusted gases. After this release, new air–gasoline mixture is sucked in (or injected), and the piston is pushed down again by the turning disc. Two-stroke engines are still in use, for example, for mopeds, mowing machines, and other garden machinery. They were even used for some cars. Most of the cars have more complicated four-stroke engines, but the principle is the same. There are various videos with moving pistons on the Internet (Wikipedia) that help with a more detailed understanding.

Rocket Engines, Jet Engines

Rocket engines are, at least in principle, the simplest of all engines. They consist in essence of a cylinder that is narrow at one end and contains a chemical that is usually solid or liquid and that can rapidly combust. The process of combustion (burning) transforms the solid or liquid into a hot gas that occupies a much larger volume. As discussed previously, a solid or liquid has a much higher density than a gas. The gas therefore leaves the cylinder under pressure with a very high velocity. To understand the consequences, we remember our discussion of billiard ball movements and collisions in Sect. 2.1.2. There we have learned that the total momentum of all billiard balls is conserved and must, therefore, stay the same in all processes that might happen. We treat now all the atoms of the rocket as billiard balls. To simplify the problem, we assume that the rocket is somewhere in outer space with no forces acting on it. Before the ignition of the fuel, the rocket stands still which means that the total momentum is zero. After the ignition gas streams out of the rocket with high speed in the backward direction as shown in Fig. 2.29. We chose the direction of the out-streaming gas as the negative direction of our coordinate system. The gas has, therefore, a negative momentum $-M_G v_G$, with M_G being the mass of the expelled gas and v_G being its velocity. The rocket has a different mass M_R, and its velocity at the beginning is zero. However, we can calculate its velocity v_R after ignition from the fact that the total momentum must always stay zero. Therefore, if after some time all the exhausted gas has a momentum of $-M_G v_G$, we must have

$$M_R v_R - M_G v_G = 0, \tag{2.68}$$

where $M_R v_R$ is the momentum of the rocket at that time. From this we immediately obtain

$$M_R v_R = M_G v_G \tag{2.69}$$

and

$$v_R = \frac{M_G v_G}{M_R}, \tag{2.70}$$

Fig. 2.29 Principle of a rocket engine: A sturdy steel cylinder is filled with a chemical that creates highly pressurized hot gas when burned. One end of the cylinder has an opening through which the gas is expelled

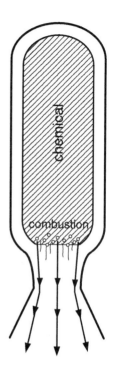

which means that the rocket moves with the positive velocity v_R, i.e., it moves in the positive x-direction opposite to the exhaust gas. Actual rocket science is based on this principle of momentum conservation. The art of rocket engineering is, of course, to get the rocket exactly where one wants it to go, as, for example, into an orbit around the earth. One therefore needs to include the gravitational forces and accelerations and thus needs to solve Newton's equations. Such a solution, including the gravitational forces of earth, other planets, moon(s), and sun can only be accomplished by use of computers.

The workings of jet engines are a little more involved than that of rocket engines. The jet engine principle is illustrated in Fig. 2.30. Jet engines are used to power airplanes and use fuel that is similar to the gasoline that is used in cars. In fact the fuel is usually a liquid called kerosine or a kerosine–gasoline mixture. The kerosine–gasoline mixture does need the oxygen of air in order to burn. The air is supplied in jet engines by a fan (the fan on the left side shown in Fig. 2.30). That fan sucks air in and compresses it into cylinders that resemble rocket engines, two of which are also shown in Fig. 2.30. These rocket-engine cylinders have an opening through which the pressurized air enters. There is no solid rocket fuel inside these cylinders. Instead, the gasoline–kerosine mixture is injected through a pipe as also shown. The air–gasoline–kerosine mixture is then ignited (e.g., by electric spark), and the combusted hot gas is propelled toward the right big opening of the cylinder and hits there a second fan and drives this fan to high rotational speed. The second fan is, in turn, connected by a metal rod (center axis) to the first (left) fan that sucks the air in.

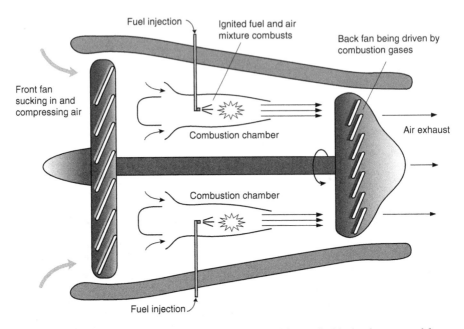

Fig. 2.30 Principle of a jet engine: A front fan is connected by a cylindrical rod to a second fan at the engines back. The front fan takes air into the combustion chamber and compresses it. Fuel is injected into these combustion chambers and ignited causing a jet of compressed air and combusted fuel to accelerate the back fan that in turn by the connection rod drives the front fan

All air is subsequently blown out at the back after being fully combusted. As is well known, such jet engines are very powerful and can lift huge airplanes. The engineering principles, that make such engines work, are the physics principles of the rocket engine (momentum conservation) plus another principle that we have not discussed yet. This second principle is the feedback between the front (left) and the back (right) fans. The back fan drives the front fan that sucks the air in and in this way pressures air into the rocket cylinders that pressure combusted air onto the back fan. The action of the two fans feedback on each other. Feedback as it is commonly termed is very important in many engineering applications. It is also important that the feedback can be controlled so that the desired power can be reached. This control is exerted by the amount of fuel that is injected. That amount is, in turn, controlled by the pilot of the plane who operates the levers that control the fuel and therefore the thrust of the engines.

2.3.4 Refrigerators and Nonideal Gases

Steam and combustion engines work at high temperatures, and this usually means that the gases that are involved behave like ideal gases. The reason is that, at high temperatures, the atoms or molecules move very fast and spend only a short time close to each other. Their interactions are thus limited to collisions of short duration. This type of interactions randomizes the direction of the velocities. The molecules and atoms behave as if they would be fully elastic billiard balls.

At lower temperatures, the velocities of the atoms and molecules are much smaller, and atoms and molecules reside next to each other for longer time periods. As a consequence, the interactions between atoms and particularly between molecules, influence the behavior that is seen by the outside world. The case of steam is a good example. If one lowers the temperature, steam condenses into water and becomes a liquid instead of a gas. Liquids are essentially not compressible, and their volume stays almost constant. Equation (2.66) is then not valid. However, the equation that describes the gas becomes already nonideal before the gas liquifies, because of the stronger interactions of the molecules at the lower temperature. These molecular interactions are a very important component for many processes in nature. They explain why geckos can run on ceilings and why water moves up through the wood of trees, from the roots to the leaves. We describe these interactions therefore in some detail and first explain how refrigerators work on the basis of these molecular interactions. Subsequently we give a brief overview of how these interactions form the basis of many processes in biology.

The physicist Johannes van der Waals was the pioneer who discussed the molecular interactions in gases. We mention his name to facilitate Internet searches for those who wish to know more about this area. The forces that are the origin of these interactions can be explained well by the example of the water molecule H_2O. Both oxygen and hydrogen are, when viewed from some distance, electrically neutral. This is because hydrogen consists of one negatively charged electron and one positively charged proton. Oxygen possesses 8 negative electrons and 8 positively charged protons in its atomic core or nucleus. When hydrogen and oxygen approach each other closely a molecule of water forms as indicated in Fig. 2.31.

The formation of molecules is described in detail in Sects. 2.4.2 and 2.5.3. For now it is sufficient to know that oxygen "likes" the electrons more than hydrogen. The consequence of this fact is shown in Fig. 2.31: the side of the molecule that contains the oxygen nucleus is more negatively charged and the side with the hydrogen more positively charged. One says that such a molecule is a dipole, meaning it has a positive and a negative side. Such a dipole likes to attract other dipoles in the way shown in the figure with positive sides pairing with negative sides and vice versa. This type of attraction and the corresponding forces, are a typical example of the forces between molecules that van der Waals had in mind. He knew that it takes energy to break up such clusters of molecules that are connected by the dipole forces and to get them farther apart. We now discuss an example of the consequences of such a breakup of van der Waals forces.

Consider the metal cylinder shown in Fig. 2.32. The wall in the middle has a small hole through which high pressure water vapor (moist air) or steam is supplied, while to the right, an outward-moving piston lowers the pressure considerably (creates a vacuum). The pressurized water vapor expands as soon as it enters the right half of the cylinder. Therefore, the water molecules move away from each other. To do so,

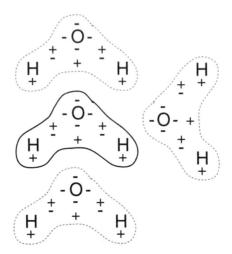

Fig. 2.31 Schematic drawing of the charge distribution of water molecules. The oxygen O "hogs" more of the negative charge of the H_2O molecule than the proton of the hydrogen H atom does. The water molecule has a shape similar to that shown in the figure. There are more negative charges "swarming" close to the oxygen, while the hydrogen side is more positively charged. If other water molecules are present, as they are shown (*dashed*) in the figure, then these molecules are attracted in such a way that the negative oxygen stays close to the positive hydrogen of the other molecules

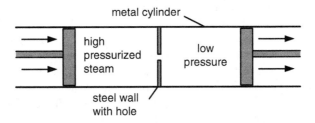

Fig. 2.32 Cylinder with piston (*left*) pressing water vapor through a small opening. A second piston moves to the *right* faster than the first one and thus lowers the pressure in the right half of the cylinder. The pressure lowering leads to a cooling of the right side. This type of cooling effect is used in refrigerators

the van der Waals forces that attract the molecules need to be overcome, and that does take energy. This energy can only be taken from the heat content of the water vapor (steam). Therefore, the expansion of the gas slows the water molecules down, which means the water vapor is cooled. One can achieve the same effect with gases other than water vapor or steam. One can use, for example, propane gas and start with pressurized propane at room temperature. Then, the expanded propane cools below room temperature. You can observe this effect if you have a gas grill supplied by a propane container. Open the valve on top of the propane container and ignite the propane as usual to operate the gas grill. The fire heats now the top grill and the whole equipment gets a little warmer. However, just at the top of the propane container, where the gas streams out and expands, you can clearly feel a significant cooling effect. This is the principle of how a refrigerator works, the only difference being that the gases of the refrigerator are not ignited anywhere, and gases other than propane are used that have a higher efficiency for cooling.

In this way cooling can be achieved in a closed space (e.g., the inside of a refrigerator). The gas itself is warmed up somewhat by the process of cooling the refrigerator's inside and then transported to the outside. There the gas is compressed again which warms it up. Fans or some other cooling equipment bring it back to room temperature. Then again, the gas is decompressed (expanded) in the tubes inside the refrigerator and cools them down. The net effect is that heat is transported out of the refrigerator to the kitchen. In addition work needs to be performed (energy used) for the compression process that also heats the kitchen. This process is called the refrigeration cycle. The inverse process is also possible. One can cool down (as just described) some space outside a house and heat up the inside of the house. A machine that accomplishes that is called a heat pump. Heat pumps are more efficient in energy use than furnaces that just burn gas, wood coal, or oil (or anything else). The reason is that a heat pump moves the heat and uses only the energy necessary for that moving. The process is of importance for possible reductions of energy usage and a way of heating that is environmentally friendly.

Thus, the physics of gases is important for a number of applications including car engines and refrigerators. At the same time one can learn a lot about atoms by considering the detailed mechanisms of these machines. It is quite common in the history of science that engineering applications that derive from science have a positive feedback and enhance and extend the scientific work of a given area. Van der Waals' work represents a great example. He was mainly concerned with deviations from the equation for the ideal gas. However, the forces between the molecules are of great importance in many areas of science and engineering.

Interactions between positive and negative charges were already well known before van der Waals' work. He pointed toward a new phenomenon: even when molecules are overall neutral, i.e., have the same number of positive and negative charges, considerable interactions between them are still possible because of the so-called dipole forces that are illustrated in Fig. 2.31. These forces are strong over short distances and act only if the molecules are close to each other. Forces of this kind also occur between liquids and solids and between two solids. A very important effect of this kind is the so-called capillary action that is illustrated in Fig. 2.33.

A capillary is just a very thin tube. When such a tube is immersed into a liquid, two effects can happen. If the liquid and the tube material attract each other, because of the existence of electrical forces as we just have described them, the liquid "creeps" up the tube walls and is therefore pulled up higher. If the tube is very thin the liquid can move up many meters (yards) high. This is actually one of the physical mechanisms that helps plants transfer water from the ground to their leaves. Repulsive forces between liquid and capillary wall are also possible and also illustrated in the figure. Such forces are, for example, observed between mercury (which is a liquid metal at room temperature) and glass. The explanation of these repulsive forces is not as straightforward as the explanation

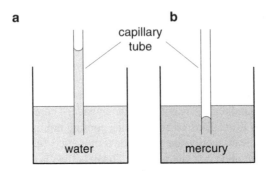

Fig. 2.33 Behavior of liquids in a capillary (thin tube). Part (**a**) shows the result if an attractive force exists between liquid and the inner wall of the capillary tube. Such an attraction exists between certain materials (e.g., glass) and water. The water is then pulled upwards. For very thin tubes, the water may be pulled up several meters (yards). Part (**b**) shows the result for a repulsive force as it occurs, for example, between mercury and glass: the mercury is pushed down. Note that the curvature of the liquid that is shown within the capillary occurs also at all other liquid-glass boundaries but is not shown

of the van der Waals forces was. The general theory that explains all of these forces, attractive and repulsive, was pioneered by Hendrik Casimir. Van der Waals and Casimir forces are of central importance in nano-science and nanoengineering, because the attraction and repulsion of nanostructures is of great importance for the workings of nanomechanical machines. Miniature laboratories made on "chips" may have movable parts that can get stuck because of attractive van der Waals forces. The interested reader is referred to Sect. 3.4 and to the Internet.

We finish this section with the following remark. Electrical forces such as van der Waals forces and Casimir forces are much stronger than gravitational forces such as the attraction of massive bodies by the earth. This is why water can be transferred against the gravitational pull to the top of a high tree.

2.3.5   The Random Motion of Atoms and Laws of Thermodynamics

Heat-related phenomena have a great significance for both science and engineering. From the engineering point of view, it is important to obtain the utmost mechanical energy from steam and combustion engines with the least amount of fuel. It is clear that we wish our cars to drive the largest distance with the greatest possible flexibility and power. The achieved miles or kilometers per gallon (or liter) of gasoline become of greater concern the more cars we use and the more gasoline is consumed. Gasoline is usually made from oil, and the oil has formed over millions of years from life-forms such as algae. Oil and gasoline have, therefore, received the name "fossil fuels." Fossil fuels are not easily replaceable, because their formation takes so long. Mankind has learned to create oils, alcohol, and other fuel types more quickly by using plants, algae, and bacteria that grow now and have not just grown in the past. These fuels are called renewable fuels, because they can be generated over and over, as our needs require. As these lines are written, however, the renewable fuels are more expensive to generate than fossil fuels are, and fossil fuels are therefore predominantly used, in spite of the fact that they cannot be replaced. We return to this topic in Sect. 4.1.4, and just note here that the supply of sufficient fuel energy to the population of the world is a very complicated and important problem. Therefore we must conserve fuels as much as possible and use engines (motors) that have the greatest mechanical energy output for the least fuel input. If we look at the engines that we have discussed above, however, we see that these engines are far from ideal. The two-stroke engine, for example, exhausts even fuel that is not completely burned, and all of these engines exhaust some hot gas, gas that carries still a lot of energy with it. The question therefore arises whether more ideal engines can be built, and engineers are continually thinking about this question. For example, there are investigations in progress about whether we can do better with electric motors. This is a great area for future STEM experts.

There is a fundamental limit to the efficiency of all engines that use heat in some form. This limit was investigated by the French physicist Nicolas Carnot. The limit is in essence based on the fact that heat is the random motion of atoms. Carnot found that we are never able to utilize all the energy stored in heat, because of this randomness of atom motion. What we wish to produce with all engines is, at the end, some mechanical power, as that of a moving piston. The pistons of engines move in a certain direction, and the gas that drives the jet engines moves also in a given direction, while heat is based on the random movement of atoms in all directions. In all the heat-based engines, we always lose some energy to the random motion of atoms and molecules. This fact is expressed by one of the basic laws of physics that is called the second law of thermodynamics (the dynamics involving heat). The second law says it is not possible to construct an engine (such as the engine of a car) that has no other effect than creating mechanical motion, while deriving the energy by cooling down some substance.

It is thus impossible that a stone just cools down and flies up into the air by using its own heat energy. This is, of course, rather obvious to us. Note, however, that the second law of thermodynamics has been checked and proven experimentally over and over and is considered as valid as the first law. The first law of thermodynamics has been discussed already above in various ways. The first law states, in essence, that heat is just a form of energy and that the sum of the various forms of energy, electrical, mechanical, and heat energy, stays constant in any closed system. This first law tells us, as we have learned on other occasion, that energy is conserved. It is therefore impossible to run a machine without supplying the energy that is produced by the machine in a different form (often mechanical or electrical). Usually the energy is supplied to the machine through some form of fuel.

It is very important to realize that we need machines to sustain human life in modern ways. We need therefore energy in the form of fuel or electricity or heat. It is also important to realize that running these machines always involves creation of some random form of energy. It is impossible for us to just cool down a heat reservoir and draw mechanical energy and do nothing else to the environment. The second law tells us that, if we wish to live and use energy, we must perturb the environment by creating heat and exhaust gases. All we can do is search for the most efficient machines that have minimal negative influences to our environment. This is a very complicated problem indeed, and one worthy for scientists and engineers to investigate probably for as long as humans populate the planet.

2.4   Chemistry and Quantum Mechanics

Chemistry and quantum mechanics are sometimes seen as separate subjects. However, chemistry had a unique influence on the discovery of quantum properties, properties that are connected to the fact that the world that surrounds us is made of elementary building blocks such as atoms, electrons, and protons. These building blocks are typically encountered and measured as a whole a so-called "quantum" and cannot easily be divided into smaller entities, although such a division is possible for the atoms and, as we will see, even for protons. Chemistry has discovered and researched basically all possible types of atoms and the way these atoms form bonds and molecules. All the materials that we encounter on earth can be understood that way, and chemistry is therefore a central science of great reach and importance. Quantum mechanics developed from the findings of chemical science and came up with a method to compute all the properties of atoms, and molecules, particularly the important energies that are involved in molecule formation. Very important steps were added by the Austrian physicist Erwin Schrödinger who found a wave equation that accomplished all of this quantitative understanding of molecules. Quantum mechanics developed then further and gave new meaning to the particles of chemistry. In particular, quantum mechanics tells us that all the particles that we can identify in nature have also properties that are typical for waves. There are some scientists who call all these entities "wavicles," and wavicles make up our world.

These introductory sentences, however, have brought us far ahead of the historic developments that have first provided an understanding of the atomic composition of our surroundings. We start the next section, therefore, with the discoveries of chemistry that have shown that the materials around us consist of different atoms or compositions of atoms (molecules), and then explain the properties of these atoms and of important molecules in terms of relatively simple rules. Only the last section deals with quanta in a more general way. There we discuss photons, the quanta of light, the emission and absorption of photons by atoms, and the energy range that is characteristic for the light emitted from atoms and molecules (the so-called spectrum). The principles of quantum mechanics are also used to derive the energies of electrons and the characteristic energies for the bonds between atoms in molecules.

1 H							2 He
3 Li	4 Be	5 B	6 C	7 N	8 O	9 F	10 Ne
11 Na	12 Mg	13 Al	14 Si	15 P	16 S	17 Cl	18 Ar

Fig. 2.34 The first 18 elements (atoms) of the so-called periodic system of elements, starting from the lightest (hydrogen) and ending with the heaviest (argon). Atoms arranged in the same column (such as H, Li, Na, or C, Si) have similar chemical properties. For example, He, Ne, Ar are called noble gases, have low chemical reactivity, and are odorless and colorless. The names of the atoms corresponding to the symbols are (H) hydrogen, (He) helium, (Li) lithium, (Be) beryllium, (B) boron, (C) carbon, (N) nitrogen, (O) oxygen, (F) fluorine, (Ne) neon, (Na) sodium, (Mg) magnesium, (Al) aluminum, (Si) silicon, (P) phosphorus, (S) sulfur, (Cl) chlorine, and (Ar) argon

2.4.1 Elements and Atoms

Considering what we have learned about gases, it is very surprising that only about one hundred years have passed since the raging debate between Boltzmann and Mach about whether or not atoms exist. It was clear already some time before Boltzmann that our surroundings are composed of a number of so-called "elements." One such element was hydrogen, another oxygen, both known to be gases but also known to occur in liquid and solid form such as in water and ice, respectively. The number of known elements was increasing over time. There were only 40 elements known to the famous French chemist Antoine Lavoisier. As these lines are written, we have 117 known elements with the possibility of creating still (very few) more by involving so-called nuclear chemistry (see radioactivity).

The Russian chemist Dmitri Mendeleev is credited with showing that many elements behave chemically in a very similar fashion to a few others. He created a list of elements in which the columns contain the chemically similar ones. The first 18 elements, listed that way, are shown in Fig. 2.34. More complete lists, called the periodic system of elements, can be found on the Internet.

Hydrogen and oxygen like each other and form water molecules when they interact or, as one says, chemically react. We can deduce from Fig. 2.34 and Mendeleev's rules that lithium and sulfur will also like each other and react similarly. To notice all these chemical similarities took, of course, a lot of time and detailed knowledge. Helium, neon, and argon have in common that they do not readily react with any other element on the list. They are singled out by this distinction and called the noble gases. Their opponents on the left side of the periodic system, hydrogen, lithium, and sodium, on the other hand, readily react with all the other elements on the right side of the table, except, of course, the noble gases. Lithium and sodium are metals and conduct electricity. Hydrogen is a gas that normally does not conduct electricity but also becomes a metal under extremely high pressure and is thus similar to lithium and sodium. Chemical compounds of the first and the seventh column are salts.

The salt that we use for cooking consists of the two elements Na and Cl. These stick closely together and form a "molecule," a new entity that behaves chemically different than its constituents. The attraction of Na and Cl arises from the fact that the sodium does not "hang on" to one of its electrons, while the chlorine not only likes to hang on to its own electrons but also takes easily an additional one. Thus the Na atom loses its electron to the chlorine and becomes positively charged, a so-called positive ion denoted by Na^+, while the chlorine becomes a negative ion denoted by Cl^-. Positive and negative charges attract each other as we know from Sect. 2.2.1 and therefore stick together. This type of bonding of atoms is called ionic bonding.

It turns out that the formation of molecules due to attractive electrical forces is a general feature that is the basis for all the chemical bonding of atoms that form molecules. The atoms on the left side of the periodic system tend to give electrons away and those on the right side tend to accept electrons. However, the degree of electron donation and acceptance varies significantly depending on the participating atoms. The forces of attraction can therefore range from reasonably strong, as for NaCl, to much weaker as for the van der Waals forces discussed in Sect. 2.3.4. The atoms in the middle of the periodic system, C (carbon) and Si (silicon) in Fig. 2.34, have a very special status. They like to donate electrons as much as they like to accept them. In other words they tend to share electrons in various ways and thereby form so-called "covalent" bonds with other atoms. These other atoms could also be carbon or silicon. For example, coal consists largely of carbon, and we are used to associate the word carbon with coal. However, under high pressure carbon atoms form a special (called "covalent") bond to each other. Indeed chemists have been able to fabricate diamonds out of pieces of coal by using very high pressure equipment. Carbon also forms bonds with other carbons and hydrogen, oxygen and nitrogen. Molecules of that kind are the nucleic acids and proteins that make up our body. It is, of course, the goal of chemistry to understand the details of the forces that hold all of these atoms together and lead to the formation of so many different substances. Chemistry can then use this understanding to produce molecules that are useful for humans in a great variety of ways, ranging from the salt in the kitchen to medicines for diseases. This is the reason for the great importance of the periodic system of elements and its understanding. A rough understanding can be gained by the electrical forces that we have just discussed and by the addition of a few rules. These rules deal with integer numbers of electrons that are preferably exchanged and shared between the various atoms of which the molecules consist. Modern quantum mechanics gives an explanation of these rules, and we will give a brief description of this explanation below and a more detailed one in Chap. 5.

2.4.2 Composition of Atoms and Rules for Molecule Formation

We have explained in Sect. 2.3.2 that the physical properties of gases follow from their atomic structure. There is a very large number (see Avogadro's number) of atoms in any larger volume (such as one liter) of a gas. The atoms themselves are very small, and it took, in the past, difficult experiments to determine their size. Nowadays we can make atoms visible with atomic force microscopes as described in Sect. 3.4. "Seen" with such a microscope, the atoms look like small spheres with a typical diameter of the order of 0.1 nm (10^{-10} m). The air that surrounds us is, of course, such a gas consisting of atoms. However, air appears to us like something continuous. For example, when wind blows in our face, or if we wave around with our hands, the air is everywhere. From such experience we could not possibly conclude that 10^{23} atoms of oxygen and nitrogen are impinging on our face or our hands and we could not possibly conclude that the space around us is almost empty except for the very small atoms. Yet, this is exactly how it is as we can see by use of powerful microscopes.

Thousands of years ago, when Democrit (Democritos in Greek) talked about atoms, he did have the idea that the matter surrounding us is composed of small particles that cannot be further divided. He had the right ideas of how things might be, but there was no way to prove any of his ideas. It took more than 2,000 years to find out that atoms consist again of much smaller entities called electrons, protons, and neutrons. We know now that the diameter of a proton or neutron is of the order of 10^{-15} m, and that of the electron is even smaller. One can, of course, not strictly speak of a diameter of an electron, because no microscope exists that would show us some sphere corresponding to an electron, neutron, or proton. We also know now, from detailed measurements and theory, that the neutron and proton have structure and consist of still much smaller entities, the quarks and gluons described in Sect. 5.4.

To understand the chemistry of atoms, it is most important to understand the interactions of electrons and protons and their consequences. Electrons are negatively charged and swirl around the positively charged protons with enormous speed and frequency of oscillations and turns. As we will see when we discuss quantum mechanics in more detail, no distinct pathways of the electrons can be determined. However, there are patterns to this swirling of the electrons that can be compared to the patterns of standing waves of strings and other objects as we have discussed them in Sect. 2.1.5. The patterns are, in general, three dimensional meaning that if we introduce coordinates (x, y, z), all three directions are of importance. Several forms of these patterns are shown in Fig. 2.35.

Protons and neutrons have a mass that is about 2,000 times the mass of the electrons, and are located in a very small volume, around the zero point of the coordinate system Fig. 2.35. The collection of neutrons and protons at the center of the atom is called the nucleus of the atom. Some of the electrons of atoms swirl around the nucleus in s-patterns that have the form of a sphere, as shown at the top of Fig. 2.35. The electrons can be anywhere inside the sphere and even outside the

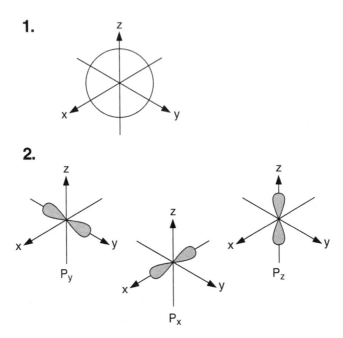

Fig. 2.35 Patterns of electrons "swirling" around the atomic nuclei that consist of protons and neutrons and are located in the center of the shown coordinate system. The *top* pattern (1) shows a circle that symbolizes the symmetry of a sphere and is called the s-type standing wave pattern or s-state of an electron. The probability of finding an electron is largest at the center of the sphere and diminishes rapidly away from the center. The *bottom* (2) shows the more complicated so-called p-type standing wave patterns. They have the form of the ∞ symbol rotated around the axis that is shown. The probability of finding an electron is again largest at the center of the shaded volumes and diminishes rapidly away from the axis around which the shaded volumes are wrapped. Other even more complicated swirling patterns (d, e, f) are also possible

sphere, and they propagate through the nucleus without any interaction other then the electrical attraction. The electrons also have a characteristic energy. We denote this energy here just by an integer number. For example, 1s means the electron exists in an s-type standing wave pattern and has the lowest energy. 2s means the electron is also in an s-type standing wave pattern but now at a higher energy with the label 2. We can also have 3s, 4s, etc. It is important to note, that the electron cannot get "stuck" in the nucleus because then it would loose its kinetic energy and such processes are not possible and energy must be conserved. Therefore, the electron must just continue in its swirling pattern forming a stable atom while moving and moving. There are also other patterns that electrons follow around the atoms. Three such patterns of the so-called p-type are also shown in Fig. 2.35. Again these patterns can correspond to different energies, and we then write 1p, 2p, 3p,... etc. There exist more complicated patterns that are important for the atoms beyond number 20, but we will not discuss them here because we wish to discuss only the major principles.

Two important rules that we need to remember are the following:

1. Each atom has an equal number of electrons and protons. This number equals the number given to the atom in the periodic system (hydrogen has the number 1, helium 2, lithium 3, etc.).
2. Each standing wave pattern with a given energy (labeled 1, 2, 3,...) can accommodate exactly two electrons. Energy 1 (the lowest energy) has only s-pattern standing waves, energies 2 and 3 have the possibility of both s- and p- pattern standing waves. The fact that each pattern with given energy can accommodate precisely 2 electrons, follows from the Pauli Principle named after the Austrian physicist Wolfgang Ernst Pauli and is related to the so-called "spin" of electrons discussed in Sect. 5.3.2.

Using these rules, we can find the electron patterns of the atoms of the periodic system of elements. We assume that the electrons will be in their lowest form of energy which is normally the case. Hydrogen (H) has only one electron. Therefore the hydrogen electron has a 1s pattern. Helium has two electrons and therefore two 1s pattern electrons according to rule (2). Because helium is a noble gas, this means that two electrons in the 1s pattern form a very stable and nonreactive atom. Lithium has two 1s electrons and one 1p electron. We note that lithium is very "willing" to give away the 1p electron because then it is left with the two 1s electrons exactly like a noble gas. We make now a jump and go to Neon. Neon has 10 electrons and therefore two electrons of each of the patterns $1s, 2s, 2p_x, 2p_y, 2p_z$. Neon is also a noble gas, and the pattern of Neon must therefore be a stable and unreactive one. Sodium (Na) has 11 electrons and therefore two electrons of the patterns $1s, 2s, 2p_x, 2p_y, 2p_z$ and one 3s electron. Again, in chemical reactions, sodium likes to get rid of the one 3s electron and then to assume the pattern of the noble gas neon. We make another jump to chlorine. Cl that has 17 electrons, two of each of the patterns $1s, 2s, 2p_x, 2p_y, 2p_z, 3s, 3p_x, 3p_y$ but only one more electron of the p_z pattern type because Cl has 17 electrons. The next higher noble gas is argon (Ar) with 18 electrons and therefore two electrons of each of the patterns $1s, 2s, 2p_x, 2p_y, 2p_z, 3s, 3p_x, 3p_y, 3p_z$ which again is a most stable and unreactive configuration. Therefore when compounds of chlorine are formed, Cl likes to obtain the one electron to approach the configuration of argon with 18 electrons.

We can now see how these rules explain the formation of the NaCl molecule. The sodium donates its one electron to the chlorine and the positively charged sodium has then the configuration of Neon, while the chlorine takes an electron, becomes negatively charged, and ends up with the configuration of argon. Thus we end up with a molecule of atomic constituents that do not like to react chemically anymore, because they are similar to two noble gases. In addition, the two atoms of the molecule stick together, because of the electrostatic attraction of positive and negative charges. This type of bonding is known as ionic bonding. Molecule formation can be generally explained based on the basis of these principles, even if the molecules are composed of many atoms. There are, however, complications when it comes to how the electrons are actually shared by the atoms. The donation of the electron to the other atom is in a way the simplest form of sharing. However, this type works only for the elements from the first and the seventh column. The sharing between two or more carbon atoms, that form a so-called covalent bond, can be very complicated. It took many decades of chemical research to find out how this sharing works in all the interesting chemical materials, starting from salts like NaCl and going to DNA molecules with millions of atoms. In the following, we discuss important molecules and explain how one can understand their formation from the properties of the constituent atoms and their place in the periodic system. We also introduce a shorthand way of writing or drawing the chemical composition of molecules that permits us to see how electrons are shared.

The way to write or draw the chemical composition of molecules is based on the three following facts: (a) the electrons of atoms form s- and p- type standing wave patterns corresponding to energies 1, 2, 3..., (b) the s- and p- standing wave patterns can accommodate each 2 electrons, and (c) the electrons of the various atoms are shared in molecules in such a way that the electron numbers of noble gas atoms are approached. For all molecules that we discuss here, this means that we will have constituent atoms with either two s-electrons corresponding to the lowest energy labeled by 1 (as in helium) or eight electrons, two s- and 6 p-type, as in neon (energy 2) or argon (energy 3).

The sharing of electrons by the atoms of the molecule is usually indicated by a line that represents two electrons. The line is drawn between the atoms that share the electrons. Sometimes electrons are simply represented by one dot for each electron. Expert chemists use lines for shared electron pairs and dots for the unshared electrons. The corresponding symbols are called Lewis structures after the chemist Gilbert N. Lewis. We do not emphasize such details of meaning when discussing chemical symbols here, because we just like to drive home the fact that, when molecules are formed, atoms attempt to achieve the noble gas configuration (configurations with eight electrons for all cases that we consider). If the sharing of electrons is important for the explanations, we will emphasize this in the text. Here are a few examples.

A gas of atomic hydrogen H is not stable because the hydrogen atoms like to form pairs with two shared electrons, one from each atom. The corresponding symbols for the hydrogen molecule are H–H or just H_2. One line corresponding to two electrons is called a single bond. The hydrogen molecule has thus two shared electrons for each hydrogen atom and, therefore, resembles the very stable helium atom (2 electrons of the 1s type) that does not tend to undergo further chemical reactions.

The air we breath consist mostly of oxygen O and nitrogen N. Atomic oxygen is also not stable and forms the pairs $O = O$ or just O_2 that are now held together by so-called double bonds, 4 shared electrons of energy 2. Each oxygen atom of the molecule has, in addition, 4 more unshared electrons of energy 2 that are often not indicated (symbol $O = O$), but could be by writing the symbol as $::O = O::$. Together, the total number of unshared and shared electrons of energy 2 is thus 8, exactly the number of the noble gas neon. Nitrogen forms molecules with three electron pairs shared by each atom plus two unshared electrons, resulting in $N \equiv N$ or $: N \equiv N :$ or just N_2. In this way, we have again the 8 electrons of energy 2 for each nitrogen atom, just as many electrons with energy 2 as neon has.

As we know already, for the combination of Na and Cl, we have no sharing of electrons; the electron simply moves from the sodium to the chlorine to form positively charged Na^+ and negatively charged Cl^- ions (each resembling a noble gas). Na^+ and Cl^- stick together by the electrical attraction and form NaCl molecules.

General molecule formation is a very complex problem and its corresponding chemistry requires significant expertise. For example, the electrons of oxygen can be shared in a different way then for O_2, and the earth's atmosphere contains small quantities of O_3, also known as ozone. The interested reader is referred to descriptions on the Internet. With more than three atoms and different types of sharing or nonsharing (as in NaCl), one deals with an enormous amount of different ways to form molecules. The formation, fabrication, and use of these molecules are the essence of the art of chemistry. We are not explaining how to actually fabricate molecules from their chemical constituents, but just introduce a number of molecules and their properties as well as their importance for humans.

2.4.3   Important Molecules: From Salt to DNA

No attempt is made to present a complete list of important molecules; there are simply too many. The molecules chosen are just representative examples to show how relevant chemistry is for our life and are supposed to stimulate interest. The true chemist and student of chemistry needs, of course, access to a lab and needs to perform experiments. There are also chemistry kits available for home use, and the Internet lists many of them.

Molecules in our Household

We have already discussed the chemistry of salt and have seen that the bond between Na and Cl arises from the electrical attraction after the electron is transferred from Na to Cl. This ionic bond is characteristic for all salts, also those that are not used for cooking such as LiF or MgF_2. You can construct all possible salts from the periodic table and from our rules. They always involve a metal such as Li, Na, or Mg on the left side of the periodic table and other atoms such as F or Cl on the right side. The metals like to donate their electrons, and F or Cl is happy to accept the electron, and the ionic bond is formed. The resulting ions must have the configuration of the noble gases, and this rule tells us how many ions we need, for example, two fluorine ions for one magnesium ion or just one sodium for one chlorine. As you may have noticed, we have not included into this discussion molecules that involve hydrogen such as HCl. Indeed such a molecule has features that are very similar to NaCl, and the attentive student may ask the question why we have not listed HCl as another salt? Indeed HCl is a very important molecule. However, there is a big difference between a sodium atom and a hydrogen atom that both have lost one electron: the sodium ion (atom minus one electron) has a noble gas configuration while the hydrogen ion (hydrogen minus one electron) is just a proton with no electron at all. This fact has the following consequences.

HCl molecules form a gas that can be dissolved in water. This solution is a strong electrolyte, meaning that the H and the Cl are completely ionized. In contrast to a solution of NaCl, however, the HCl solution is not just a salty liquid but is very reactive and attacks many substances. It is called a strong acid. Why is HCl much more reactive in certain ways than NaCl? The reason is the following. Hydrogen does indeed donate its electron and the chlorine takes it, just as we had it for Na and Cl. However, once hydrogen has donated its electron, there is no electron left, just the positively charged proton. The positively charged Na^+ owns still electrons and thus resembles the noble gas Ne. The proton does not compare to any noble gas. It actually does like to react with other atoms, unlike the noble gases do and therefore protons lead to acidity. In fact acids are just molecular entities that can supply protons. It is interesting that the name proton is of rather recent origin; it was introduced by Rutherford in 1919 during his investigations of atomic nuclei. Rutherford's discussion and naming of protons was seen at the time, and still appears to some, remote from the experiences of daily life. Yet, humans have tasted protons since ages in their drinks. There exist great varieties of acids. Weak acids are in many of our drinks such as lemonade or Coca Cola. Watery solutions of HCl form strong acids. Nevertheless HCl resides in our stomach. It helps turning solid food into liquid and starts the digestion process. HCl also destroys bacteria and therefore protects us from illness. Nevertheless, sometimes our stomachs contain too much acid and then the stomach walls are attacked and suffer, leading to stomach pain and ulcers. The medicines that are then prescribed are proton inhibitors, substances that inhibit the supply and availability of protons in our stomach.

The acidity of food needs to be controlled very carefully. For example, if we buy chicken soup, it has to be very close to "neutral" and not acidic at all. Food science and engineering require, therefore, a very careful control and measurement of acidity. The acidity has its own scale like temperature and is measured by so-called pH values. The symbol pH comes probably from "power of hydrogen." Pure water is said to be neutral (not acidic or tasting sour) and is assigned a pH value of 7. A lower number means that the substance is acidic, 1 being very acidic. The pH scale is a logarithmic scale which means in essence that every point moves you a factor of 10. A pH value of pH = 1 means that the substance is 10 times as acidic as pH = 2 and 100 times as acidic as pH = 3. A pH value above 7 means that the substance does not like to provide protons but instead likes to accept protons. Such substances are called bases. An example for a strong base is the molecule Na^+OH^- because it likes to accept the proton in the chemical reaction

$$Na^+OH^- + HCl \rightarrow Na^+Cl^- + H_2O; \qquad (2.71)$$

in words, the base NaOH added by the acid HCl results in salt and water. Salt is still a little corrosive but not as bad as the acids or bases are. If the simple ionic bond can already have so many consequences for the science of our food, imagine what

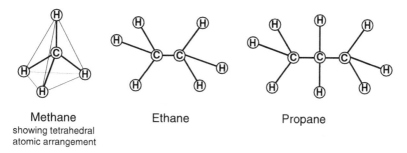

Methane
showing tetrahedral
atomic arrangement

Ethane

Propane

Fig. 2.36 Three so-called hydrocarbon molecules consisting exclusively of hydrogen and carbon. The nearest neighbor atoms of each molecule are arranged in the geometrical form of a *tetrahedron*. The illustration attempts to show this 3-dimensional arrangement

effect all the other forms of chemical bonds have, especially covalent bonds and the participation of a multitude of atoms.

Of particular importance are the following carbon-related molecules. Carbon (C) is a very special element because it is the first element that is located in the center (4th) column of the periodic system. It has therefore altogether 6 electrons, two each of energy 1 s-type and energy 2 s-type as well as energy 2 p-type. Thus it has 4 energy 2 electrons and needs another 4 energy 2 electrons to achieve the chemically inactive neon configuration. There are many different ways of receiving these 4 additional electrons by sharing with other atoms. The most straightforward way for carbon to get 4 additional electrons is to share electrons with 4 hydrogen atoms. CH_4 molecules are the molecules of the gas methane. The next more complicated way of sharing is to have two carbon atoms that share two electrons with each other, and each of the two carbons shares electron pairs with three hydrogen atoms. C_2H_6 is called ethane, and one can go on like this and find molecules of gases and liquids that all burn and combust well and are therefore of use for gas grills or to drive combustion engines. Gasoline is made of such molecules that are called hydrocarbons. Figure 2.36 shows the first three hydrocarbons. Note that each carbon is "connected" to 4 lines, with each line (or two dots) representing a pair of shared electrons and thus altogether 8 electrons.

The true shape of the molecules is, of course, three dimensional, and some attempt is made in Fig. 2.36 to indicate the three-dimensional arrangement of the atoms. It is important to remember that carbon likes to arrange its neighbors in the form of a tetrahedron: the four lines of shared electron pairs point to the corners of a tetrahedron as indicated in the illustration. In general, it is difficult to plot molecules in three dimensions. Some texts for chemistry try to introduce certain ways of shading the atoms to indicate in which direction they are pointing. We have not attempted to include any special way of 3-dimensional rendering. Many Internet sites have moving or rotating pictures of important molecules so that you can get a better 3-dimensional feel for them.

Adding different kinds of atoms such as oxygen (O) brings us to another important type of hydrocarbons. We know that the atom combination –O–H

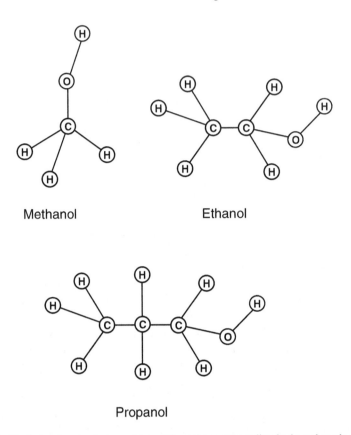

Methanol   Ethanol

Propanol

Fig. 2.37 Alcohol molecules that are obtained from the corresponding hydrocarbons by replacing one H atom by -O-H

"desires" to share an additional electron. Adding hydrogen that contributes one electron for the shared pair, one obtains the water molecule H–O–H or H_2O. We can also insert the –O–H group instead of one of the hydrogens in the molecules of Fig. 2.36 to arrive at new molecules that are shown in Fig. 2.37. These molecules are called alcohols. The second, ethanol, is especially well known as a component of drinks such as wine and liquors. It is also combustible and is often used as fuel for racing cars. Ethanol, particularly mixed with gasoline, is also frequently used as fuel for ordinary cars.

The molecules described above can be constructed from rather simple considerations of electron sharing. As we will see in the section on quantum mechanics, this sharing means that the electrons form something like a standing wave, an analog to the vibrational patterns of objects such as strings or drums, in between and around the various atomic nuclei. For single atoms, these patterns are, as we have learned, of s- and p-type (or d-, e-, f-type for the elements with higher numbers than those listed in Fig. 2.34). The molecules discussed above can still be explained in terms of these patterns and the tendency to form configurations that are close to that of

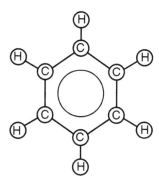

Fig. 2.38 Benzene molecule exhibiting a ring-like structure. Every *straight line* of the illustration represents the sharing of two electrons. The *inner circle* indicates six electrons that are shared by all carbon atoms and move freely between them. This special sharing of many electrons makes the molecule stable, because in this way the carbon atoms approach the noble gas electron configuration

the noble gases by sharing pairs of electrons. However, there exist many molecules for which the sharing is of different nature. An important example is benzene, a smelly liquid. The molecule of benzene has 6 carbon and 6 hydrogen atoms, and its chemical symbol is therefore C_6H_6. How on earth can the electrons be shared to approach the noble gas configuration (eight electrons of energy 2 for carbon and two of energy 1 for all hydrogens)? Well, there is a way, and it is shown in Fig. 2.38.

The first new fact that we notice is that the molecule contains a closed ring of carbon atoms. This closed ring provides, of course, new possibilities of electron sharing. We could, for example, have two hydrogen atoms for each carbon leading to a molecule C_6H_{12}. Such a molecule does indeed exist and is called cycloalkene. The benzene molecule, however, has another novelty in it. This novelty is the sharing of carbon electrons among all the carbons. We can see that ring-like structures will therefore add numerous possibilities to molecule formation. The number of possibilities becomes even larger, if we admit atoms other than carbon into the ring itself. The possibilities of sharing electrons between all of these atoms increase significantly, and the understanding of this electron sharing between many atoms including rings becomes a real art. Chemistry is therefore divided in subdisciplines such as inorganic chemistry (the chemistry of NaCl, HCl, etc.) and organic chemistry which is in essence the chemistry of carbon-related molecules. Biochemistry is related to the molecules of living beings that are carbon-related.

We finish this section by mentioning a molecule of greatest importance: glucose. The name derives from the Greek word "glukus" for sweet. Glucose is a form of sugar that our cells use as a source of energy. It exists as a string of carbons with oxygen and hydrogen atoms attached resulting in the molecule $C_6H_{12}O_6$. There exist also several variations of glucose that contain a closed ring in which one of the carbon atoms is replaced by oxygen.

Sugars are of far broader use and presence than just for the energy supply of many life-forms including humans. Sugar molecules can be put together in long

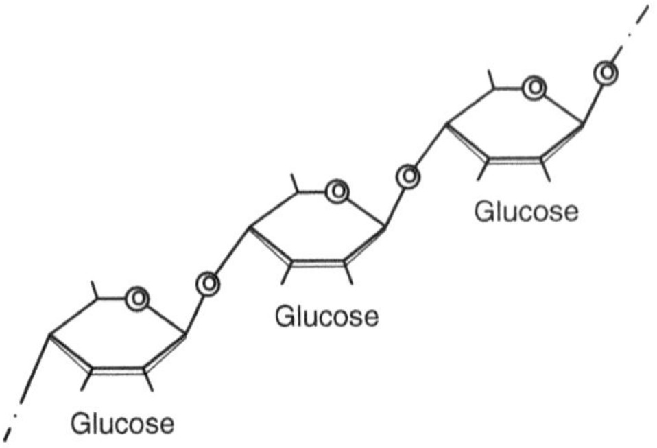

Fig. 2.39 Schematic of a
single strand of cellulose, a
very long molecule. Many
such intertwined strands form
the basic material for plants

Glucose

Glucose

Glucose

chains that consist of up to 10,000 glucose molecules connected by oxygen. These chains then form cellulose, a very important basic material. The long molecules intertwine with other long molecules to form the stable material of which green plants are made. Such a material made of intertwined chains is called a polymer. Poly is the Greek word for many. Plastic materials are polymers and are very important for our daily life. Figure 2.39 shows a single strand of cellulose.

You can imagine the enormous possibilities of constructing and using molecules, be they chain-like or ring-like and polymers of them that form three-dimensional structures and materials of all kinds. This is naturally great fun for the real chemist. Chemists analyze molecules and also put them together (synthesize). They have synthesized enormous numbers of useful molecules including the most significant and important of all: DNA. DNA is what we discuss next.

Molecules We Are Made of and Molecules that "Make" Us

The molecule that is central to life, including human life, is the DNA molecule. This is a giant, long molecule like cellulose. Remember that cellulose consists of smaller sections of sugar-like molecules that are repeated and repeated all over and connected by chemical bonds. DNA has a "backbone" or "skeleton" (also "spine") made out of repeated sugar-phosphate units (see Internet for the phosphate molecules) to form long strings of repeated molecule sections. However, the really important ingredients of the DNA are chemicals that are denoted by the letters A (adenine), T (thymine), C (cytosine), and G (guanine). Early chemical analyses by Erwin Chargaff had shown that each giant DNA molecule contains equal quantities of adenine and thymine and also equal amounts of guanine and cytosine. This was an important clue for the two young scientists, James Watson and Francis Crick, who came up with the precise structure of the DNA in 1953. What does DNA actually do and how does this molecule look like?

DNA contains a code that dictates how proteins, the materials that make up our bodies, are produced. Proteins, in turn, determine the general properties and functionality of our bodies, such as muscular strength. The code of the DNA is similar to any code that can be expressed by a few symbols, such as the binary codes that form computer programs and are based on the symbols (bits) 0 and 1. In the case of DNA, we deal with the four symbols A, T, C, and G that stand for the corresponding molecules. A message of the DNA is therefore equivalent to a string of these symbols, as for example

$$\text{ATCGATTGAGCTCTAGCG.} \tag{2.72}$$

According to Chargaff's rule, one is forced to assume that the DNA that contains the above string will also contain a second string of code given by

$$\text{TAGCTAACTCGAGATCGC,} \tag{2.73}$$

because we need to come up with equal numbers of (A, T) as well as (G, C), respectively. Thus, the code of the DNA must occur in two strings, the second of the two being obtained from the first by exchanging adenine (A) with thymine (T) and guanine (G) with cytosine (C). Only in this way can we end up with a given arbitrary code and equal quantities of adenine and thymine as well as guanine and cytosine. The two lines above look like the "positive" and the "negative" of film. At the time when film was still very important, the words positive and negative reminded everyone immediately of making copies of photos. Nowadays, with digital cameras that work without paper, different thoughts are connected to the processes of making copies. Nevertheless, it was an important clue for the discoverers of the DNA structure that the process of copying DNA molecules that we describe below, was somehow already contained in the structure of DNA and could be, and as we now know is, the basis of the reproduction of life. The actual DNA molecule is indeed made of two strands of code, like the ones shown above only much longer. These two strands are mounted on a sugar-phosphate skeleton, and the skeleton and strands are twisted to form the so-called "double helix" that is shown in Fig. 2.40.

The word double helix indicates that there are two spirals nested in each other. The outer boundary of the spirals is formed by the sugar-phosphate skeleton, and the spirals are bound to each other, because the A, T molecules and the G, C molecules match up and stick together. This sticking together is accomplished by hydrogen-sharing bonds, as we have discussed them in connection with van der Waals forces and the sticking together of water molecules that was illustrated in Fig. 2.31. The A, T, G, C molecules contain rings of carbon with nitrogen and oxygen substitutions. The hydrogen bridges are located between the nitrogen atoms or between nitrogen and oxygen atoms and are shown together with the molecules in Fig. 2.41.

Any given cell of any living being, for example, a bacterium or a human being, contains at its core such DNA molecules. The number of A, T and G, C pairs, that a cell contains, depends on the sophistication of that being. For humans, the number of pairs is about 3 billion. The first artificial DNA, as synthesized by G. Craig Venter and colleagues, was for bacteria and contained "only" about a million pairs. In spite of the rather small number of (A, T) and (G, C) pairs of this artificial DNA, the experiments of the Venter group have a great significance. The artificial DNA was introduced into lifeless bacterial cells that did not contain DNA. These cells started

Fig. 2.40 A section of a DNA molecule. The winding "tapes" symbolize the sugar-phosphate skeleton. A code of molecular sequences of (A, T) and (G, C) pairs is also shown. Notice that we are attempting to show a 3-dimensional molecule. Shorter molecule "connections," as, for example, the =T A= at the *top* of the figure, indicate only that the connection points out of the page, while the longer connections are parallel to the page

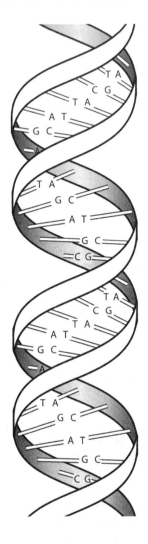

then to divide and multiply just as natural living cells do. This means that the science of DNA is a step closer to a predictive and experimental repeatable understanding of bacterial life-forms. The process of DNA division and replication and multiplication is very similar for all life-forms and works as follows.

In the first step the DNA must be unwound and "unzipped" into two strands, a "positive" and a "negative," as discussed above. This is accomplished by the presence of a chemical (enzyme) called helicase. When this chemical is supplied, the cell starts to divide. Then the hydrogen bonds are broken and the two strands of the double helix are pulled apart forming two single strands. In the next step these two strands are immersed into a different "soup" that contains another enzyme called DNA polymerase as well as a sufficient supply of A, T, G, and C molecules, and each molecule of each strand will pick up its exact partner. On the other side

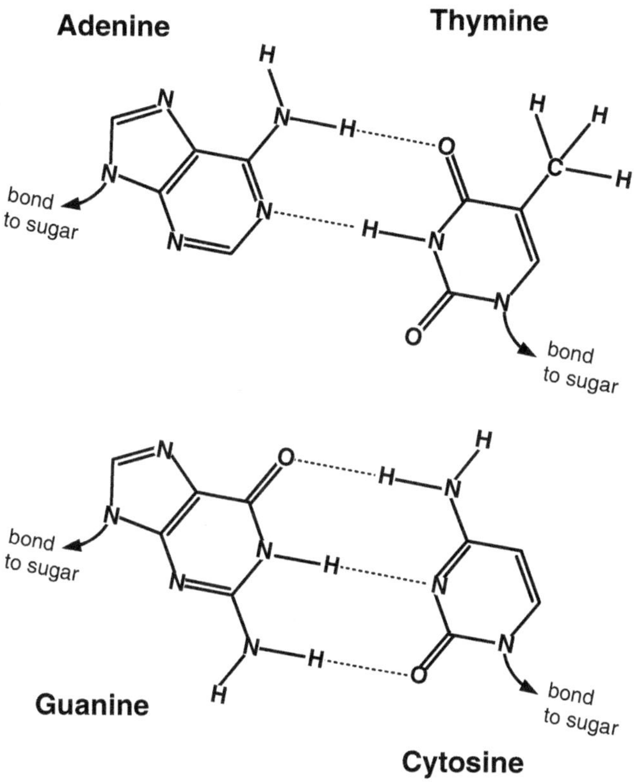

Fig. 2.41 The molecules adenine, thymine, guanine, and cytosine contain rings that include carbon (not indicated but located at *ring corners*) and nitrogen (indicated by N). Adenine and thymine are connected by two hydrogen bonds, guanine, and cytosine by three. The bonding to the sugar-phosphate skeleton (not shown) is indicated by *arrows*

of these new partners, the sugars and phosphates join together and form the new skeleton, and now you have two new DNA double helix strands. The double helix strands are the core material of the so-called genes that contain all information that makes our cells "tick," the information and properties that we inherit from our parents, and that we hand down to our offspring.

Of course, something needs to be done with the information that the DNA provides. Some materials need to be formed, in order to end up with new beings, be they bacteria or humans. This process of transforming the code of the DNA into actual materials is by now reasonably well understood. The materials, that are built by the coded information, are the so-called proteins. Proteins are the stuff we are mostly made of; they form our organs and muscles, and are polymer materials (like cellulose, the basic materials for plants). Proteins are composed of only 20 different chemicals that we either obtain from food or that our body produces. These 20 different chemicals are called amino acids. The formation of any given amino acid takes triplet DNA code sequences, such as CAA or CAG, that are called codons. Each of the elements of a triplet may be one of the 4 basis A, C, G, and T. Therefore, there exist $4 \cdot 4 \cdot 4 = 64$ different code triplets that lead to the production of the amino acids. Because we have only 20 amino acids and 64 codons, the coding is redundant. For example, both CAA and CAG lead to the production of the amino acid called glutamine. A string of 300 such codons instructs the cell to build a protein consisting of 300 specially arranged amino acids. All the complicated proteins, that our body needs and consists of, can be generated by codon strings. It is interesting to note that almost all life-forms, from algae to humans, use the same codons, hinting that all these life-forms have developed in similar ways. This fact is one of the cornerstones of the theory of evolution teaching the evolution of more complex life-forms from simple ones.

The details of this story, how cells multiply, how information is transmitted from the parents to offspring, and how all the materials that are involved are coded and produced, all of these, are currently explored in highly interesting and complex research projects. These projects are relevant to everything that concerns us humans, from health to illness and to hereditary traits. The projects encompass the composition of the materials that make our bodies and extend to complicated personality properties such as a hot temper. There is hope that all of these important problems related to how we live and die will one day be understood as well as Euclidean geometry is understood now. This is a wonderful playground for those interested in STEM and, particularly of course, those interested in biochemistry.

Crystals: Metals, Semiconductors, and Insulators

Cellulose and DNA are giant molecules. Many intertwined cellulose strands form leaves and stems of plants and can form all kinds of 3-dimensional shapes. The DNA in the human genes contains billions of A, T, G, and C molecules and would be several inches long if stretched out. Characteristic of both types of giant molecules is a lack of regularity. This lack of regular arrangements of atoms arises, in the case of cellulose, from the intertwining of the strands that is typical for polymers. Polymers are, thus, not very "regular." In the case of the DNA, it is the codes that can be all different, and the lack of regularity is grounded in the very nature of DNA and its complicated coding system. We can compare this lack of regularity with that of books: there is some regularity in the arrangement of the letters and words, but the text is naturally all different and describes different things. Crystals, on the other hand, are giant collections of atoms (or molecules) with a completely regular arrangement of the atoms (or molecules). The Internet teaches you how you can easily grow salt crystals of centimeter (0.01 m) size. Such a crystal contains more than 10^{23} Na and Cl atoms that are all arranged in a regular pattern as shown in Fig. 2.42. The NaCl crystal is, electrically speaking, an insulator, meaning that it does not conduct electrical currents. The Na^+ and Cl^- ions are fixed in the crystal and cannot be moved by normal electrical forces such as available from a battery.

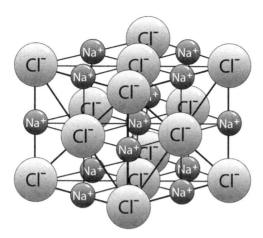

Fig. 2.42 Geometrical arrangement of Na^+ and Cl^- ions forming an NaCl crystal. Notice that the arrangement of the atoms is cube-like as indicated by the lines. This cubic shape is also maintained for very large crystals. The atomic configuration of a large crystal is obtained by simply continuing the shown pattern to all sides in such a way that any given "face" of the cube becomes the first layer and face of the next cube. All such cubes then flawlessly assemble to form the big crystal

The electrons of the two ions are also localized with these ions and cannot easily move to neighboring ions. Ordinary electric fields are simply to weak to move the electrons from one such ion to the other.

Sodium (Na) by itself can also form a crystal, and such a crystal has also the symmetry of a cube. However, the sodium crystal is a metal and an excellent conductor of electricity. The reason is simply that sodium likes to give away one electron when connecting with other atoms, in order to achieve the neon configuration. Therefore, all Na atoms of the crystal give away and share the outermost (energy 3 and s-type) electrons of the single sodium atoms. Because these electrons are shared by all atoms, they can freely move in between them, and thus sodium crystals conduct electricity. The bonding of the sodium atoms by all the shared electrons is called metallic bonding. This type of bonding is relatively weak and is characteristic for all metals (this is why gold can be formed into jewelry).

Crystals made of elements from the 4th column have different and very special properties. Carbon atoms can form a variety of crystals, depending how they are arranged and share electrons. Graphite is one of these carbon crystal types and conducts electricity as metals do. Graphite consists of sheets (thin two-dimensional layers) of graphene that sit on top of each other and are only loosely held together by weak bonding forces. Graphite is relatively soft, because the two-dimensional layers can slide on top of each other. They also can be pealed off, and graphite is, therefore, used as the writing material in pencils. The single graphene layer sheets themselves, however, are much more robust and show that the graphene bond is much stronger than that of metals and is based on a special sharing of the electrons. The best known carbon crystal, with very strong bonding, based on another similar type of special electron sharing, is diamond.

Fig. 2.43 Arrangement of carbon atoms in a diamond crystal. Notice the tetrahedral arrangement of the nearest neighbors and the resulting shape of a cube. A large crystal is obtained, as in the case of NaCl, by continuing the pattern to all sides

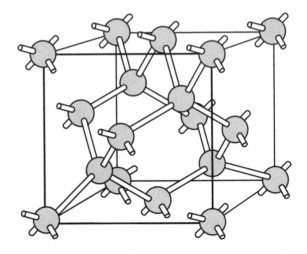

Diamond is known to be a very tough substance that cuts glass, and pure diamond is an insulator, that is, it does not conduct electricity. This fact indicates that the bonding of carbon atoms in diamond is not based on the metallic sharing of electrons. The carbon atoms of diamond are located in a tetrahedral arrangement of the nearest neighbor atoms as shown in Fig. 2.43. As also shown in the figure, the atoms still fit in the form of a cube with every face of the cube containing atoms at each corner and one in the center. Such a crystal lattice (the set of all dots that symbolize the atoms) is called a cubic face-centered crystal lattice. The bond between the carbon atoms is called a covalent bond, meaning that basically only neighbors share these electrons.

The lack of electrical conduction in diamond crystals is only observed for "pure" crystals. If we introduce into the diamond small amounts of atoms of a different kind, such as the right-hand neighbors (in the periodic system) nitrogen (N) or phosphorus (P), then diamond does conduct electricity. The reason is simply that N and P have one more electron compared to carbon, and they like to give away this electron to the carbon atoms. This additional electron can then be shared by all carbon atoms and can therefore contribute an electric current. Such additional atoms are called "donor atoms," because they donate electrons to the insulator and make a conductor out of it. One cannot put arbitrary amounts of donor atoms into a crystal, because this would either ruin the crystal or lead to the formation of a different crystal. Usually one can only introduce as much as about 1 % of the donor atoms relative to the crystal atoms. Therefore the conduction of such crystals with donors is not as good as the conduction of gold or the Na crystal that are highly conducting metals. Crystals like diamond, particularly when supplied with donors, are therefore called semiconductors. The important factor for semiconductors is that the conduction can be controlled and therefore "engineered."

There is another very important factor that makes semiconductor materials so important for electrical engineering. This is the possibility of introducing so-called acceptors such as boron (B) and aluminum (Al) atoms that are to the left of C in the periodic system. Boron has one electron less than carbon. Therefore it tries to take away one electron from the rest of the crystal so that it would be configured similar to the carbon atoms. Because an electron is taken away, a positive "hole" (lack of negative electron) is created in the carbon crystal. Other electrons of the carbon crystal can jump into the hole. Such a jump creates another hole, and therefore positive holes can move and conduct electricity as the shared electrons do. A hole is thus a *missing* electron and is shared by all crystal atoms. It can be regarded as a freely moving positive charge that contributes to conduction. The number of holes can be chemically controlled and engineered.

As we will see in later chapters, the conduction by negative electrons and positive holes is the secret of all of modern electronics. Electrons and holes in semiconductors form the basis of our computer revolution. You might say now: "yes, but are diamonds not too expensive for electronics?" The answer is, of course, yes, they are! In addition, diamond crystals are much more expensive if they are bigger, like the ones in the crowns of kings and queens. Fortunately, there are more inexpensive crystals that can be used for electronics. Silicon is in the same column of the periodic system as diamond and has therefore very similar properties. The electronics industry uses silicon to grow their crystals. Silicon crystals are easier to grow and do not require the high pressure that diamond formation needs. Silicon crystals can be grown with diameters up to around 0.6 m. These are really giant crystals that can then be sliced into thin sheets called wafers. Wafers are used to produce the electronic chips that are in our computers, in cell phones, and in almost all instruments of our households. More of this will be discussed in Sect. 3 on engineering and technology.

2.5 Energy of Atoms, Electrons, and Photons

From the above description of atoms, of molecules, and of the electromagnetic forces, it follows that almost everything that surrounds us, all that we can touch, smell, and see, is either made out of atoms (positive nucleus plus negative electrons) or can be described by electromagnetic waves including visible light. This section describes what we know about the nature of electrons and light and some of the important equations for the energy of these entities in atoms, in molecules, and in free space. The classical description of physical phenomena is usually given in terms of either particles or waves. Billiard balls are particles, and the wireless communication discovered by Hertz is accomplished with electromagnetic waves. As physics evolved into its modern form, it was found that nature does not present us with a clear distinction between particles and waves, certainly not when we deal with the very small, with atoms, electrons and molecules. As science penetrated down to the atomic scale, it was found that the clear distinction between particles and waves is lost and the classical laws need to be replaced by an "amalgamate" of wave and particle descriptions. This amalgamate is known as the quantum theory of matter and is described in this section in a phenomenological way and in more mathematical detail in Chap. 5.

2.5.1 Light: Waves and Particles

Newton thought of light as particles. Maxwell came to a different conclusion as discussed in Sects. 2.2.4 and 5.3. His wave equation for electromagnetic phenomena definitely suggested that light is an electromagnetic wave. Hertz followed Maxwell's theory and created such waves in the laboratory and demonstrated the possibility of wireless communications. The problem was not settled, however, and investigations of the emission of light from heated and glowing bodies could not be explained by a theory based only on waves.

Photons

Max Planck studied the available data for light emission as a function of temperature and came to the following conclusion that he presented to the German Physical Society:

"We found—and this is the essential point—that the energy E of light must be thought of as being composed of a given number of equal parts. The energy E of these parts can be determined from the equation:

$$E = h\nu, \tag{2.74}$$

where h is a constant given by $h = 6.626\,10^{-34}$ Joule seconds (J s) and ν is the frequency of the light or any other electromagnetic radiation (such as infrared)."

It is pretty amazing how clearly Planck expressed himself. He talked to a room of experts and what he said is, in its essence, understandable to everyone who can read a simple equation. Nevertheless, it was Einstein who made Planck's statement entirely clear. Einstein stated that these "equal parts of energy" are the particles of light. He further postulated that the energy shown in the above equation is the energy of one such light particle. Particles of light are nowadays called photons, and Eq. (2.74) is now called the Einstein–Planck equation. The mathematical-physical thought-process of Planck and Einstein, which took them from the continuous light waves of Maxwell to the postulate that these waves were actually made up of particles (photons), is called quantization. Remember the digital representation of the grey shades and colors of pixels. This representation of shades and colors by binary numbers is also a form of quantization, a quantization that is completely man-made and is introduced only because electronic handling of photos works nicely that way.

The quantization of electromagnetic fields as formulated by Planck and Einstein is, on the other hand, a quantization that we have to introduce, because otherwise we cannot explain the experiments with light. We can see from Eq. (2.74) that these equal parts of energy, the energies of the photons, are extremely small quantities, because Planck's constant h is extremely small. The unit of h is the unit of what one calls the action, which is energy multiplied by time (expressed above in Joule seconds). This complicated unit is needed because the above formula tells us that if we multiply Planck's constant by the frequency ν which has the units $[\frac{1}{s}]$ (one over seconds), then we obtain the energy. The frequency of visible light was discussed in Sect. 2.2.4 and is around $6 \cdot 10^{14}[\frac{1}{s}]$. This means that the visible photons have an energy of about $4 \cdot 10^{-19}$ J, which is a very small number. The energy radiated by the sun onto one square meter of the earth during one second is about 100 J. This means that about $\frac{100}{4 \cdot 10^{-19}} = 2.5 \, 10^{20}$ photons hit every square meter of the earth during one second. This is an enormous number. As a homework problem, you can calculate how many photons the whole sun emits during a second. For this calculation you need to imagine a sphere with a radius that equals the average distance of the sun from the earth and you need to calculate the surface area in square meters. It is because of the fact that the numbers of photons is so large that we really do ordinarily not realize that photons behave like particles. This fact can be compared to the case of the pressure of gases that is caused by very many atoms whose single effects and impacts we do not feel at all. We also cannot see or feel single photons. We know, however, that in large numbers, they follow exactly the wave equation and the rules of Maxwell. Therefore photons must be some strange mixture of wave and particle that we called previously already a wavicle. We will see below that such a mixture of wave and particle properties also describes electrons, protons, atoms, and everything else we know. They all behave like waves, yet when we actually measure some quantity, then this gives always the result that the total energy must be thought of as being composed of an integer number of equal parts, just as we would expect for particles.

We know from Sect. 2.1.5 that any wave has a wavelength, the distance between two valleys or two maxima of the wave. Therefore we need to be able to associate a wavelength with electromagnetic waves such as light or radio waves. This is indeed easy to do with radio waves because we can measure the electric field at a number of points in space and what we get is indeed the form of a wave. This type of measurement is already difficult for visible light, because the wavelength is small (of the order of 10^{-6} m) and becomes even more difficult for X-rays that have a very very small wavelength. However, it still can be done because of the phenomenon of diffraction that was explained in Sect. 2.1.5. We have discussed this phenomenon in some detail, because it is at the heart of the wave- and particle-like nature of everything we know, at least as far as I understand it. Diffraction shows that light and all electromagnetic radiation behave like waves do. At the same time we know from Planck and Einstein that such radiation always comes in "lumps" of energy $h\nu$. These facts seem to be in contradiction to each other. But wait, it gets even more interesting when we look at electrons and other "particles."

Wavicles

Diffraction is a general effect for all wavelike phenomena and works for water waves, radio waves, light, and any other wave. The amazing thing is now that diffraction effects can be measured also for electrons, protons, and all other building blocks of our universe, because everything we know is made out of "wavicles"; all constituents of our world behave part time as particles and part time as waves. As strange as this may sound, this is a basic truth of our world, and because it is so basic, we need to discuss it in more detail. Before we do this, we note that the exact calculation of what happens if photons hit gratings (one-, two-, or three-dimensional gratings including crystals) must include all possible pathways the photons can go and not only the two pathways that we have shown in Sect. 2.1.5. This calculation is indeed possible and was pioneered by Richard Feynman. Feynman also showed that this type of calculation works for all wavicles that we know, and he designed a general method that always can be used. This method is called Feynman's path summation or path-integral method. This calculation with all its bothersome details is painfully complicated. Fortunately it can often be replaced by simple rules that one deduces from the addition of a few such photon paths, and there are also modern computer software packages that can deal with it.

2.5.2   Electrons: Particles and Waves

Electrons were at first considered to be particles, in the sense of macroscopic bodies like billiard balls, that had concentrated their mass in a certain volume and also had some charge concentrated essentially in a point. This incorrect, or at least incomplete, analogy received a fatal blow when Louis de Broglie suggested that electrons may also behave like waves, at least partly so. Experiments, that involved crystal materials that can be viewed as natural gratings with a spacing of the lines (atoms) of about 10 nm or less, did confirm the wavelike properties. Electrons were reflected by the crystal gratings exactly as if they had a certain wavelength. This wavelength depends on the momentum $p = mv$ of the electrons, where m is the electrons mass and v its velocity. Louis de Broglie suggested the following equation:

$$\lambda_{dB} = \frac{h}{p}, \tag{2.75}$$

where h is Planck's constant. This means that we should imagine that the electron is something like a very small particle-like "blob" that behaves also like a wave. The size of the blob is about a few wavelength λ_{dB}. If we wish to calculate that wavelength, we need to know the velocity of the electron, which may, of course, greatly vary depending on circumstance. As is known now from many experiments, the electron does not really stand still under any practical circumstance. In fact, in atoms, the electron velocity is of the order of $3 \ 10^6 \frac{m}{s}$, which results in a momentum of $2.73 \cdot 10^{-24} \frac{\text{kg m}}{\text{s}}$ because the mass of the electron is $9.1 \cdot 10^{-31}$ kg. From Eq. (2.75) we obtain then a de Broglie wavelength of about $\lambda_{\text{dB}} = 0.24$ nm. This is a very rough estimate for the approximate value of the de Broglie wavelength that is encountered in atoms, molecules, and crystals. Thus we can imagine the electron in an atom to be a "blob" of a size around 0.2 nm having some wavelike structure with that small wavelength. This is for all practical purposes then a very small "particle" and the wavelike nature can only be observed with gratings (or any kind of structures) that have a very small spacing (around 0.2 nm). This explains why the electron has been considered in the past to be a very small particle and not a wave.

Only after de Broglie had the idea that the electron may also have some wave properties, did scientists perform measurements with such small gratings or structures. They found that electrons indeed show a wavelike behavior and exactly as if they were waves with the wavelength given by de Broglie. How did de Broglie get this great idea? Well, this was a stroke of genius based on Einstein's theory of relativity and will not be described here. Here we just describe the consequences of this idea. One consequence is that the wavelike nature of electrons is very important for atoms. Atoms have nanometer size and the electrons around atoms cannot be regarded as particles but must be treated as wavicles. It is unfortunately not easy to give a picture of what happens to the wave of electron(s) around the atomic nucleus (proton in case of hydrogen). The reason is that the electrons around a nucleus do not have a constant velocity and therefore do not have a constant wavelength. Playing with a lot of different wavelengths to imagine how an electron might behave around a nucleus or proton is a complicated project, more complicated than visualizing a cell phone antenna surrounded by electromagnetic waves. As for antennas, what one needs is a mathematical equation, the wave equation of Maxwell (see Chap. 5), which provides us with the actual result for patterns of such complicated waves. Such an equation has been found for electrons and other wavicles by Erwin Schrödinger, and we present this equation and its solutions for the electron energies in Chap. 5. De Broglie's idea of using waves for the electrons in atoms came only to full fruition after Schrödinger found his great wave equation.

2.5.3 Electrons and Atoms: Standing Waves and Energy Spectra

We do know that every atom has a very small nucleus consisting of the relatively heavy protons (a proton has about 2,000 times the mass of an electron) that also carry a positive charge. There are also neutrons in the nucleus, but we can disregard them for the present discussion. Each atom contains an equal number of negatively charged electrons and positively charged protons and is therefore overall neutral (without net charge). If we consider a hydrogen atom only, then we deal with exactly one proton that forms the nucleus and one electron. The massive proton can be regarded as the center of the atom. The electron swarms around the proton and is attracted by the proton's electric charge but otherwise does not interact with the proton at all. The energy of the total system is always conserved. Therefore, if the electron is far away from the nucleus, it has a low kinetic energy but a high "potential energy." The potential energy is the energy the electron can potentially obtain by being accelerated due to the electrical attraction toward the nucleus. At the place of the nucleus, the electron then has a very high kinetic energy. Does it hit the nucleus and destroy it? No, the electron does not interact with the proton, it goes right through it and toward the other side. There it loses kinetic energy because the proton now attracts the electron against its motion. In this way the electron oscillates or "vibrates" around the nucleus trillions of times per second. Together with these vibrations of the electron, we also have rapid fluctuations of the electromagnetic fields of electron and proton, because of the change of the position of their charge. It appears therefore that there is much change going on in an atom while at the same time atoms are stable and, if not disturbed, exist forever. This reminds us of the discussions of ancient philosophers.

Description of Atoms: Analogies and Probability

Thousands of years ago, the Greek philosopher Parmenides claimed that change is impossible and existence is timeless, while Herákleitos of Ephesus maintained that everything is in flux and you "can not step twice into the same river." Perhaps nature favors a mixture of the two possibilities as it is described in Conrad Ferdinand Meyers poem "The Roman Fountain":

> Up shoots the beam, and falling fills the marble basin's round,
> That veils itself and flows into a second basin's ground.
> The second gives, it grows too rich, the third its waving crests
> And each of them just gives and takes and flows and rests.

The scientists who founded quantum mechanics have also merged these seemingly contradictory views of change and stability by giving the electrons and other particles wavelike properties. Waves mean that there is change. However, the waves that make up the atom are thought to be standing waves, and standing waves are stable, such as the vibrations of a guitar string, and can last for long time. They may last forever and ever because there is no loss of energy in an atom (unlike the loss of energy of the guitar string to the guitar body and the surrounding air). The dance that the electrons, the protons, and the electromagnetic fields perform in an atom is stable and appears constant and unchangeable to the outside world. Indeed, to disturb that "dance," one needs a significant amount of energy. To remove the electron from the hydrogen atom, we need the energy of 13.6 eV. This large energy is usually not available. Therefore, we can think of the atom as a stable unit, representing some form of standing wave of the involved wavicles (electrons, protons) and the electromagnetic field. We know that this unit will stay stable as long as less than the critical energy is supplied, for example, by shining light onto the atom.

A word needs to be said here about oversimplified pictures of atoms that still persist in the literature and on the Internet. These pictures go back to the work of Rutherford and describe atoms analogous to our solar system: the atomic nucleus corresponds to the sun and the electrons to the planets. It is usually stated that one really should not think like that, but then, the names that are used to describe atoms are still reminding everyone of the solar system. For example, the electron patterns around the nucleus are called orbitals or shells. Remember that the path of a satellite is called the orbit, and ancient superstition had the planets move in crystal spheres. This analogy is now known to be totally incorrect, and we therefore have avoided its use. We use only the term standing wave pattern to describe the electrons swarming around the nucleus. How false the planetary analogy is becomes clear when one realizes that, for the s-type standing wave pattern, the electron is frequently near the center and propagates through the nucleus. In the planetary picture this would mean the planets would dash frequently through the sun. The electrical force is also trillions and trillions of times stronger than the gravitational forces between sun and planets. We therefore ask the reader to stay away from the planetary analogies when talking about atoms. Bohr taught us that new thinking is necessary when we discuss the physics and chemistry of atoms. He suggested the following words of wisdom. We are now in a new field of physics, and we know that here the old concepts do not work, because otherwise atoms would not be stable and exist forever. However, when we wish to speak about atoms, we must use words, and these words can only be taken from old concepts, from the old language. Therefore we have a dilemma.

Bohr's words are, of course, of great wisdom. For example, how should one understand and describe that indeed electrons, photons, and all other particles are reflected (or diffracted) by gratings exactly as waves would be, and then, when an actual measurement is made, the detector instruments give a single click and thus detect single particle-like entities? As far as I understand it, this strange fact lies at the heart of all phenomena that deal with the very small, with atoms, photons, electrons, neutrons, and protons. This represents a "Gordian knot" that is very difficult to untangle or even to speak about. There is one elegant way to cut the Gordian knot, and this is the modern way of seeing the waves. This way goes back to the concept of probability. When we throw a coin up in the air, we know that so many things can happen to that coin that we really cannot tell whether it will land on one side or the other. We express this by saying that there is a certain probability that the coin will land on heads or tails. For example, a so-called fair coin will fall with the same probability on heads or tails, meaning that if we throw the coin a very large number of times, the number of heads will equal the number of tails; then one says the probability to show either heads or tails is $\frac{1}{2}$. If the coin is not fair, for example, because it is heavier on one side, then we might have a probability of $\frac{1}{4}$ for heads and $\frac{3}{4}$ for tails, meaning that if we throw the coin 4,000 times it will fall about 1,000 times on heads and 3,000 times on tails. Note that we need to make sure that the sum of the probabilities adds up to 1 because we can only have either heads or tails. To make the long story shorter, the probabilities can assume rational and even real numbers between 0 and 1, while the outcome is then just either heads or tails.

If we label one side of the coin by 0 and the other by 1 then the outcome is one of the natural numbers 0 or 1. This gives us an idea how to handle the mechanics of atoms and photons or any quanta. We can describe these quanta by probability amplitudes, the amplitudes of some waves characterized by a continuity of real numbers, while the measurement outcomes are denoted by 1 when a particle detector clicks, and thus a particle is detected and by a 0 otherwise. In other words, the waves are just a "guide" for the particles, and their amplitudes just indicate where the particle is most likely to be. This interpretation of the wavicles is a very successful one and is used by all physics texts. There is a consequence to this interpretation. As long as we describe quanta-like electrons or photons in this way, we cannot exactly determine the location of these quanta. We can only say that there is a certain probability that we find an electron at a given location by some type of measurement. The electron can be in any volume in which the wave resides. Only if the wavelength is zero would we be able to know the exact location.

Uncertainty Principle

There is some very deep truth to this uncertainty of particle location, and this truth is usually formulated as the uncertainty principle that is attributed to Werner Heisenberg. This principle can be stated in a variety of ways. One useful way to get a feeling for principle in one dimension (only x-direction) is the following. The location of any quantum particle is uncertain, and one can only determine that a particle will be found in a range Δx of the x-axis. This range depends on the uncertainty Δk with which the wave number $k = \frac{2\pi}{\lambda_{\text{dB}}}$ can be determined. Assuming that the location of the electron cannot be determined within one wavelength, one obtains

$$\Delta x \, \Delta k \geq 1. \tag{2.76}$$

Δk enters because the de Broglie wavelength is crucial for describing the spacial extension of the wavicle. If the wavelength goes to 0 which means k becomes infinitely large, then we may have $\Delta x = 0$ and thus know the exact location. If on the other hand Δk becomes very small and close to 0, because the de Broglie wavelength becomes very large, then Δx must also be large, and we cannot determine at all where the particle is actually going to be found. If we use the equation for the de Broglie wavelength Eq. (2.75) together with Eq. (2.76), then we obtain

$$\Delta x \, \Delta p \geq \frac{h}{2\pi}, \tag{2.77}$$

where h is Planck's constant. The precise mathematical derivation of Heisenberg, based on his special quantum mechanics that uses the mathematics of matrices, results actually in

$$\Delta x \, \Delta p \geq \frac{h}{4\pi}. \tag{2.78}$$

This equation expresses the fact that wavicles with very precisely known location must exhibit a very large uncertainty of the momentum. On the other hand, if we try to determine the exact momentum p, then the location must become completely uncertain. This all may sound a bit strange and indeed is not anything that we see in our daily life. The reason is simply that the de Broglie wavelength is very small to start with, and Planck's constant h is very small, and therefore the uncertainties become significant only when one approaches very small Δx. Then, however, this uncertainty becomes an important law of nature!

As far as I understand it, one of the reasons for this mysterious uncertainty is the following: Whenever we try to measure some quantity involving electrons, protons, and photons, we need some special instrument. Any instrument is, however, at least as "big" as an electron, and we therefore cannot perform any measurement on the electron without disturbing it with the instrument. It is like measuring mosquito positions with a flyswatter. However, there is more to the uncertainty principle than this analogy. We can see that from the appearance of Planck's constant, the explanation using the de Broglie wavelength and the result of Heisenberg are given in Eq. (2.78). All measurements have confirmed the uncertainty principle with great accuracy, and all investigations point to its deep significance. In fact the principle can be generalized (with some caution) for any product of physical quantities that have the same units as Planck's constant, such as energy E times time t. One can thus obtain an energy-time uncertainty principle:

$$\Delta E \, \Delta t \geq \frac{h}{2\pi}. \tag{2.79}$$

This means that the energy of a particle becomes totally uncertain if we try to pin down the precise time of measurement, for then we would have $\Delta t \rightarrow 0$ and $\Delta E \rightarrow \infty$.

Knowing all of this, what is it that we would find out if we were able to somehow make a measurement of the position of an electron that swarms around a proton? One can indeed make such measurements by using so-called atomic force microscopes. These microscopes are described in Sect. 3.4 and consist of a very fine "tip" like the tip of a needle, only much finer. The tip end is in essence a single atom. With this tip one comes close to the atom that one is interested in and one measures the force with which the atom repels the tip. Another way of measurement would be to shoot electrons toward the atom and measure how these electrons are scattered away by the atom. The electrons of the atom repel, of course, the incoming electron because both have the same negative charge. From the repulsion and scattering one can then estimate the location of the atoms electrons. If one performs such experiments with hydrogen, then one finds that the electron of the hydrogen is with highest probability at the center of the atom and with lesser probability at larger distances from the center. If one plots surfaces of equal probability to find an electron, then one finds spheres, with the probability decreasing as the spheres increase in size. This is exactly what we expect for an s-type standing wave pattern that we already plotted in Fig. 2.35. In Fig. 2.44, we have plotted the probability of

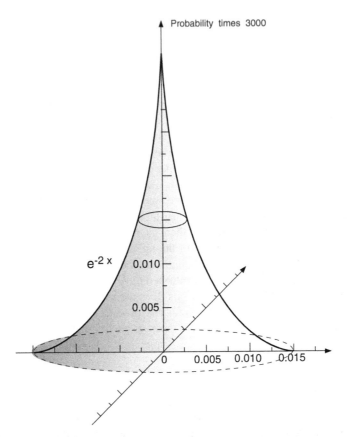

Fig. 2.44 Probability to find an electron around the nucleus for the s-type standing wave pattern of energy labeled 1. The scale of both x- and y-axis is given in nanometers. The x-axis is horizontal (*right arrow*), and the y-axis is perpendicular to the page

finding an electron in the volume of a small cube with the length of one side of 0.005 nanometer and one corner located at $z = 0$ and any given x-, and y- coordinate. The nucleus is the origin of the coordinate system. This is again for the s-type standing wave pattern of the energy labeled 1. Only this time we have plotted the value of the probability as a function of x, y for $z = 0$, while Fig. 2.35 just shows the spherical symmetry of this probability. Note that the probabilities of finding the electron at any given point are rather small. It is only certain (probability 1) that we find the electron somewhere around the nucleus.

The patterns of probability to find electrons are different for different energy numbers. For the lowest energy that we denoted as energy 1, the probabilities to find an electron are highest in the center of the atom where the proton resides. We can excite the electron of the atom to higher energy by somehow transferring energy to this electron and move it up to energy 2. This can, for example, be done by shining light of a certain energy (to be discussed below) onto the atom.

Then we obtain similar patterns for the probability to find the electron, but now this probability shifts away from the center. This is also what one finds, for example, if one investigates the s-type electron of energy 2 in carbon and other atoms. If we investigate the atoms with p-type standing wave patterns that we have described in Fig. 2.35, then the probability to find such an electron has exactly this p-type shape: the electron is never found at the center, the nucleus of the atom, but can be found in the "lobes" extending in the x-, y-, or z-direction as they are shown in the figure. These probabilities to find electrons around atoms correspond to 3-dimensional standing wave patterns, as one can show by solving the Schrödinger wave equation (see Sect. 5.3). A two-dimensional analogy to such patterns would be, for example, the standing wave pattern of a drum. The locations of large drum-vibration amplitude correspond to the location in the atom where the probability to find an electron is highest. If you hit a drum in the center, then the center amplitude is largest with a lesser vibration toward the boundaries of the drum. This corresponds to the s-type standing wave pattern of an atom.

As mentioned, one can make atoms and their standing wave patterns visible by using atomic force microscopes. The visualization of atoms and the handling of atoms by the tip of atomic force microscopes is very important in modern nanostructure science and engineering. However, for chemistry, it is even more important to know the energy of the electrons in atoms and molecules. It is this energy that determines the energy of light that an atom can emit or absorb, and it is this energy that determines whether a chemical reaction will release energy when it happens (e.g., if gasoline is burned) or whether energy is needed for a chemical reaction to occur. Because of the conservation of the total energy, chemical reactions can only take place if the energy of all the ingredients (including released or absorbed heat) stays the same before and after the reaction. Therefore, if we can understand and compute the energies of the standing wave patterns of electrons around atoms and molecules, we can basically understand all of chemistry. It was the Austrian physicist Erwin Schrödinger who figured out the equation that allows us to calculate these energies. His equation and a way to solve it is presented in Sect. 5.3. The solution of the Schrödinger equation becomes increasingly difficult for atoms with larger numbers of electrons and for molecules with larger number of atoms and will still be an interesting STEM problem in years to come.

Energy Levels, Spectra, Molecule Formation

When sunlight shines on a fine grating such as that represented by a DVD, the light is decomposed into all the colors of the rainbow. This range of colors is often called the spectrum of the sunlight. In general, one calls a range of frequencies of electromagnetic waves generated by some source a spectrum. A careful examination of the sunlight, usually performed with special gratings that can resolve the detailed structure of the spectrum, reveals that the continuous "bands" of the colored light are interrupted by many dark lines, so-called absorption lines. These dark lines are related to the bright lines that one can see, again by use of a special grating, when analyzing the light emitted by atoms or molecules. For example, if you sprinkle table salt in a gas flame, or simply the flame of a candle, then you see a bright yellow light. This is also the yellow known from streetlights that use a gas containing sodium. The spectrum of light emitted by sodium atoms consists mainly of two narrow frequency ranges or "spectral lines." In general, atoms of any element can be excited by heat or electricity to emit light. They always emit just a sequence of narrow lines each corresponding to a narrow range of frequencies. For our purposes, we can assume that we are dealing with ideal lines of a given frequency that exhibit some broadening arising from the uncertainty principle. Below, we explain the reason why atoms emit and absorb light at frequencies that are characteristic for the given atom type. We also explain the broad rainbow frequency bands contained in the sunlight and similar bands emitted by liquids and solids. This relation of atom and molecule type to the emission and absorption of frequency bands permits the investigations of atoms and molecules by studying the light that they emit. Such investigations are of great importance for science and tell us, for example, the details of the composition of stars. The spectra also inform us about the energies that are involved in atom and molecule interactions, and that are involved in chemical processes.

Spectral Lines

The physicist Niels Bohr had the fruitful idea that the electrons in an atom are in "quantum states" or just "states" that correspond to certain given energies, and the emission and absorption of photons with given energy is then linked to transitions of electrons between these quantum states. The states correspond, of course, to the standing wave patterns and the energy of these standing waves that we have mentioned repeatedly. A precise mathematical definition will be given in connection with the solution of the Schrödinger equation that is presented in Sect. 5.3. It is an amazing fact that Bohr was correct and that electrons swarming and tumbling around protons have very precise energies. It was known for a long time that the energy and frequency range of vibrating strings and drums is characteristic for the particular vibrating object. The mathematicians Sturm and Liouville had even presented methods to solve equations and calculate such vibrational energies. They called this type of mathematical problems eigenvalue problems. The vibrations are only possible for a sequence of energy values and frequencies that can be enumerated by integer numbers. These integer numbers 1, 2, 3,... correspond to those that we had used before to label the energy of the standing wave patterns of atoms. One says now that the electrons are in quantum states 1, 2, 3,... that have a definite energy each. The energies corresponding to the s-type patterns of the hydrogen atom are plotted as lines in Fig. 2.45 in units of eV. Remember $1 \, eV = 1.602 \, 10^{-19}$ Joules.

We can read from this figure several important facts. The lowest energy of the electron, its so-called ground state, is $-13.6 \, eV$. One counts the energy negative because the electron is attracted to, and bound by, the proton, and it therefore takes

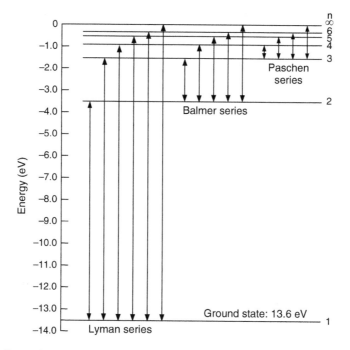

Fig. 2.45 Energy levels of the hydrogen atom in units of eV. The spectral lines of light emitted or absorbed by hydrogen atoms correspond to energy differences of these lines

energy to get the electron away from the proton. If one pulls the electron very far away (infinitely far in theory but only a few nanometers in practice), then one needs the energy of 13.6 eV. The other energies of the quantum states, also called energy levels, are obtained by dividing by the square of the energy number:

$$E_n = \frac{13.6\,\text{eV}}{n^2}. \tag{2.80}$$

Thus we obtain, for example, $E_2 = \frac{13.6\,\text{eV}}{4} = 3.4\,\text{eV}$. This simple equation lets us calculate the spectral lines, that is, the dark absorption lines or bright emission lines, that atoms show when they absorb or emit light, respectively. If a photon is emitted, because an electron had been excited to energy 2 and then drops back to energy 1, the energy of the photon is $h\nu = E_2 - E_1$. In general, if the emission of a photon is due to a transition from energy E_m to E_n we have

$$h\nu = E_m - E_n = \frac{13.6\,\text{eV}}{m^2} - \frac{13.6\,\text{eV}}{n^2}. \tag{2.81}$$

If p-patterns or other types of standing wave patterns are involved, then the energies change slightly from that for the s-type, and all these energy levels and transitions between them are by now well known and understood. The photon emission and absorption energies are characteristic and different for each atom. Each atom and also each molecule have characteristic and different standing wave patterns and emit therefore photons of a characteristic energy sequence. This sequence is as typical for the atoms and molecules as the DNA sequence is typical for living beings. For example, we can determine from the spectral lines of the sun, and from all stars similar to the sun, the amount of hydrogen, helium, carbon, and any other elements that are present on their surfaces. Spectra give us therefore an idea of the chemical composition of the universe, far beyond our solar system. Energy spectra are also of greatest importance on earth to help in identifying atoms and molecules. The calculation of these spectra is one of the great achievements of quantum mechanics.

Molecules, Spectral Bands

The energy levels of molecules, and even giant molecules such as DNA or three-dimensional crystals, can also be calculated from the Schrödinger wave equation. This calculation is much more difficult than the corresponding calculation for single atoms and, with a few exceptions, requires our largest computers to be accomplished. Even the largest computers have difficulty calculating the energy levels of large molecules with large numbers of atoms, particularly when the atoms are not regularly arranged. This is an area of research that, as far as I understand it, will be ongoing and will not be solved entirely for a long time (a playground for future STEM experts). There are a few facts, however, that one generally finds in such computer calculations and that we list here because they are important for many applications, such as designing electrical devices, ranging from LEDs to lasers and from transistors to computer chips.

If two atoms form a molecule by sharing electrons, then each energy level of the two atoms splits into two levels that are usually closely spaced. This is shown in Fig. 2.46. The top of the figure shows two well-separated independent hydrogen atoms with the corresponding two very closely spaced energy levels that are still about at the energy E_1 of single hydrogen atoms. Also shown are the probabilities of finding electrons around the protons, and these too correspond to the probabilities to find electrons for two single hydrogen atoms.

The lower part of the figure shows the two protons much closer together, as they indeed are for the real hydrogen molecule. The energy levels of each hydrogen atom split now into two well-separated levels. The figure shows the lowest energy level E_1 and its splitting into two levels denoted by E_1', E_1''. The probability to find an electron that occupies the E_1' energy level is also indicated. Note that the probability to find the electron in between the two protons is considerable. This leads to an attraction of the two hydrogen atoms because the negative "electron cloud" in between the protons attracts the two protons. Thus the two hydrogen atoms are bound together by the sharing of their two electrons. The fact that E_1' is lower than E_1 means that by the sharing of the electrons one ends up with a state

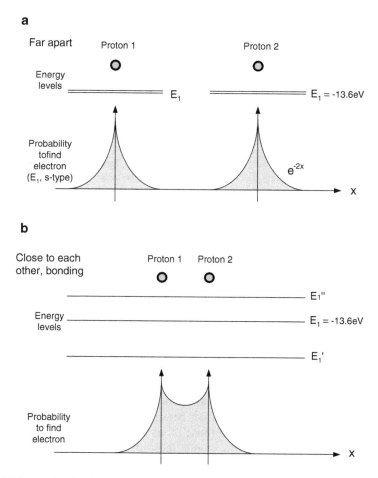

Fig. 2.46 Energy levels of two hydrogen atoms. At the *top* of the figure, the hydrogen atoms are assumed to be very far apart and separate. The lowest energy level E_1 does split into two very closely spaced energy levels that are still located at an energy of about -13.6 eV, the value of the lowest energy level of a single hydrogen atom. The bottom of the figure shows the energy levels for hydrogen atoms that are in close vicinity to each other, as they indeed are for a real hydrogen molecule. In this case, the original lowest energy level E_1 (still shown but not existing anymore) splits into two levels E_1' and E_1'' that are separated by a significant energy

of lower energy. In other words energy can be gained by the pairing of the hydrogen atoms. Therefore, if we put atomic hydrogen into a container, the hydrogen atoms will pair, and energy will be freed, for example, in the form of heat. The bonding energy of a pair of hydrogen atoms is 4.52 eV. If we pair a typical number of atoms that we have in a container, say $2 \cdot 10^{23}$ hydrogen atoms, then we will obtain the considerable energy of $4.52 \cdot 10^{23}$ eV corresponding to $7.24 \cdot 10^4$ J.

The splitting of the energy levels of electrons for neighboring atoms is a general effect that always occurs. Any given energy level of an atom splits into as many

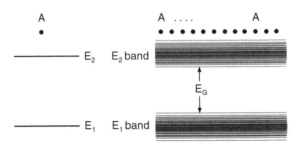

Fig. 2.47 Energy levels of a crystal. There are as many energy levels as there are atoms in the crystal. These many energy levels are closely spaced and form so-called "bands." The broad bands correspond to the original energy levels of the atoms and are separated from each other by so-called gaps that do not contain energy levels. The spacing of the single levels of the bands varies and is typically highest toward the center of each band. The energy levels may be so dense that they totally overlap; this is indicated by the *darker shading*

energy levels as there are atoms once the atoms come close. Close means as close as a few de Broglie wavelengths, which is around 0.2 nm. For the case of a crystal, we have many many atoms closely spaced together, and the energy levels of the atoms split, therefore, into a large number of energy levels that resemble "bands" and are also called energy bands. This is shown in Fig. 2.47 for an arbitrary atom type denoted by A.

The light spectra of crystals also do show bands and, in fact, often very broad bands of frequencies in both emission and absorption. This reflects the fact that the transitions of the electrons that emit and absorb light now occur between the bands of densely spaced energy, and not between well-separated discrete energy levels as occur in atoms. Such occurrence of very broad absorption and emission bands, in place of the very narrow atomic lines of the atoms, is typical for any situation where atoms are closely spaced as they are in crystals and also in glasses and even in liquids. Therefore emission and absorption of light from bodies with high density material such as stars always contains such broad spectral bands. The sunlight appears to us, therefore, more like a continuum of frequencies, and this is exactly what we see if we look at the reflection of a DVD or a crystal. Then we see a continuum of rainbow colors.

Insulators and Metals

The splitting of energy levels in crystals and glasses, and generally in solids and liquids, provides also an explanation of their electrical properties. Solids can be insulators, semiconductors, or metals, depending on how well they conduct electrical currents. The explanation of these properties involves the knowledge of how many electrons are actually contained within the bands of energy levels of the solid or liquid.

Depending on the atom type and on how many electrons each atom has, some of the energy bands may be filled with electrons and others may be empty. Materials consisting of atoms that lead only to either totally full or totally empty bands are called insulators, because they do not conduct electricity. It is easy to see that empty bands will not conduct the current. The fact that entirely full bands do not conduct either is more complicated to explain, and we ask the reader to just accept this fact. The full bands are separated from the empty ones by an energy range that does not contain any energy levels at all. This range is called the energy gap. If this gap is very big, then the solid is an excellent insulator. If it is small, the material does not insulate as well. The reason for this is that at normal temperature, the electrons can gain energy because the whole crystal vibrates and vibrates more if the temperature is higher. Then electrons can gain enough energy to go to the next higher band, with the consequence that then one band is not entirely full and the other is not entirely empty. The material becomes slightly conducting and is called a semiconductor. We have already discussed semiconductors in Sect. 2.4.3. There, we discussed another reason (not higher temperature) for semiconductors to conduct electrical currents: the use of donor and acceptor atoms.

Still other materials have just enough electrons to fill all bands, except for the highest band that is only partly filled with electrons. These electrons in the highest band are then shared by the whole crystal and therefore can contribute to electrical currents. Such materials usually conduct electricity very well and are called metals. Metals are used as connection lines from power plants to the cities and from our power outlets to the appliances, etc. A very good conductor is copper, and copper is most often used for such purposes. Silver is an even better conductor for electricity but too expensive to be used for ordinary electrical connections. Some expensive cables, however, do even have golden parts, although gold does not conduct as well as copper. Gold does, however, not corrode as easily as copper does and is therefore used for the connections of expensive equipment. A typical example would be the input and output plug of a cable for high definition TVs.

Chapter 3
Engineering and Technology: Math and Science Meet Creativity and Design

As we did in Chap. 2 on science, we ask ourselves now what we mean by "technology" and "engineering." Technology is a term referring to the state of the art in the whole general field of "know-how" and "tool use" at any given time. We will discuss, as an example of such tools, the personal computer and, on a more elementary level, all of its parts, the processor, and memory chips as well as the devices (transistors) and the materials (mostly silicon) that make up the computer chips and make the computer tick. There are, of course, many other tools of great importance. For example, if we were stranded on an island, it would be very important for us to know how to light a fire, how to feed ourselves, and how to create tools to catch fish. The author just hopes that none of the readers will ever be stranded on an island. Therefore, we have covered instead machines such as the two-stroke engine, the jet engine, the electromagnetic motor, and wireless communication tools, and we will also discuss, in addition to the personal computer, a few other tools such as disc recordings, lasers, displays, and cameras.

The next question is: What is engineering? Again we can find a straightforward definition on the Internet:

Engineering is the discipline, art, and profession of acquiring and applying technical, scientific, and mathematical knowledge to design and to use materials, structures, and machines to realize a desired invention.

Next I tried to find out what design was, and the Internet gave the answer: No generally accepted definition of "design" exists, and the term has different connotations in different fields. Therefore, I will try to explain the word design by the examples below and only as far as it is directly related to the engineering of a product and not to its beauty. For example, record discs for playing music where preceded by cylindrically shaped objects. Disc recordings replaced the cylinders because they could be more easily mass produced. Of course, the design beauty is a major factor for marketability and therefore very important. STEM, however, deals more with the engineering and technology principles, at least as far as I understand it.

K. Hess, *Working Knowledge: STEM Essentials for the 21st Century*,
DOI 10.1007/978-1-4614-3275-3_3, © Springer Science+Business Media New York 2013

Since ancient times, great engineering tasks were performed by ingenious people. The machines they designed were mostly for use in wars and battles at land and at sea. For example, Archimedes constructed machines to attack ships and according to legend even set ships on fire using mirrors to concentrate the sun-light. Whether or not this is true, Archimedes was certainly one of the greatest scientific geniuses of his time and applied his own most modern scientific ideas. He was an engineer because he designed "engines," a term that referred in the past to military machines. On another level, engineers operated these engines and needed to understand them very well. Therefore, those who were called engineers covered a wide range from very basic scientific invention to the construction, design, and operation of machines, in the past mostly mechanical machines. In modern times the machines are still in some instances of military use. However, the most important machines are those that serve the broad interests of humans. It is fairly clear that a music playing disc is useful to many more people than a torpedo that can sink ships. Engineering therefore has in modern times received a label related to mass production, and it was the particular designs that enabled mass production of products that were of greatest importance, ranging from vacuum cleaners and TVs to washing machines, cars, telephones, and personal computers. The companies that have produced and are producing all of this instrumentation that serves us and (hopefully) improves our lives, have managed to do so by supporting great research laboratories. The success of these research laboratories is based, on the one hand, on a symbiosis of science and engineering that furthers invention and innovation and, on the other hand, on their dedication to designs suitable for mass production of products useful and (or) desirable for existing or possible future customers.

3.1   Engineering Design: From Edison's Recording Cylinder to the DVD

Thomas Alva Edison's phonograph, an instrument to record sound and to subsequently play the recording, is one of the most successful inventions of all time. The subsequent developments of Edison's original invention and design show how far engineering principles can be pushed. We therefore present the principle of the phonograph and some of its engineering advancements as the prime example of engineering design and technology at its best.

3.1.1   Recording Sounds

Edison started with scientific knowledge about sound. He knew that sound was nothing but wavelike vibrations of some medium. Sound in air is caused by waves of pressure that modulate the density of the air, compressions followed by dilatations and dilatations followed by compressions. Therefore sound can be represented mathematically as a wave (see Sect. 2.1.5). The amplitude of that wave corresponds then to the top air pressure. Louder sound corresponds to larger air pressure and larger amplitudes. The pitch of sound depends on the frequency of the wave. Higher pitch corresponds to higher frequency. A low humming corresponds to about 50 oscillations of the air pressure per second, and a higher tone like the screeching of tires to about 1,000 oscillations per second. Edison knew also, that the sound vibrations of the air can be transferred to solid bodies such as thin metal plates. If you speak toward a thin metal plate, such as the bottom of a cooking-pan, then you can feel vibrations when you touch the pan. One says that the metal plate is a "resonator" and resonates with the pressure variations of the air due to the sound. This is the scientific background of knowledge that one needs to have in order to understand Edison's invention. The background of why an invention was needed by people was the following. At the time of Edison, it was not possible to record discussions in any other way than writing down what was said. Therefore, expensive secretaries were needed who had the capability to write and type very fast. Court stenographers are performing such functions to this date. Not always, however, are discussions important enough to employ specialists to record them, yet one still may wish to hear later what was said. Therefore, there was (and still is) a need to be able to record and to replay sound. Edison had knowledge about sound, he had knowledge about this need, and he had an idea.

What if one puts a needle in the middle of a small and thin metal plate or any type of membrane such as used for drums? Then, this needle will vibrate as the metal plate or the membrane vibrate. Therefore, the vibrations of the sound in the air will cause the needle to vibrate. If a needle vibrates, then if we bring some deformable soft material close to it, the needle will make stitches into this material. We are now almost at the main point of the invention. All we need to do is move this deformable material with constant speed, and then the needle will stitch a wave into the material that has exactly the form of the vibration of the metal plate and the wave of the sound in the air that is exciting the vibrations of the metal plate. So this is very simple, we just move a metal plate (or membrane) with a needle over a soft material and speak. Our speech is then transferred as a wavy needle-stitch onto the soft material. Then, if we want to hear our words again, we take the plate with the needle and move it over the material that was used to record the sound. Then the needle will go up and down exactly in the pattern that it has stitched into the soft material. That will, in turn, make the metal plate, and with it the surrounding air, vibrate. Thus, we will be able to hear again what we have recorded. A great idea of a great inventor. However, it took a long time to bring this invention to fruition and even longer to perfect it. Actually, we still may not have perfected it yet, as we will explain below.

The realization of the idea had many obstacles to overcome by clever engineering design. Edison first used tin foil and then a form of wax as the soft recording material. The material also had to move, so he mounted it onto a cylinder that could be rotated. The metal plate (membrane) with the needle also needed to be moved along the cylinder as the cylinder rotated. If you have difficulties imagining this, you need only look at some of the photos of this machine that can be found on the Internet. There is a definite design problem with the soft recording material. It must be soft enough so that the needle can stitch the wave amplitudes into it, but the recording material must be strong enough so that you can play the sound by moving the needle over it. The needle tends, of course, to scratch the material when used often, and then the recording is ruined. Edison managed to design a cylinder with wax that could nicely record and play quite a few times before being damaged by needle scratches.

As it turns out, Edison's phonograph had a different major application than first intended. It was soon realized that one could record music and bring the music of orchestras and singers to everyone's home. Many people wished to buy recordings of famous singers and play them at home. Therefore it was necessary to find a method of mass production of such recordings. If you have to record a new cylinder for each customer, then the singer must sing again and again for every customer. This would not work. Therefore what one has to do is somehow copy what is on the cylinder. It turns out that this is not easy with cylinders, and one needs a different engineering design: the record disc. Consider the following. If we record onto a disc instead of onto a cylinder, we can then (just for the principle of explanation) pour over this disc some material that is first liquid and fills all the vibration patterns on the disc. Then we somehow solidify this liquid material (e.g., by cooling it down) and have now a "negative" copy of the recording. This negative copy can be used as a stamp, and one can stamp the pattern into a softer material, again and again and again. This is a method of mass production of the recording, and this is what was done with music recordings that are still available and sometimes used. Thus we have followed the invention of sound recording and some of the engineering design to the actual possibility of mass production. Are we at the end of this invention process? Of course not! All the improvements of material and production did not help to remove one problem: if one plays the record over and over, it deteriorates. It did not help much to replace the needles by diamonds, that just made the recording tip (now a diamond instead of the needle) last longer. After playing a record many times it always lost quality. So how could one get around that? This step took major design changes and a number of additional inventions.

3.1.2 Digital Recording Using Laser Diodes

The main idea for this new design was to use light to read what was recorded on the disc. Light does not destroy material even when used over and over. However, light also does not precisely follow the form of the stitches of a needle, and the light reflections would, therefore, not follow the amplitude of the sound with precision. It took another major step of engineering design and ideas to overcome this problem. This major idea was to change the information of sound vibrations and amplitudes into digital information. We know that louder sound causes larger amplitudes of needle vibrations. Information of this type is called analog information, meaning that louder sound causes deeper engravings . We have learned in Sect. 1.3.4 and

Fig. 1.23 that we can transform such analog information into bits or numbers, that is into digital information. In the case of Fig. 1.23, we had given numbers to the grey shades or color shades of a photograph. We can apply the same procedure to sound by giving a number to the pitch (sound frequency) and another number to the loudness or amplitude of the sound. One needs a machine to do that. That machine is called an "analog-to-digital converter." Such machines have been developed, and this development has had wonderful consequences. Having such machines, we do not need to make deeper and shallower stitches to record different sound. One only needs some kind of mark, for example, a tiny hole or just reflecting surface, to read from a light beam the information bits 0 (no light detected) or 1 (light detected), respectively. For example, any groove will reflect much less light than the unperturbed surface of the disc, and can represent a zero. Such marks can be burned into the material directly, one groove after the other, by use of powerful laser light (see below). Burning one mark after the other can be done relatively quickly, and that is the way it is done when you burn a so-called compact disc or CD on your personal computer. As everyone knows, CDs can store many songs. It is, of course, possible to store in this way any information whatsoever not just sound. We know from Sect. 1.3.4 how to turn photographs and thus movies into digital information.

Photographs have a lot of pixels and movies are made of very large numbers of photographs. The marks that are used to store movies therefore need to be very small and one needs special discs, so-called digital video discs or DVDs to store movies. For mass production, the process of burning such discs may be too slow, because it takes a computer several minutes to write a DVD that requires billions of burned markings. If one wishes to distribute many DVDs, as companies that sell movies do, then one needs again to write the information on a type of stamp and produce the copy DVDs by just pressing the stamp into some material. The technology to write markings on a DVD either by use of a laser or by just using a stamp, has been developed to a high level. This means in essence, that suitable materials have been found in which the markings can be made reliably and of very small size. It is easy to estimate that the markings need to be of micrometer (10^{-6} m) size. Then one can fit a gigabyte of information on a disc that is about 10 cm in diameter. A byte represents in its most common definition just 8 bits. The estimate of how many bits or bytes fit on a DVD is actually a nice homework problem to think through. With a storage capability of several gigabytes, you can store more than 100,000 low-resolution photos and therefore whole hours-long movies.

For high-resolution photography, the markings need to be made still smaller. This also means that the wavelength of the laser light that reads the markings needs to be smaller. Remember, if one reads information with light, the wavelength needs to be about as small as the markings are. Therefore, one needs blue lasers that have indeed a small wavelength to produce and read smaller markings. This has led to the name of Blu-ray technology. How can we push the technology further to get higher resolution and even better picture quality? We just need to make the markings smaller and smaller and use ultraviolet lasers to read the information. Currently no such technology exists that is inexpensive and lends itself to mass production.

Do we actually need such a technology? Well, think of the following problem. A photographer likes to photograph animals that move with high speed such as a leopard following his prey. The photographer likes all photos in very high resolution which means you need about 10 Mbit of information or more per photo. The photographer also wishes to look at the photos after the shoot in slow motion, because she wishes to show the leopard's movements in detail. This means that the camera needs to take about 1,000 photos per second or 10^{10} bits of information per second. For about 20 min of photography, that gives about 10^{13} bits of information or 10 Tb of information. DVDs can currently not handle such a flood of information. The hard drives that are discussed below in Sect. 3.2.3 can do this job. Hard drives, however, are still more expensive than DVDs and cannot be copied by using the stamp method. So we see, there is still a lot of work to do if we wish to fulfill the above photographer's wishes.

3.1.3   Recording with Atoms

How far can we push all of this? As we will see in Sect. 3.4, engineers are now able to write information by moving single atoms. 1 bit can then be represented by a single atom (the "1") or no atom (the "0"). The distance of atoms in solids is about 0.5 nm. This means that a line of 8 cm or 0.08 m, which is the length of a credit card, can hold $\frac{0.08}{0.5\cdot10^{-9}} = 1.6 \cdot 10^8$ bits of information. The width of a credit card is about 5 cm, and therefore a credit card can hold $\frac{0.05}{0.5\cdot10^{-9}} = 10^8$ such lines of atoms. This means that we can store altogether $1.6 \cdot 10^{16}$ bits (the product of lines and atoms per line) on a credit card. Therefore, you could store about 16 million ordinary movies on such a card, or 1,600 high-speed high-resolution movies (such as 20 min of the hunting leopard). You could also store on such a card the whole Library of Congress with relative ease. Can we really do this? Well yes, at least in principle! As mentioned, there exist atomic force microscopes that can move single atoms. You can also write the letters of your name with atoms. However, the writing process with atomic force microscopes is very slow. It would take many years to write a high-speed movie. The price of this work would be exorbitant. There is no simple stamping process that can deal with single atoms yet. So you can see, we already have a proof of concept about what really can be done, but the engineering design and inexpensive mass production would require some very new ideas that we are currently lacking when it comes to such high densities of information. We can see, however, that nanoscience and nanotechnology have an enormous potential to fulfill many of our future needs and dreams.

The storage and retrieval of bits of information brings us in a natural way to the personal computer, because this is what is needed to accomplish computation. All numbers that a computer deals with can be represented by a switch or a gate that can be on or off. If the switch is on, it can transmit an electrical signal that represents a "1" while if the switch is off, that represents the bit "0." We have illustrated and discussed this in connection with Fig. 1.4. The switch that is shown in this figure, could of course be created mechanically. We could, in principle, have such a gate made out of metal that we could open and close by hand. Using such mechanical switches, we could build a mechanical calculator and that is indeed close to what Charles Babbage designed more than 150 years ago. His machine was only recently built by admirers; it was too expensive and complicated at Babbage's time. Looking at this machine on the Internet, tells you immediately that this machine, however admirable, is very slow. All these mechanical parts with the large mass of the metal gears can not be moved quickly. To access and store bits of information much faster, the method of burning marks and reading these marks with light is already a big improvement. Such a burned mark involves the movement of only a few billion atoms and that can be done faster than the movement of a gear with 10^{25} atoms. However, electrons can be moved around still much faster than atoms. Compare the mass of an electron of $9.1 \cdot 10^{-31}$ kg to the mass of an atom. Even the lightest atom, hydrogen, is 2,000 times heavier. Newton's law of acceleration, as given in Eq. (2.38), tells us that lighter objects are accelerated to high velocities in less time by use of the same force. Therefore a switch based on the movement of electrons will beat any switch that has to move atoms. We like to have gigabit operation speed corresponding to dealing with 10^9 bits per second (gigahertz). Modern computer chips, based on semiconductor technology, do indeed operate that fast, and we wish to have even faster ones. The key to their fast operation is indeed the use of electronic devices, i.e., devices that work with electrons and move electrons but do not move any mechanical parts or atoms. The most efficient of such devices currently known are transistors. Transistors and their use for computer memory and processor chips are discussed next.

3.2 Semiconductor Chip Technology

My own involvement with semiconductor devices started when I began working at the University of Illinois. There was a great excitement about semiconductors and transistors there. One of the inventors of the transistor, John Bardeen, had founded laboratories that worked on the cutting edge of semiconductor technology including work on transistors and LEDs. I remember graduate students of the University of Illinois who were looking for a job and were received by the research directors and even the CEOs of companies like Intel, Texas Instruments, and Hewlett-Packard. It is very telling, that these industry leaders took their time to pick their top employees and future leaders. This is how technology keeps its momentum, and this is how industrial research laboratories worked worldwide at that time. They knew and hired the most knowledgeable STEM students.

In the great industrial laboratories, one had the feeling that science and engineering was forming a symbiosis and followed one purpose only: to improve or renew the design of the companies' major products. The leading minds had a clear understanding of the stumbling blocks, and they knew who else in the world was working on the problems and also knew about the best students at the various universities. After receiving a Ph.D. in Austria in applied physics, I had studied in the USA with John Bardeen and with his associates. Before I returned to Europe (as it turned out only for a short time), I had already a letter of invitation to visit the head of the Siemens research laboratories in Germany, Heinrich Welker. I visited him right after landing in Europe. There was a lot going on in his office, and one of his assistants ushered me in at an opportune moment. "So, what is new with transistors," he asked me? "What did you find out from your research in the USA ? " "I investigated the effects of high electric fields," I said. "We know about high fields since a long time," he said, "you did not have to go to the USA to learn about that." "Well, all the work at Siemens is limited to high electric fields at one contact only," I said. "The smaller transistors have high electric fields everywhere." "Hmmm, that is new, and what does this mean for the device design and for the customer?" he asked. "The transistor current is lower, yet the reliability is reduced," I said, thus summarizing my one year of work. Welker was happy, he had heard what he needed to know, and I was dismissed with a friendly smile. Compare this story to that of the executive of Acme Electronics from Section "Aim of the Book" of preface, who thought nothing new could be learned about TVs and discontinued much of the Acme electronics research. I am sure he did not even know what a transistor was, much less how it functioned. I have told this story because it illustrates why we need STEM education to perpetuate technology innovation. We also need industrial leaders who understand STEM and pay attention to it. I am now going to present some facts of the history of the transistor, again to give you an idea what engineering invention and design is all about. On the way, I will also try to tell you how such a transistor works. The goal is to give a typical introduction to the machinery of engineering and technology and to set the stage to discuss then all the details about one of the most important machines of our time, the computer. The story of the personal computer started early, already with the mechanical machines like the difference machine of Babbage. However, modern computers work with electrons, and therefore we start with the history of the transistor.

Among the first fast switches based on electrical currents were the so-called vacuum tubes. Although these tubes were invented with great ingenuity and indeed switched electron beams with very high speed, they have all but vanished from existence. The reasons for their demise were mainly their size, poor reliability, and large power consumption. A typical vacuum tube was the size of a fist and contained a glowing wire that used as much power as a light bulb. The first computers built with these tubes filled and heated rooms, and, in spite of this high energy use, had very minor capabilities as compared to the smallest personal computers that we have now. These tubes were not only used for the first electronic computers but also for switching telephone connections. The researchers of the American Bell Telephone Company had a great desire to have smaller switches with much less power consumption, and this is what started the invention of the transistor. The vice president of Bell, Mervin Kelly, had the great foresight that one should replace the vacuum tubes by devices made out of solids, in particular out of semiconductors. Kelly hired a dream team to do just that. The leader of this dream team was William Shockley, and its two most distinguished members were John Bardeen and Walter Brattain.

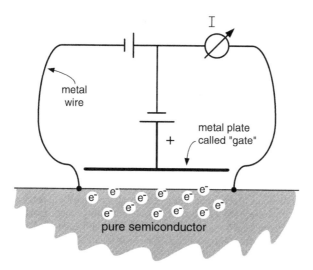

Fig. 3.1 A schematic of Lilienfeld's and Shockley's basic idea of a solid-state switch, a transistor. A metallic plate, called the gate, is mounted above a semiconductor crystal. Wires are connected to the crystal surface to the left and right of the gate. Batteries are included in such a way that the metal gate becomes positively charged. This positive charge on the gate attracts electrons (labeled by e^-) to the crystal surface. The electron accumulation at the surface causes an electrical current I to flow. The current can be measured by an instrument as indicated. If, however, a negative voltage is applied to the gate, then all electrons are repelled from the surface and no current flows. We therefore have a switch, a resistor that can be controlled electrically, a "transfer resistor" or "transistor"

3.2.1 The Transistor

The team followed an idea of Shockley. Unbeknownst to Shockley, this idea was not entirely new and had been patented about 20 years earlier by the Polish physicist Lilienfeld. It is illustrated in Fig. 3.1.

Shockley's design involved a pure semiconductor crystal. Bardeen and Brattain used germanium crystals because it was known how to produce them in very pure form. Nowadays silicon is the semiconductor of choice. As shown in the figure, a metal plate that has the function of the gate is mounted right above the semiconductor. Wires are connected to the semiconductor to the left and right of the gate. In addition we have batteries in the wire loop. One of them causes a current to flow, if the semiconductor contains mobile electrons to support this current. A pure semiconductor does not contain many mobile electrons that can conduct a current and is therefore almost insulating. The essential idea of creating an electronic switch is now that, if another battery is introduced that causes the gate to be positively charged, the gate will attract or "induce" some mobile electrons toward the semiconductor surface, and a current I will flow and be indicated by the instrument (labeled by the current-symbol I in the figure). Walter Brattain, the experimentalist of the transistor team, tried to make such a structure.

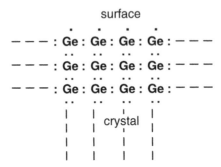

Fig. 3.2 Two-dimensional rendering of a germanium crystal and the sharing of electrons that "bind" the germanium atoms together. Shared electron pairs are usually indicated by a *line*. Here we have plotted the single electrons as *dots* for clarity. Germanium was the first transistor material. Silicon behaves exactly the same way, because it is also in the 4th column of the periodic system and is the material of choice for computer chips. The electrons are shared by neighboring atoms in such a way that the eight electrons of the nobel gas configuration accomplish the bonding within the crystal. Notice, however, that at the surface, one partner atom is missing, and therefore only seven electrons are available to be shared. Therefore, any additional electron that comes to the surface will be captured (to accomplish the noble gas configuration) and cannot contribute to a current I at the surface

However, no matter how hard he tried, there was absolutely no effect of the gate shown. No current was flowing no matter how the gate was charged. There was an essential flaw with Lilienfeld's and Shockley's idea, as Bardeen found out. One problem was that a pure semiconductor has no mobile electrons, so where do the electrons come from altogether? Well, they can be pulled out of the metal wires. So this lack of electrons in the pure semiconductor could not be the reason for measuring no effect at all. What was the reason? Bardeen figured it out, and it is shown in Fig. 3.2.

This figure shows the sharing of electrons between the atoms, as we know from chemistry. The actual crystal is, of course, three-dimensional, and there is the typical tetrahedral arrangement of the nearest neighbor atoms. There are exactly eight electrons shared by each atom to accomplish the bonding, just as shown in the two-dimensional rendering of Fig. 3.2. The situation is different as soon as one comes to the surface of the crystal. Then the top neighbor is missing because the crystal ends, and therefore there are only seven electrons available. This is unsatisfactory from the viewpoint of chemistry. The surface atoms "desire" to have one more electron for the bonding. If a gate with positive voltage attracts additional electrons toward the surface, then they are immediately used for the bonding to complete the number eight of shared electrons. These electrons can therefore not contribute to the electrical conduction. This is why Shockley's idea did not work, and this is what Bardeen realized.

During their intensive work, Bardeen and Brattain invented another type of transistor, the point-contact transistor. This type of transistor was mass produced and used for several years. Its interesting properties and small size led to much additional research. In the course of this additional research, it was finally found

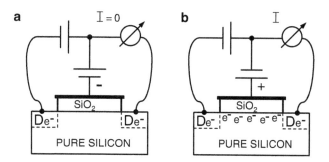

Fig. 3.3 A transistor as derived from Shockley's idea and improved by the use of silicon as the semiconductor as well as a layer of the insulator silicon dioxide on top of the silicon. The gate is deposited on top of the silicon dioxide (SiO_2). The atoms of SiO_2 supply the electrons to complete the noble gas configuration of the atoms in the top silicon layer. Donor atoms are introduced into the silicon to the left and right of the gate in the regions labeled by "D." The remainder of the design is just as in Fig. 3.1 with (**a**) showing the transistor current switched off and (**b**) switched on. The electrons of the "on" state, located below the insulator at the top of the silicon crystal, originate from the "D" regions. They are simply pulled out of the "D" regions by the positive charges of the gate

that the idea of Shockley and Lilienfeld could be made to work after all. Their idea was realized by using the semiconductor silicon. The problems related to the surface could be "cured" by oxidizing the silicon surface and thus "growing" the insulator silicon dioxide (SiO_2) on top of the silicon crystal. Silicon dioxide is a very stable and durable material and also highly insulating. It is most commonly found in nature as sand or quartz and is used in the production of glass.

The silicon dioxide that is used for computer chips is of greatest chemical purity. Chip producers fabricate extremely thin layers (about 1–2 nm) of SiO_2, or similar insulators, on top of silicon by using "recipes" that are still company secrets. A highly conducting gate material is deposited on top of the silicon dioxide to form the transistor gate. Arsenic or phosphorous donors are introduced into the silicon to the left and right of the gate, in order to create conducting layers that supply the electrons that are needed when the transistor is switched on. The completed transistor is shown in Fig. 3.3.

The figure also illustrates the on- and off-switching of the transistor by application of a positive and negative gate voltage, respectively. The silicon that is used in the illustration is pure and therefore practically insulating and does not conduct a current. The left "D" region is called "source" region, and the right "D" region is called the drain of the transistor. If the gate is charged negatively, the electrons stay in the "D" regions because the negative gate repels them. If a positive voltage is applied to the gate, then electrons are pulled out of the "D" regions and distribute themselves under the gate below the silicon dioxide. This layer of electrons conducts, and the current I is switched on. Of course, not every transistor on a chip has its own battery. These are just included here for the purpose of illustration. The voltages that supply and switch the transistors on a chip are provided through metal wires as discussed below.

Transistors switch extremely fast. The reason being that electrons have a very small mass and can be accelerated quickly. The transistors of today's technology are also very small. The distance between the two "D" regions (called source and drain) may be just a few nanometers. The electrons can be pulled over this short distance in picoseconds (10^{-12} s) or less. This enormous switching speed is the key to the success of solid-state electronics and semiconductor chip technology.

For completeness we note that the transistor described above is called a MOSFET (metal-oxide-semiconductor field-effect transistor). The pulling in of the electrons by the positive gate is called the "field effect." The semiconductor that is actually used in chip production contains acceptor atoms such as boron. The effect of these acceptor atoms has been described in Sect. 2.4.3. Acceptors supply positive background charges in the semiconductor that make it easier to switch the transistor off. The reasons for this effect will not be explained here, and the interested reader is referred to the Internet. The layer of electrons that forms when the transistor is switched on is then called an inversion layer, because the charge is inverted to be negative while it was positive before the switching. A transistor designed like that is called a "n-type inversion layer MOSFET"; the "n" stands for the negative electrons in the on state. Another transistor type that switches on when the gate is negative uses acceptors instead of the donors in the "D" regions. In other words the "D" regions are replaced by "A" regions while the pure semiconductor is now supplied with donors instead of acceptors. Such a transistor is called a p-type inversion layer MOSFET. The "p" stands for the positive charges that conduct the current in this type of transistor. The technology that uses both types of transistors with inversion layers is the most important technology for computer chips today and is called CMOS. Any detailed understanding of this technology requires Internet investigations and additional work. Some explanations that are necessary to understand inversion layer transistors are given in the section on diodes below.

3.2.2 Memory Chips and Transistors

There are several ways to use the switching properties of a transistor to create a computer memory, and there are several types of memories. The main difference in the properties of these memories is given by the length of time that these memories can actually store and retain information with and without a supply of power. Think of a standard cell phone, for example. You have stored all the telephone numbers of your friends, and you like to keep all these numbers stored even if the phone is switched off or its battery becomes fully discharged. Therefore, the memory of the phone that stores these telephone numbers needs to keep that information for a long time, for many years. This type of memory is called non-volatile memory since it persists even when no power is supplied. The memory sticks you carry around in your pockets also contain nonvolatile memory. Volatile memory, on the other hand, must be constantly supplied with power, or it loses its information. This disadvantage is outweighed by the fact that information can be stored and retrieved

Fig. 3.4 Four transistors and capacitors of a "dynamic" memory. Each transistor is connected to a capacitor denoted by the symbol −||−. The capacitor can be charged and discharged by switching the transistor on. The voltage for switching the transistor is supplied by a metallic line (*horizontal*) called the word line. The charging and discharging of the capacitor, as well as the "reading" of its charge, is done through a second (*vertical*) metallic line, the bit line. All capacitors are connected to a common electrical contact the so-called ground

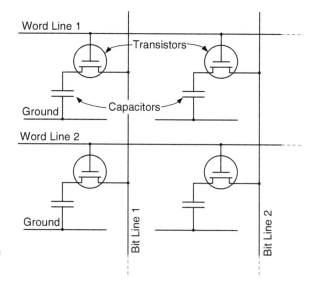

much faster from volatile memory than non-volatile memory. There's a very good chance that the STEM students reading this book will someday develop a new type of memory that combines non-volatility with high-speed operation. For now, however, today's high-speed computers utilize fast, volatile memory. There are two major volatile memory types that work that way. One is called static memory and the other dynamic memory. Static memory retains the information continuously and will not lose its bits as long as the power is on. Dynamic memory, on the other hand, would lose its bits after a short time, even when the power is on, and needs frequent refreshing of its contents. This type of memory is used in current computers for many purposes. The reason, why the author has chosen to explain just this memory type, is because it is the easiest to explain on an elementary level. The principle of this memory type is shown in Fig. 3.4.

The actual device that stores the bit is a capacitor. We have described how a capacitor works in Sect. 2.2.2. It consists of two metal plates that can be charged by applying a voltage. In the case of the memory cell of Fig. 3.4, the capacitor can only be charged (or discharged) when the transistor is on and then carries an electric current that handles the charge. The transistor is shown here by the symbol that is used in electrical engineering. This symbol shows the connections to the gate as well as the two other contacts (left source, right drain) within a circle. To address a particular transistor and charge the connected capacitor works in the following way: consider the transistor in the right lower corner of the figure. To switch that transistor on, we have to apply a positive voltage to the wire (a thin line of metal atoms) that is called word line 2. Then we can charge the upper plate of the capacitor positively by applying a positive voltage to the wire called bit line 2. If we like to read the information, we switch the transistor on by charging again the word line positively and then just read the voltage on the bit line. If it is positive, the capacitor was charged. In this way we can find every transistor and capacitor and check out the charge of the capacitor by just knowing the number of the word line and the bit line that connect to the transistor. If you are not yet used to electrical circuits, think of the whole system working with water pipes. The capacitor is then just a container that can be filled with water by a pipe that can be turned on and off (the transistor switch). This analogy also lets us understand that the switch that supplies and drains the water can be a little leaky, and so we can lose water. This leaking makes it necessary to refresh (refill) and explains the dynamic nature of this memory.

The marvelous thing is now that one can mass produce such memories with millions and millions of transistors and capacitors both inexpensively and reliably. This is done in the following way. One starts with a silicon crystal. We have described how to produce NaCl crystals in Sect. 2.4.3. Growing a silicon crystal is a much harder job to do. One needs to grow it out of molten silicon at very high temperature. The material also needs to be very pure. The semiconductor materials science is different from the material science of metals or insulators, because small additions of atoms (other than the crystal atoms) can change the electrical properties very significantly. We know the effects of donors and acceptors already from Sect. 2.4.3. Therefore additions of atoms other than silicon need to be very carefully controlled. One foreign atom, in a million silicon atoms of a crystal, can already cause unwanted and damaging effects. The silicon crystal also needs to be big for reasons of inexpensive mass production. Very pure and large cylindrical crystals can nowadays indeed be grown. These cylinders are then sliced like sausages into very thin circular slices called wafers. From each of these wafers one can then create many "chips" formed by repeated photolithography (see below) on top of the wafer (see Fig. 3.6) and later cutting it into rectangular pieces. Each chip, in turn, contains millions of transistors.

How can we put millions of transistors and capacitors on a silicon chip by using photographic methods? We illustrate the well-established and inexpensive way that is called photolithography, by describing just a few steps of the standard chip fabrication. The particular steps we show are the introduction of the "D" regions of the transistors of Fig. 3.4. The method that is used to create all these "D" regions for millions of transistors at once is derived from photography (therefore the name photolithography). The photographic method that is used here is very special and uses ultraviolet light instead of visible light. This is necessary to create very small structures. To create such small structures one needs to use light with the smallest possible wavelength as we already have pointed out in connection with Blu-ray technology. You can easily imagine that one cannot create a small structure with electromagnetic radiation of long wavelength. Here we need even shorter wavelength and a material that changes its properties when it is exposed to the ultraviolet light. Such a material is called photoresist (PR). Photoresist comes in two varieties that are called positive photoresist and negative photoresist. Positive photoresist has the property that it can be easily washed away after being exposed to the ultraviolet light. For negative photoresist the unexposed regions are easily washed away. Using positive photoresist, it is now possible to create the "D" structures. The necessary steps are shown in Fig. 3.5.

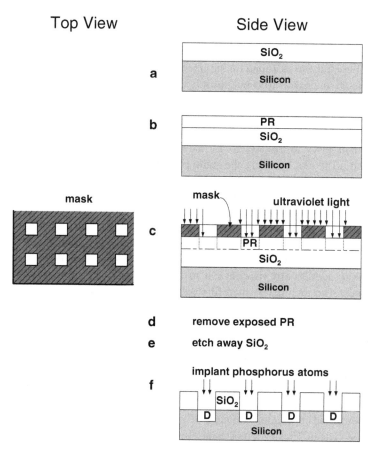

Fig. 3.5 Six steps used to create the "D" regions for millions of transistors simultaneously. Only eight regions for four transistors are shown. (**a**) Oxidize the silicon wafer (side view shown) to create an SiO₂ layer. (**b**) Apply a layer of positive photoresist (PR). (**c**) Expose PR through a mask. (**d**) Remove exposed PR. (**e**) Etch away SiO₂ in regions without photoresist. (**f**) Implant phosphorus atoms through windows

The six steps shown in the figure are for forming the "D" regions of 4 transistors by use of a photographic process with a so-called mask that is also shown. Of course, the mask can be made for many transistors and is routinely made for millions and millions and even billions of transistors just by repeating the shown pattern. In order to fit millions of such rectangles on a mask to create the "D" regions, the single rectangles need to be very small, much less than a micrometer (10^{-6} m). The process works as described in the figure. First the silicon is oxidized and a layer of photoresist is put on top of it (steps (a) and (b)). Then the mask is put on top. The ultraviolet light is blocked by the mask except in the areas of the rectangles that are transparent, and the photoresist is exposed there to ultraviolet light. Afterwards, the exposed photoresist is removed (steps (c) and (d)).

Finally the SiO_2 is etched away in the windows where the photoresist was removed, and phosphorus (or arsenic) donor atoms are "implanted" into the silicon in the open windows so that the "D" regions are formed. The implantation process is not described here but can be found on the Internet by searching for "ion implantation." In principle all one has to do is put phosphorus or arsenic into the windows and raise the temperature to "diffuse" these atoms into the silicon.

To finish a working memory chip, one also needs to create the capacitors and all the metal wire interconnections of all the transistor. This is considerably more complex than just the process to form the "D" layers that we have discussed. However, it all works by using the same photolithographic principle and method. This method makes it possible to create millions of transistors and capacitors in a number of steps of applying photoresist and masks. More of these steps are necessary to finish the memory chip. Of particular importance are the metallic interconnections, as for example, the word and the bit lines. These interconnections may be formed by use of many layers of metal lines embedded in insulating silicon dioxide. In the case of the memory chip that we have discussed, these metal lines enable one to address any random transistor and capacitor by just using its word and bit lines. The memory is therefore called a dynamic random-access memory (DRAM).

Other types of semiconductor memory are produced in similar fashion. An important variation is the flash memory chip, which is a form of nonvolatile memory. Flash memory does not use capacitors (that lose charge and need refreshing) as the DRAM does. Instead it stores the electric charge in a more permanent way in a so-called floating gate, a small volume of silicon encapsulated in the insulating silicon dioxide. Flash memories need no power to maintain their stored information. They are therefore used in memory devices of various kinds, for example, for cameras and memory sticks. Flash memories are in essence used the way hard drives (see below) are used (e.g. for iphones, ipads). Their advantage is higher speed and the lack of any mechanical parts.

Memory chips and all other types of semiconductor chips are made on large silicon wafers that are cut from cylindrical silicon crystals. Each large wafer contains many chips that are all created by the same photolithographic process. A typical example of such a wafer is shown in Fig. 3.6.

One of the current guiding principles (industry benchmarks) for chip technology is to double the number of transistors on a wafer of a given diameter every two years. This is also known as Moore's law, named after Gordon Moore who was one of the pioneers of chip technology. Moore's law is not a law for the ages such as Newton's law for force and acceleration. It represents rather a very important rule for mass production and has led to much research and innovation. Doubling the number of transistors on a wafer necessitates the shrinking of transistor dimensions by the square root of two. In doing so every two years, industry has reduced some of the transistor dimensions to the length of a few atoms. Naturally, these dimensions cannot be reduced much further, and Moore's law may come to a halt. Designing and arranging billions of such minuscule transistors, as well as connecting them in a meaningful way, requires advanced levels of training in all areas of STEM.

Fig. 3.6 Silicon wafer with many chips (*rectangles*) on it (photo courtesy of Prof. Joseph Lyding, University of Illinois, Urbana)

The accomplishments of transistor downsizing have brought us desktop personal computers, laptops, and smart phones, and many more devices and instruments that are improving our lifestyle.

3.2.3   The Personal Computer

The development of semiconductor devices including transistors and memories, has led in a natural way to inexpensive and mass producible computers, so-called personal computers, that everyone can afford. To appreciate this fact, we need to know what a personal computer actually does. This is not so easy to understand these days, because the personal computer (PC) seems to do anything and everything. The PC is balancing your check books, it is transferring money from your bank to the cable and power companies, it lets you look at your photos, send e-mail, write letters, and the list could go on and on. In principle, however, all that a PC does, no matter how complicated the task, is processing digital information, digital numbers. We have discussed in Sect. 1.3.4 how the computer can handle digital photographs by just crunching binary numbers. The computer can also represent the ABCs as binary numbers, and therefore lets you write and correct letters. Naturally, the computer works with numbers when it balances your bank account, and it certainly works with numbers when it does mathematics.

Use of Semiconductor Memories and CPUs

It turns out that we can understand the major principle that is involved in the workings of a personal computer as soon as we understand the few lines of code that we have discussed in Sect. 1.1. We reproduce these lines here and refer to them as a typical code or "application" (Apps) code:

```
begin
M = 1000
m = 1
A = 0.0
Label[1]
r = m · m · m · m
s = 1/r
A = A + s
m = m + 1
If m = M Goto Label[2] Else Goto Label[1]
Label[2]
Display or Print A
end.
```

This code, that could be used for actual computers, calculates the sum $s = \sum_{m=1}^{1,000} m^{-4}$ for M terms, and M was chosen to be 1,000. Modern computers use a different computer language than that shown here, for example, Java. We use the above language for educational purposes, because it is easier to understand. We could do the whole calculation of the code above by hand. However, this would take a day or so, depending on accuracy. A computer does this in fractions of a millisecond. You can try this by using MATHEMATICA and increasing M from 10 to 10,000, and you will not see a difference in the calculation times because they are less than a second. The reason for this enormous speed is that each transistor switches extremely fast.

The above code contains simple arithmetic operations such as addition and multiplication. The semiconductor chip that accomplishes such tasks is called a processor chip. Processors are built by using photolithography in exactly the way that we have described for the memory chip, but processors are a little more complicated. The processor chip fulfills also other tasks than the arithmetic operations of addition, subtraction, multiplication, and division. It can perform certain operations such as going to a specified address (location) of the memory chip (e.g., Label[1] in the above code), and it can store numbers in, and retrieve numbers from, that memory location. The chip that accomplishes all of this, elementary arithmetic operations as well as processing instructions, is called the central processing unit or CPU. The code above represents a typical example of what we may wish the computer to process and execute. It is a particular application (App), that solves a particular problem and performs a certain task.

What happens then if we "mouse click" the icon of a particular App on our PC, such as the simple App shown above? The CPU will then process all the steps starting with "begin." First M is stored to be the number 1,000, next m as the number 1, and next $A = 0.0$. The label $Label[1]$ corresponds to a binary number that marks this particular location of the code. The next step uses the arithmetic capability of the CPU and calculates the 4th power of m by multiplying it three times with itself and setting r in the memory equal to the result. Next, again, arithmetic is performed and $s = 1/r$ is calculated and stored in the memory. In the next step a new A is calculated by adding s to the previous A and storing the new A in the memory. Now m is increased by 1 and in the next step the CPU compares m and M and goes to Label[1] if they are not equal. This starts the calculation all over resulting in the next A and m. If $m = M$ then the CPU directs the process to Label[2] and the final value of A is printed which gives us the whole huge sum $s = \sum_{m=1}^{1,000} m^{-4}$ that was desired. Of course, this App is rather simple, and all it does is calculate a sum of numbers. Mathematically speaking, this accomplishes already a lot, because to calculate such a sum is all you need to calculate the sine and cosine functions or to calculate any area of any geometrical object (see integration in Sect. 5.2.2). General Apps do a lot more and work not only on your PC but also may involve the Internet and other computers to accomplish very big and often complicated tasks, such as finding your favorite restaurant on maps, shop in a music store, or display your newspaper.

Use of Hard Drives

This brings me to another topic of importance. On any personal computer, we have many Apps and not just one. The semiconductor memory, big as it is and even if it holds a gigabit or 10^9 bits, will not be big enough to hold all the possible applications that you ever may wish to have. Semiconductor memories, such as the DRAM that we discussed, have the great advantage that any transistor of them can be randomly chosen and operated, they also have the great advantage that they are extremely fast, but they have a major disadvantage: when the computer is switched off, their memory is lost. There are semiconductor memories that do not lose the information. However, all semiconductor memories are also a little more expensive than other memories that can store humongous amounts of data. These other memories are slower, and the information can not be accessed randomly. They function similarly to the DVD discs; however, they store the information not by burning holes but by magnetizing tiny metal pieces. These magnetic memory devices, called "hard drives," store many terabits, 10^{12} bits, of information and are ideally suited to store all your Apps, your music, your photos, and whole movies. Because of their importance, I will give a short description of how they work. The storage capacity of hard drives is often not designated by the number of bits but by "bytes" instead. 1 byte corresponds to 8 bits. 1 Tb means then that you can store about 10^{12} characters (such as letters), each corresponding to 8 bits (0 or 1) of information.

The principle of operation of a hard drive is shown in Fig. 3.7. It consists of a circular disc that is driven by an electric motor and can rotate with high speed. This disc is called the platter, and it has on its surface a very thin layer of magnetic

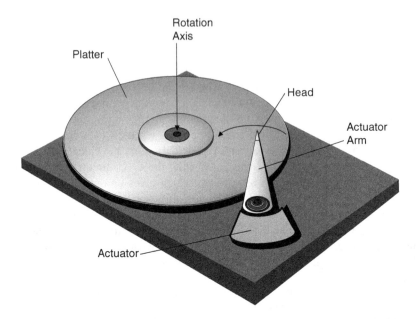

Fig. 3.7 A hard drive consists, in principle, of a rotating disc, the platter, that is coated with a very thin magnetic material. The information is written and read by devices at the end of an actuator arm. The tip, or head, of the arm contains a tiny coil that generates a magnetic field and magnetizes a small area at the surface of the platter. A second little device on the arm reads the magnetic field on the platter

material (magnetic materials are derived from iron or cobalt). This layer is thinner than $\frac{1}{10,000}$ of the thickness of paper that is normally used for typing, meaning that it is a few nanometers thick. It can be magnetized in small areas by a very small coil that is mounted at the tip of the so-called "actuator arm." This actuator arm can be moved with very high speed and acceleration in such a way that the tip can move from near the center to the outer edge of the spinning platter, which gives it access to the entire data storage area of the platter. Therefore, every small point-like area on the platter can be magnetized, i.e., turned into a tiny magnet, as desired. The tip of the actuator arm also contains a tiny instrument that can read magnetic fields. In other words, it can read whether a magnetic field is zero or large, and in which direction the magnetic field points (north or south pole). Thus, we have an instrument that can read and write bits. If a magnetic field is on (or is pointing toward a certain direction) this is a 1, and if it is off (or points into another direction) we have a 0. We can both read and write this information. To read and write this information fast enough, the actuator arm needs to move very fast and needs to be accelerated back and forth. The actual accelerations that are reached in today's technology are more than 500 times the acceleration g due to the earth's gravity. Nevertheless, hard drives are still not as fast as the electronic semiconductor memories, because these do not have to move any mechanical parts.

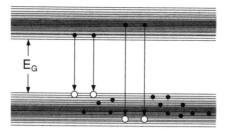

Fig. 3.9 Two bands of closely spaced energy levels of a crystal, just as in Fig. 2.47. Full and empty circles are added to indicate the presence or absence of electrons in the respective energy levels. Emission of photons follows a transition of electrons from full higher levels to empty lower levels

Light Generation by Diodes (LED's)

We have not yet discussed important facts that are connected to the destruction of electrons and holes, except that the energy of electrons and holes differs by about E_G. What happens to that energy? We know that energy conservation is the most basic law of physics and the energy can therefore not vanish. We also know from Sect. 2.5.3 that electrons making a transition between the energy levels of atoms emit electromagnetic radiation. That is also what happens in some crystals when electrons and holes destroy each other. They emit electromagnetic radiation. In principle, the electron occupies an energy, level at higher energy and the hole is nothing but an empty energy level at lower energy, and thus we have a transition of the electron from higher to lower energy and emission of radiation. The energy levels of a crystal already have been shown in Fig. 2.47 and consist of groups of closely spaced levels separated by the energy E_G. We show the energy levels of a crystal again in Fig. 3.9 with a slight modification. The novelty of the crystal energy levels as compared to the levels of atoms is that the energy levels are closely spaced in two or more ranges (so-called bands) and transitions can only occur if electrons occupy higher energy levels and if some lower levels are empty, i.e., if holes are there. We have indicated that situation in Fig. 3.9 by plotting electrons in the higher levels as dots and holes in the lower level as empty circles. Note that the higher energy levels are mostly empty while the lower ones are mostly full. Another new feature of crystals (as compared to atoms) is that both electrons and holes can move freely through the crystal while both the empty and filled states of the atom are localized to the place where the atom is located.

The frequency of the radiation arising from electron–hole destruction can be calculated from the Einstein–Planck relation $E = h\nu$ and is close to E_G as can be seen from Fig. 3.9. Several semiconductor materials emit infrared to visible light. GaAs crystals emit infrared. More complicated chemical compounds, such as AlGaAs, emit red light and some, such as GaN, even blue light. Not all energy is converted to light when electrons and holes destruct. Some of the energy is converted into heat, meaning atomic vibrations. The ratio of energy that is converted to light to the energy that is converted to heat depends on the particular crystal material. In silicon crystals, almost all the energy is converted to heat, while in AlGaAs, almost all energy is converted to light.

Diodes that emit light are extremely useful and are ubiquitous in semiconductor electronics and optics. The diode used in this way is called a light-emitting diode, or LED. We already have explained that the difference of electron and hole energies is, in some crystals, turned into light. We also have explained that the electron–hole destruction stops because of the positive and negative fixed-charge regions that develop in the middle. How can one continue the light generation? This is also indicated in Fig. 3.8. One simply applies an external voltage to the crystal in such a way that electrons are pushed to the right and holes to the left and a diode current flows. This process can go on continuously because the metal wires (shown in Fig. 3.8) supply electrons from the left and extract electrons at the right (and thus create new holes). In this way the energy from the battery is turned into light. Remember that in some crystals the process goes on without any significant generation of heat. The energy of the battery is just used to supply the electrons and holes that emit the light with virtually no excitation of atomic vibrations (heat). This makes LEDs more efficient than ordinary light bulbs that generate not only light but also much heat that represents wasted energy.

It is important to realize that heat production can be reduced to very low levels for LEDs by choosing the right material. Light bulbs, as invented by Edison, always produce heat because their light originates from glowing wires. Currently, LEDs are used in a variety of ways. We encounter the red light of such diodes on a variety of computer equipment, household equipment (washing machines, flashlights, and recently in TVs). LEDs are also used for cars, most prominently for the red brake lights. Here we have an additional advantage when using LEDs. These diodes switch on basically without delay when the driver hits the brake, because the light is generated instantly without any heating of wires that takes time. Light bulbs have a delay of around 0.3 s that it takes for the metal filament to glow brightly. This 0.3 may be valuable time when the brakes are hit in an emergency situation. At 60 miles per hour a car moves about 26 feet (more than 8 m) in 0.3 s.

LEDs are also used increasingly to illuminate residences and business offices. However, here they have not had the breakthrough that they deserve because of their greater efficiency with respect to energy use. One reason is that there exist currently no implementations that replace the light of bulbs with its warm tones. The white light of diodes that is currently available is often perceived as "cold" because of its greater content of bluish colors. Indeed it is generated without heat, and it is not surprising that its spectrum is different from that of bulbs. However, this matter is only a question of material choice and should not be an insurmountable problem. There is good hope that diodes will replace bulbs, even the more efficient gas discharge bulbs, in due course.

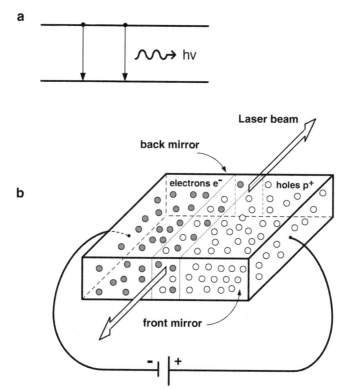

Fig. 3.10 Principle of stimulated emission: (**a**) Electrons are shown to reside in the higher energy level of an atom with an empty lower energy level below. The transition of the electrons to the lower energy level occurs spontaneously if no light is present, and is stimulated (i.e., enhanced) by any existing light. (**b**) Electrons and holes in a diode. If they "meet" in the center of the diode, the electrons destroy the holes by emitting light. This destruction of electron–hole pairs that generates light is also stimulated by any preexisting light. Mirrors are attached to the front and back of the diode

Lasers

There is only one major step necessary to turn LEDs into lasers. The word laser stands for light amplification by stimulated emission of radiation. The principle of stimulated emission was established by Einstein and can be described as illustrated in Fig. 3.10.

Figure 3.10a shows two energy levels of an atom. The upper level contains two electrons that are indicated by dots. Normally the two electrons would reside in the lowest energy level. Two electrons with opposite spin can occupy each energy level as we have learned in Sect. 2.5.3. Because the lower energy level is empty, the electrons from the upper level can transit to the lower level by loosing energy in form of electromagnetic radiation. This radiation is indicated in the figure by the wavy arrow toward $h\nu$ that means that a photon with the energy $E = h\nu$ is emitted by each electron. This emission occurs "spontaneously" if no light is present to start with. The word "spontaneously," in connection with the emission of light by an electron, indicates that the electron makes the transition without any discernible reason much like the atomic nucleus of a radioactive substance may fall apart while no "reason" for this process can or could be found. Typically such a spontaneous transition happens in a time period that depends on the light frequency ν (actually on ν^3), and that can be as small as microseconds or even nanoseconds. Einstein found out that if light is present to start with in the surroundings of an atom, then this "stimulates" the electron to make the transition even faster, say in picoseconds. Thus the present light can be seen as the "stimulating" reason for the transition of the electron. That transition creates then an additional photon, additional light. This process gives us the possibility to engineer a feedback mechanism: we start with light present and therefore create quickly, by stimulated emission, new light that stimulates even more and creates more new light and on and on, until an enormously strong beam of light, a laser beam, emerges.

How can this be done? All we need is a LED with mirrors on two of its sides. These mirrors are made, for example, by using a thin metal layer. If this layer is thin enough, then some part of the light gets through the layer to the outside, while most of the light is reflected and stays inside of the diode. Thus, the light that is generated by the electron–hole transitions creates more light, that creates more transitions and more light and so on. The light becomes so strong that also the small fraction of the light that goes through the mirror becomes very intensive, and therefore a beam of intensive light, a laser beam, emerges from the diode. In this way, one can create very strong beams. Both electrons and holes need, of course, be resupplied in a LED by the application of a voltage. Thus, to fabricate a diode laser, all we need to do is put mirrors at the end faces of the diode as indicated in Fig. 3.10b.

Lasers and Fiber Communication

Diode lasers are used for many applications, for example, to read and write the bit-information stored on DVDs and CDs that let us watch movies and listen to music. Usually red light lasers are used for this purpose. For highest quality (high-definition movies) laser diodes with shorter wavelength are necessary, as we have explained above. The shorter wavelength corresponds to blue light, and the technology is named Blu-ray. There are numerous other applications of laser diodes. The laser pointer that is used by speakers during slide presentations is well known. Lasers also play a major role in optical fiber communications.

An optical fiber is a thin glass-like material that is very transparent. These fibers are cylindrically shaped and often carry light signals over many miles. They transport light signals very efficiently and can carry many gigabits of digital information per second. The reason for this high-speed capability is the fact that light itself is an electromagnetic wave of very large frequency ν (see Sect. 2.1.5) and can therefore be modulated (switched on and off) extremely fast. Think of an electromagnetic wave of a certain wavelength and frequency. You can not switch that wave on and off faster than it oscillates itself. Therefore, if you have a radio wave of one megahertz, you cannot switch it on and off faster or more frequently than one million times a second. Thus you cannot transmit more than one million bits per second. Light has a typical frequency $v = 6 \cdot 10^{14}$ Hertz (oscillations per second). Therefore, we could, at least in principle, transmit 10^{14} bits per second, with each bit (light pulse) still containing six oscillations. Laser-driven optical fiber transmissions are therefore used when the transfer of enormous numbers of data are needed, and the "backbone" of Internet connections uses optical fibers for this reason. Such fibers are even used to connect continents, for example, Europe and America. Charles Kao is credited with pioneering transmissions of signals through optical fibers. Optical fibers are also used in medicine, and you can find bundles of fibers, with light emerging from their endpoints and creating sparkling effects, even in toy stores.

Solar Cells

Light-emitting diodes and laser diodes use electrical power (voltage and current) to produce light. Inversely one can use light and its energy to produce electrical power by using diodes as so-called solar cells. Consider the diode of Fig. 3.8. The center of the diode contains no mobile charge, only fixed positive donors left from the center and fixed negative acceptors to the right. If one shines light on such a diode, the light creates electron–hole pairs by lifting electrons from the lower band of states to the upper band (see Fig. 3.9). This is the inverse process of light emission that is indicated in the figure by downward arrows. The negative electrons are then pushed away by the left negative fixed charge and attracted to the right by the positive fixed charge. Inversely, holes are attracted to the left by the fixed negative charges left to the center and pushed away by the positive fixed charges at the right. Because both electrons and holes are mobile, this results in a voltage and an electric current, i.e., in electrical power that can be used for a variety of purposes. Best known are solar-driven garden night-lights. The electricity of solar cells is used during the day to charge a battery, and the battery discharges at night to produce the light. Solar electricity is a candidate for clean energy production. However, the fact that the electricity is produced only in bright daylight and needs to be stored for use during night (in a battery or by other methods), makes this type of energy currently still expensive as described in more detail in Sect. 4.1.3.

3.3.2 Cameras and Displays

Electronic Cameras

Assuming that everyone knows how a camera with film works, the electronic camera is easy to describe. Those readers that are not familiar with ordinary cameras are encouraged to do some research on the Internet. The essential point of the workings of a camera is the creation of a two dimensional image of whatever is photographed by use of a glass lens. A light-sensitive film is placed at the location of that image in conventional film cameras. Electronic cameras use a light-sensitive micro-chip instead of the film (photo paper).

The principle of how a light-sensitive chip works is as follows. We know from our discussion of solar cells that light generates electrons and holes in semiconductor materials. Consider now an area of a semiconductor that is covered with metal gates and spaces between them. We know from Sects. 3.2 and 3.2.2 how to produce many gates on top of a semiconductor. We also know that such gates can be used as capacitors because they are just metal plates on top of an insulator; this works also when the metal is on top of a semiconductor. These gates can be arranged in rows and columns that form a so-called "array" of gates. Assume now that we have a positive voltage on all the gates of the array and shine light on it. Where the light is bright, many electrons and holes are created, and the negative electrons are attracted to the positive gates and accumulate under them. Each such gate corresponds then to a pixel, a picture element. If there are electrons under the gate, that means that light was shining there, the brighter the light, the more electrons. These electrons can now be used to create an electronic signal. They can be detected by a transistor that is mounted close by the metal gates that are used as capacitors. We have discussed such a structure in Sect. 3.2.2. Thus, with little changes, we could use a structure similar to the memory chip to create a light-sensitive chip for an electronic camera. The detected light is stored in the form of electrical charge under gate capacitors and that charge can be read by transistors and word and bit lines. Actual camera chips have variations of this principal way of functioning. These variations are CMOS and CCD camera chips. CMOS stands for complementary metal-oxide semiconductor, and complementarily refers to a special use of both electrons and holes. This technology is the "workhorse" of modern electronics and is also used in processor chips. CCD stands for charge-coupled devices. These devices are in essence just gate arrays and the charge that is generated by the light under the gate is just transferred to external detectors by applying suitable voltages to the gates. This works in principle like the bucket brigade of firemen. The gates with the electrons can be compared to buckets of water (the electrons), and each bucket is poured into the next neighboring bucket and in this way electrons are transported to the end of the array and detected. The first cameras of this kind were created at Bell Laboratories by Willard Boyle and George Smith.

This way of transforming light into electrical signals that can be stored, for example, on a hard drive, is the modern way of forming and storing images. There is much less photo paper in use since the advent of electronic cameras. The images are stored electronically on DVDs, hard drives, or flash memories. From there, they can be transferred to printers or to the computer screen or TV.

Liquid Crystal Displays

We have talked about what makes computers tick and about a number of electronic devices that create light such as LEDs. We did not yet explain how displays, such as your computer screen work, and add here a few remarks. Modern displays are mostly based on so-called liquid crystals (LCDs stands for liquid crystal displays). These are liquids that change their optical properties, such as letting light go through them, when a voltage is applied to them. The way this is achieved, is by use of transparent electrodes. We are used to metal electrodes, for example, made of copper, that do not let light go through. However, there exist also transparent electrode materials, and these are used for displays. The liquid crystal materials are sandwiched between them. If a certain voltage is applied to the electrodes, the liquid crystal becomes dark. Every pixel of the screen consists of such electrodes and liquid crystal material and can therefore be dark or light and thus form a black-and-white picture. It is also possible to perform the same "trick" with colors, and this requires just more pixels and special liquid crystal materials. The light must be generated separately, because liquid crystals do not emit light themselves. As these lines are written, the best displays use LEDs behind the liquid crystal materials in order to light up the screen. One says that the LCDs are "backlit" by LEDs. It has taken many years to develop displays to the perfection that we are now used to see. The principle of how they work is pretty simple. However, the development of the technology has required armies of design engineers and scientists working over many years.

3.4 Nano Structures and Microscopes

As we know, nanostructures are structures of a typical size around one nanometer. This is a very small size: one billionth of a meter! Small structures are of great importance to many applications including the chips of personal computers or cell phones. As discussed above, a chip often contains hundreds of millions of electronic devices such as transistors. Naturally, if you wish to fit a hundred million transistors onto a chip, which is itself smaller than a postage stamp, you need to fabricate very small transistors. There does not seem to be an end to the need of ever more powerful computers containing more and more electronic devices, and it is clear that we wish to make the devices as small as possible. The smallest size of any electronic device will be the size of a small molecule that contains only a few atoms; in other words it will be of nanometer size. Nanometer size structures are not only desirable in electronics, the molecules of chemistry are also of that size, and nanometer size structures are important in all areas that use chemistry, including medicine and the science and engineering of materials. The possibilities and opportunities of nanostructures are developing as these lines are written. What we are using today are structures that have nanometer size in one direction but rarely in all three dimensions. The insulation layers under the gate of transistors are made of silicon dioxide or other materials with a thickness close to one nanometer. The width and length of the gate of typical transistors, however, are a lot larger than one nanometer, because mass production has not yet been developed to nanometer sizes in all directions. Below we report the current status of technology and then give an outlook to what is possible already in principle but can not yet be mass produced.

3.4.1 Limitations of Current Miniaturization

In the middle of the twentieth century, it was just a dream to make things very small. Feynman asked the question whether it was possible to write the content of 24 books on the head of a pin and answered the question immediately. The head of the pin is about a sixteenth of an inch across (about $\frac{1}{6}$ cm). If you magnify the head of the pin's diameter 25,000 times, you end up with a diameter of about 1,500 in. (4,000 cm). If one models the magnified head of the pin as a circular disc, one obtains an area (π times the square of the radius) of about 1.8 million square inches or 13 million square centimeters. One page of a book is around 300 square centimeters, and thus, the area of the magnified tip corresponds to about 42,000 pages, even more pages than 24 books typically have. We know from Sect. 3.2.2 that, in order to write very small, one inverts the process: instead of magnifying the head of a pin, we reduce the size of the book pages so that they all fit on top of the head of the pin.

Imagine that we shrink a car by a factor of 25,000. The car would then have a length of 0.0002 m, that is about the diameter of a human hair. In order to resolve details of the car, such as the bumpers and mirrors, we would need to be able to resolve pieces that are of the length of a fraction, say about one hundredth, of the length of the car. This would mean we need to control all feature sizes down to about 0.000002 m, i.e., down to two micrometers. This is still equivalent to 2,000 nm. Structures with such length are commonplace in modern technology. Transistors produced by photolithography as described in Sect. 3.2.2, can be fabricated in mass production.

The limitations of photolithography to make things small are roughly determined by the wavelength of the light. We already have repeatedly stressed that the size of the tiny structures that we wish to make can not be (much) smaller than the wavelength of the light that is used in the photolithographic process. The connection of the wavelength λ of light to its frequency ν is $\lambda \nu = c$ (see Eq. (2.42)). Therefore, if we wish our structures to be very much smaller (symbolized by $\ll$) than one millionth of a meter, we must have

$$\lambda << 0.000001 \text{ m,} \qquad (3.1)$$

where m stands for meter. Therefore,

$$\frac{c}{\nu} << 0.000001 \text{ m,} \qquad (3.2)$$

which is equivalent to:

$$\frac{c}{0.000001 \text{ m}} << \nu. \tag{3.3}$$

Inserting $c = 3 \cdot 10^8 \cdot \frac{m}{s}$ we obtain:

$$\nu >> 3 \cdot 10^{14} \text{ s}^{-1}. \tag{3.4}$$

We know from Sect. 2.1.5 that light with frequencies higher than about $8 \cdot 10^{14} \text{ s}^{-1}$ is called ultraviolet (beyond the visible violet). It is possible to obtain a higher resolution than the one we just estimated by using laser light (e.g., excimer lasers). However, if we wish to make structures of one nanometer size, then we would need frequencies that are still a factor 1,000 higher. This is the frequency range of X-rays. Therefore, we cannot use ordinary photolithography to produce such small structures. We can, however, deduce from these considerations that photolithography lends itself to produce structures of micrometer size with relative ease. Feature sizes of $0.01 \, \mu\text{m}$ (10 nm) can be produced by using extreme ultraviolet light. Therefore, one can not only put the content of thousands of books on a chip, but one can store all of these books by using transistors. This digital and electronic storage of books, or of any type of information, has revolutionized our abilities to deal with that information and has brought us the Internet and Internet browsers.

Feynman's dream to be able to write 24 books on the head of a pin has been far surpassed by the existing silicon chip technology and its mass production. Feynman noticed, however, that the ultimate limitations of miniaturization are determined only by the size of atoms. Can we write and read information by using atomic feature sizes?

3.4.2   How Do We Fabricate Nanostructures?

There exist novel types of microscopes that can make the shapes of atoms and molecules visible and, in turn, can be used to move atoms around and to construct geometrical arrangements of atoms. For example, these microscopes can be used to write letters made of single atoms on the surface of a crystal or any solid. They also can be used to put entire molecules, in given geometrical arrangements, on crystal surfaces. An example is given in Fig. 3.11 that shows 4 different types of molecules as well as single atoms arranged in a V-shape on the surface of a silicon crystal.

The silicon atoms of the crystal surface are also clearly recognizable. Surfaces are more complicated to understand than the inner volume of crystals. At the surface atoms can deviate from the regularity of the inner crystal volume simply because there are no neighboring atoms on the outside, and there are therefore electrons missing that would result in the number 8 required for the chemical bonding (see Fig. 3.2). If no oxide or other form of atom is available on top of the surface, the surface atoms try to pair with each other in order to achieve the chemical noble gas

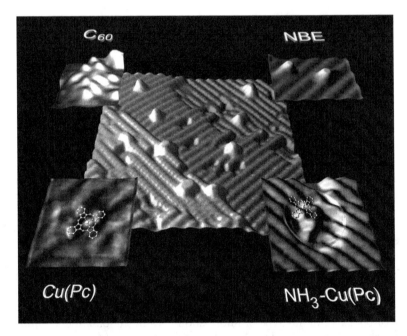

Fig. 3.11 Surface of a silicon crystal, imaged with a scanning tunneling microscope (STM), showing several flat terraces that have been exposed by cutting the crystal at a slight angle. The lines at the crystal surface actually contain pairs of atoms (or dimers) that share electrons, and the lines change direction by 90° on adjacent terraces due to the crystal structure of silicon. This silicon surface has been chemically treated with hydrogen and the V-pattern represents eight locations where single hydrogen atoms have been removed by the STM. The four peripheral images show molecules that have subsequently bonded to these missing hydrogen locations. One of these molecules, C60, is the famous buckyball structure made of 60 carbon atoms. NBE refers to the molecule norbornadiene, Cu(Pc) refers to the molecule copper phthalocyanine, and NH3-Cu(Pc) is ammonia-treated copper phthalocyanine. Incidentally, copper phthalocyanine is a dye molecule that is used to make your blue jeans blue (courtesy Prof. Joseph Lyding, University of Illinois, Urbana)

configuration. The lines that can be seen at the grey surface of the crystal consist of pairs of silicon atoms called dimers. The lines of dimers rotate by 90° (right angle) on adjacent flat terraces due to changes in the orientation of silicon atoms in the underlying silicon crystal.

Also shown in Fig. 3.11 are four types of molecules that have bonded to single silicon atoms. The figure demonstrates that we can control structures at the atomic and molecular levels. To create some electronic or chemical function of such structures is still a step further out and requires some new ideas. Such ideas for molecular electronic function are just in their infancy and will not be described here. We just describe how the geometrical arrangement of such structures can be accomplished.

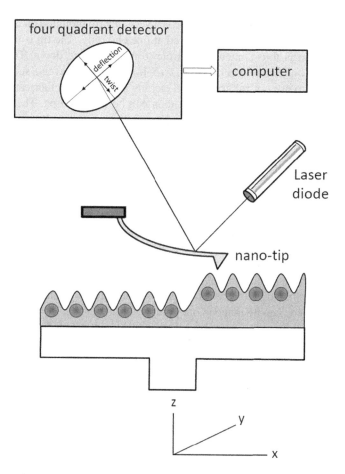

Fig. 3.12 Principle of atomic force microscopy. A tip with nanometer-scale sharpness is integrated onto a flexible cantilever. The tip is brought very close to the surface to be investigated. The sample is mounted onto a (x, y, z) scanning device so that it can be scanned beneath the tip. A laser beam shines on the cantilever and is reflected to a four quadrant photodetector, which detects the cantilever deflection by the relative strengths of the signals coming from the detector quadrants. These signals are transmitted to a computer for processing and to render an image of the surface. Note: In addition to vertical deflection, the cantilever will twist if the sample is scanned perpendicular to the long axis of the cantilever (i.e., out of the page). Such twisting is a measure of the frictional force acting between the cantilever tip and the surface, and it results in a deflection of the laser spot on the four quadrant detector that is perpendicular to the cantilever bending vertical deflection (courtesy Prof. Joseph Lyding, University of Illinois, Urbana)

Key to making atoms visible and arranging them on crystal surfaces are new microscopic methods such scanning tunneling microscopy (STM) and atomic force microscopy (AFM). The AFM was invented by Binnig, Quate, and Gerber, and its principle is illustrated in Fig. 3.12.

Key for the workings of the microscope is a very fine tip mounted on a cantilever. A cantilever is a small "beam" that is fixed at one end only. The tip is mounted on the other side and is thus somewhat flexible to move up and down. The tip apex must be of about one nanometer or less in size. It is not easy to create such fine tips, but it is possible to use crystal growth and etching techniques to fabricate pyramid-like structures on beam like objects that then can be used as tips. The tip is then used like a pointer above a rough surface. In actual experiments, however, it is not the tip that is moved, but it is the crystal surface that is subjected to very slight translations in x, y, and z directions. This movement is accomplished electronically by use of materials that change their size when a voltage is applied to them, so-called piezoelectric materials. The surface atoms exert forces on the tip as they move. For example, the electrons of the surface atoms repel the electrons of the tip atoms by the standard electrostatic repulsion. Thus the tip moves up and down depending on the presence of atoms on the surface and their position. Laser light that shines on the cantilever is then reflected in different directions depending how the cantilever moves. Very small movements of the cantilever cause much larger movements of the reflected beam that falls onto a light detector or camera that is sensitive to the position of the light and feeds the information about this position to a computer. In that way, the surface can be mapped out point by point, and the computer can store the information about every point and then reconstruct out of the data an image of the surface. It is important to note that the image that is formed is formed by the process of scanning a tip over a surface. At no point in time do we actually have an image as we are used to from optical instruments or cameras that use light. The image is created by the computer that digests the results of the scanning procedure. This image composition by computers using some type of scanning over small sections is very important and is used in variations in medicine also (see MRI, CAT-Scan in Sect. 4.3).

Atomic force microscopes can resolve feature sizes below 0.1 nm and therefore can make atoms, molecules, and their shapes visible. This is a very big achievement! You can estimate the magnification of a microscope that created Fig. 3.11 in the following way. The size of the single atoms that one can see in the figure is about 0.1 cm or 0.001 m. The actual size of the atoms is about 0.1 nm or 10^{-10} m. That means the atoms are magnified by a factor of 10^7! Imagine that a car is magnified that much; that car would be about 40,000 km or 25,000 miles long! The AFM cannot only magnify and show the shape of atoms, the tip can also be used to "roll" atoms from one point at a surface to another. That way you can write letters with atoms or molecules such as the V shape in Fig. 3.11. Therefore the atomic force microscope can be used to control structures of molecular and atomic size, and it is a great instrument to achieve nanotechnology goals. It is much more powerful than the optical microscopes. However, from the viewpoint of mass production, it does have a great disadvantage: the images and handling of atoms are achieved by scanning. Scanning is a sequential process and takes time. It is neither as inexpensive nor as fast as the process of stamping patterns into discs, as can be done for DVDs. Therefore, we can achieve the goals of nanotechnology at this time only in principle and not yet for mass production. In spite of this deficiency, nanopattern generation

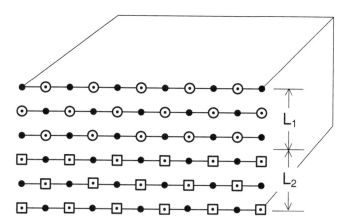

Fig. 3.13 Crystal consisting of thin layers that differ in their atomic composition. A cross section of the crystal in the plane of the page is shown. There are three different atoms involved and symbolized by full and empty circles as well as squares. The two layers L_1 and L_2 correspond to two different crystal types. In practice the full circles could stand for As (Arsenic) atoms, the empty for Al (Aluminum) atoms, and the squares for Ga (Gallium). Then the layers would correspond to an AlAs layer on *top* and perfectly matched to a GaAs crystal *below*. Laser diodes are made of such layered materials

is very interesting from the viewpoint of exploring the possibilities when things become very small. It is even possible to generate three-dimensional nanostructures as follows.

Modern crystal growth techniques have advanced significantly and it is possible to create layered crystal materials that are composed of very thin layers formed with different atoms. An example is shown in Fig. 3.13.

The figure shows a cross section of two such different crystal layers on top of each other. The crystal material can, in principle, be any combination of atoms. However, the distance between the atoms must be about the same in all layers in order to have a perfect match as shown. The distance between the atoms in the crystal layers depends, of course, on the chemical bonding and, therefore, on the types of atoms that are involved. Crystal layers of GaAs and AlAs are perfectly matched with respect to the atom distance and are commonly used to create laser diodes with light frequencies in the red. We also already have seen the use of other types of layers, for example, silicon and silicon dioxide, when describing chip-technology. The thickness of such layers can easily be controlled on the nanometer and even atomic size scales. Modern materials technology does not only use crystal layers but also layers of glasses. In glasses, there exists no completely regular distance between the atoms, so the inter-atom distances fluctuate slightly. The silicon dioxide that is grown on top of silicon in transistor technology is actually such a glass. Other types of materials such as organic compounds and polymers also show electronic function (such as solar cells and transistors) and are being developed for electronic applications. For all of these materials, the capability exists to create nanometer thick layers. The layers themselves can be patterned by photolithography down to about 10 nm as we have discussed in Sect. 3.2.2. Smaller structures can also be made with AFMs but not currently mass produced. Yet, nanometer-size patterned materials do exist in nature and nature is producing big and relatively inexpensive quantities of these materials (e.g. proteins). How does nature do it? This is discussed next.

3.4.3   How Does Nature Produce Nanostructures?

We have learned in Sect. 2.4.3 about molecules and polymer materials that are the stuff that plants or even our bodies are made of. Cellulose, the basic plant material, is a polymer composed of sugar-type molecules. Remember polymers consist of long molecules that are intertwined and thus hold together and behave similar to plastic materials. Our bodies consist to a large extent of different types of proteins that form our muscles and organs such as the liver. As we have learned, these proteins, in turn, are made of 20 different chemicals (amino acids). How is all of this, the countless number of plants, animals, and people, made and mass produced? We have seen in Sect. 2.4.3 that the key to their production is the DNA molecule, a giant molecule that contains code sequences that can be transcribed and used for the production of proteins and, of course, also plant material. Thus nature has a way to reproduce itself and to grow. Remember, the DNA consists of two strands that are wound in a helical form around each other and contain the information carrying molecule pairs A-T and G-C as shown in Fig. 2.40. These strands can separate and then generate two new strands of identical information. These two new strands can separate and form four new identical ones and so forth. This gives us 2, 4, 8, 16, 32, 64, 128,…. and very rapidly enormous numbers of new strands. Everyone knows how fast bacteria can grow and multiply (wash your hands)! Thus we do not need photolithography to create a large numbers of useful molecular structures. Our bodies, plants, and bacteria have a way to produce these structures by themselves, a self-organizing way. As these lines are written, the production of materials and structures by messenger molecules and self-organization is just being understood in greater and greater detail. Some new types of materials are called "smart materials" because they do something by themselves. For example, one can include into some materials small pellets of glue. If a break develops in these materials, glue is released from the pellets and the material repairs itself. This is just a basic example of materials that can do something more than just being there. Nature has gone far beyond such simple possibilities and repairs its materials in much more sophisticated ways. Thus, nature has found great ways of fabricating nanostructures and nano-materials. Can we expect to copy nature and create in this way electronics and the materials of the future?

Yes, we can expect that our ability to copy nature and to produce useful nanostructures will continue to develop. There exists an enormous need in medicine for such structures. Think of what it would mean if we could replace damaged human tissue, or if we could deliver safely (by way of nanostructures) chemicals that destroy tumors! Would we not like to have a whole laboratory on a tiny chip that checks the fluids of our body and sends out medicine, such as insulin controlling the blood sugar in diabetic patients, if necessary? Here, the possibilities are fantastic, and future innovations in this area are certain to occur and will be of great importance.

Electronic chips are, however, very special in one way: unlike in biological materials, the basis for the electronic function of computer chips is the movement and control of electrons. Bio-material functionality is based on electrons, protons, whole atoms, and even giant molecules. Therefore, biology deals with the motion of entities that are having much larger masses than the electrons. As we remember from Newton's law, electrons can easily be accelerated because of their small mass. As a consequence, our electronic chips are very fast and can work at gigabits per second. Single transistors can switch on and off in picoseconds! This high speed of electronic devices is of great importance for our most cherished applications. Think of just downloading a file from some location of the World Wide Web (www). Think of how long you would have to wait if the computer would be 1,000 times or a million times slower! The speed of man-made chips is unique and is the one property that exceeds everything known in nature. Thus, the known ways of self-organization to fabricate nanostructures do not have this one very desirable property of extremely high switching speed. We need to open new horizons if we wish to have it all: ultrafast nanostructures inexpensively mass fabricated. This area is wide open for future STEM innovators.

3.4.4 Nano-materials and Mass Production

As outlined above, Feynman's dreams to "write" artificial patterns of nanostructures in three dimensions can be realized. We have also learned that it is possible to produce very small electronic devices in large numbers. These ultrafast and powerful electronic circuits can be connected worldwide, with the speed of light, by using optical fiber and wireless communications. This has enabled mankind to create the Internet and the World Wide Web. Most of us have used the World Wide Web without being surprised that we can sift through gigabits of information that are located thousands of miles away from us. In fact, most of the time, we do not know where the computers that serve us are located. We can even watch remotely taken videos on portable phones, and we can download movies from stores far away and watch them on our PCs. The demands on information, on transmitting this information and on processing this information, are still rapidly increasing and will continue to do so. As a consequence it is desirable to realize the full dream of Feynman, to create three dimensional nanostructures and, in addition, to be able to make them perform all kinds of desirable functions. These functions include ultrafast electronic switching and the generation and communication of electromagnetic signals such as light (optical fibers) and radio frequency waves (wireless communications). Other functions, such as chemical and mechanical, are also of greatest importance, for example, for the medical applications that we mentioned including the lab on a chip.

The mass production of such nano structured materials is therefore, at least in the opinion of the author, one of the most rewarding areas for future innovation, invention, research, and development. The production of materials, structured materials, and smart materials with electrical, optical, mechanical, and chemical functions will control and further future progress. To finish this section we list just a few areas where progress in materials is necessary. We already have mentioned electronic applications where we have an insatiable appetite for ever larger circuits that work faster and with an increasing number of devices. Further progress is needed in fiber communication to transmit larger amounts of data in shorter time intervals. All of this needs new or at least novel structuring of materials and their mass production.

There exist a large number of important problems that we can only solve if we make progress in the understanding of materials be they crystals, glasses, or polymers. We need batteries that can hold more charge and that can be recharged more often. Consider cars that work with electric motors instead of combustion engines. Such cars can help us to a better and cleaner use of energy because electric motors are very efficient and have no exhaust of burned and unburned gases. To make such cars practical, we need batteries that can supply the car with electricity over several hundreds of miles before they are recharged. Furthermore, one needs to be able to recharge the battery several thousands of times before a new battery is needed to reduce cost. Current lithium batteries are just barely fulfilling these requirements.

We need solar cells to turn sunlight into electrical power with great efficiency and under extreme conditions of both cold and heat. Converting even a small fraction of the sunlight that hits the earth into electricity could meet all of the energy demands for the entire earth's population. Again, however, current solar cell materials do not have the desirable efficiency, durability, low cost, and insensitivity to extreme conditions. In addition, one needs to store the energy derived from solar cells for nightfall when the cells do not work. Therefore one needs inexpensive and powerful batteries or other energy storage to go along with the solar cells.

We need an electrical grid that supplies all corners of every state and that is smart enough to distribute that energy to places where it is momentarily in high demand. We may have, for example, a lot of wind energy available at one location but need more electrical energy in another because of industry demands. The materials that are used for electrical wire connection include copper and aluminum. Copper is already extraordinarily expensive and keeps going up in price. New types of materials, such as superconductors (see Chap. 5) could replace copper wires.

The list of new types of materials that are needed, nano structured or not, for the ever-increasing demands of an ever-growing number of people, is large, and we will hear more about it in the next chapter.

Chapter 4
STEM in Our Daily Lives

This chapter deals with topics that are very directly related to STEM education and in addition are of great importance for our daily life. We cannot exhaust all the important STEM topics, so we restrict ourself to a few examples. From knowing what has changed during the lifetime of the author, he can only hope to cover what will still be relevant 50 years hence.

4.1 Energy

We have learned about many forms of energy, including kinetic energy (see Eq. (2.18)), electrical energy as stored in batteries (see Sect. 2.2.1), thermal energy or the energy related to heat (see Sect. 2.3.2), and other forms of energy such as the energy released in chemical reactions as described in Sects. 2.3.3 and 2.4.3. We know from these sections that energy is something useful and that we need energy in our daily lives in all of its forms. We supply energy to our body by eating, because the food contains chemical energy for the needs of our bodies. We use electrical energy when we turn on the light, the telephone, the computer, the kitchen stove, the washing machine, the TV, and a very long list of useful machinery and appliances. Naturally we use the energy of fuel when we drive cars or fly airplanes. Energy is also needed by the farmers who work on their fields when they plow or supply fertilizer and produce the food that we later use. If one wishes to describe what we need most in our life, then it is energy, energy, and energy again in many many possible forms. In fact we need and use so much energy that we often hear that we need to look and see that more energy is being "conserved." The STEM expert notices, of course, that this is not what we need to do. Energy is always conserved automatically. The law of energy conservation is the most basic natural law that we know. Energy can neither be destroyed nor created out of nothing. The real point is that we need energy in certain useful forms such as gasoline or battery power. Other forms, such as the hot exhaust gases of cars or jet engines, are not very useful, and we need to find ways to produce the least of the useless energy and to obtain

K. Hess, *Working Knowledge: STEM Essentials for the 21st Century*,
DOI 10.1007/978-1-4614-3275-3_4, © Springer Science+Business Media New York 2013

the energy that we actually can use. All useful energy originates from the sun and either is derived directly from sunlight, or has been derived in the past from the sun in the form of the so-called fossil fuels or even further back at the beginning of our planetary system in the form of nuclear material.

Unfortunately, it is not possible to obtain and use the energy that we need in a way that is 100% environmentally friendly. We know from Sect. 2.3.5 that it is not possible to generate useful mechanical energy without generating also random heat energy. As we will outline below in more detail, we can therefore not use any energy sources without changing something in the environment. Some of you, who have already heard about many methods that supply us with energy may say: "oh, this is not true, there are some ways to obtain energy that do nothing to our environment, for example one can generate energy by using the tides."

The tides are the rise and fall of ocean water due to the gravitational influence of the moon and the sun on the oceans of the rotating earth. Thus one could store the rising waters during high tide and then, during low tide, let the waters flow back and drive an electrical generator, so-called turbine. This sounds like a real winner with no influence on the environment. However, we know that energy is conserved and can not come from nothing. Where does the tide energy come from? Energy gained from tides originates mainly from the rotation of the earth. Therefore, that rotation of the earth will be influenced if we take energy out of that system. In fact, such tidal energy consumption happens automatically but slowly. The tidal distortions of a planet or moon (even without water) lead to the so-called tidal lock which means that in the very distant future the earth will only rotate once a year around itself and thus have always the same side facing the sun. The moon is already now in tidal lock with earth, and, therefore, we can only see one side of the moon. If we influence the tides to generate much electrical power, we influence the tidal process. The forces created by influencing the tides may also shift the outer crust of the earth relative to the inner viscous or even liquid parts. Is this shift dangerous?

Here lies the crux of all environmental considerations and concerns. How do we define dangerous? If we require no change over hundreds or even thousands of years, then we have a problem, even with energy from tides. The earth rotation would change considerably in thousands of years if we take a large portion of the energy that we need from tidal power. Even a very small shift of the upper crust of the earth could change the flow of liquid lava in the hot inner earth and therefore rechannel volcanos, say those of Hawaii. The Hawaiians may not like this! It is extremely important to realize that there exists no zero tolerance solution for the environment. We cannot obtain and use energy without perturbing the environment. Even photosynthesis (see below) changes the environment and reduces carbon dioxide while producing oxygen. All we can look for and hope to find is the most inexpensive way to create renewable energy resources that disturb the environment as little as possible. In the following we describe a few possible pathways to such solutions. The author does not claim that anybody knows which procedure is going to win in the future. We only know that this is a great and important topic for future STEM experts.

4.1.1 The Warmth of Sunlight

The most elementary form of energy that humans use is the warmth of sunlight. The sun emits a broad spectrum of electromagnetic waves, ranging from waves in the radio-frequency range to the infrared, to the visible, to the ultraviolet, and even up to γ radiation. We have discussed all of these forms in Sect. 2.5. The rays of the sun are partly absorbed in our atmosphere and finally on the ground and in the oceans. Much of this energy is turned into heat that warms all of our surroundings and keeps us alive. Any heated body also emits electromagnetic radiation depending on its temperature. The earth, therefore, emits mostly infrared back into space, particularly during the night and cools down in that way. Heating and cooling during summer and winter and day and night, respectively, leads to average temperatures at the different locations on earth that determine the climate from the cool arctic regions to the warm temperatures of the equator.

The balance of heating and cooling of the earth depends on many factors including the absorption of light on the ground and in the oceans, the angle of incidence of the sunlight (that changes depending on geographic location and season), the composition of the atmosphere and its humidity, the formation of clouds, and the wind patterns and ocean currents. Some of the gases of our atmosphere, particularly carbon dioxide and water vapor, are so-called greenhouse gases. These gases lead to an increased absorption of infrared sunlight and to a decreased radiation of infrared back into space. Thus they have an extra warming effect, comparable to the warming in a greenhouse from which their name derives. Greenhouse gases are, therefore, partially responsible for the comfortable temperatures that we enjoy on earth. The earth's climate has changed in the distant past from ice ages, during which glaciers have formed (even in Hawaii), to heated periods during which the north pole was green. Greenhouse gases are, at least in part, thought to be responsible for these climate changes.

Recently scientists have become concerned that too much of a warming may occur if we generate large quantities of greenhouse gases, such as carbon dioxide, due to the exhaust of our machines and the burning of coal and oil. It has been proven, that the influence of human energy production and use has indeed already led to measurable effects on earth and that "global warming" is on the increase. Pollution can also cool the planet due to the effect of pollution clouds that transmit less sunlight to the surface of the earth. All of these conflicting and complicated effects make it very difficult to precisely predict the effects of human influence and what we should do about it. Much research is needed here, and it is imperative that present and future STEM experts explore this area in great detail. The necessities of energy use to sustain our life and, at the same time, to disturb and pollute earth as little as possible present us with two conflicting demands of greatest importance. Science has currently not progressed far enough to present us a "royal road" solution to the problem of global warming. It is clear, however, that we need to be careful and not pollute our planet beyond repair. At the same time, we need to use energy to sustain our life. These conflicting problems were the motivation for my choice of topics in the following sections.

4.1.2 Energy Sustaining Life: From Photosynthesis to Fossil Fuels

The energy that is necessary to grow food is directly retrieved from sunlight by members of the food chain, such as algae, plankton, and plants of all kind, and is used by animals, fish, cows, and finally humans. Food is a most important form of energy, and the production of food is one of the most investigated topics of human research. Food production is subject to science and engineering in progress and research, and advances are still definitely necessary. The understanding of DNA and the engineering of the molecules of biology offers here a world of possible innovations. From the educational viewpoint, we need to know about the following important processes of nature that are basic for the energy supply of life.

Photosynthesis is a word derived from ancient Greek. It means putting something together (synthesis) by use of light (photons). What is it that is put together? It is mostly sugars, such as the glucose discussed in Sect. 2.4.3. Glucose ($C_6H_{12}O_6$) is put together from water (H_2O) and the carbon dioxide of the air (CO_2) with the help of the energy of light. And imagine, the waste product of this process is oxygen! We know from Sect. 2.4.3 that sugars are very important and represent the basic molecules that build cellulose, the material that plants are made out of, and that even our DNA has a sugar-phosphate backbone. We will see immediately below how sugars are used to provide energy to our body. Thus, photosynthesis provides us with basic molecules that are necessary for our life. Nearly all life on earth depends on photosynthesis as source of energy either directly or indirectly. Plants, algae, and even some bacteria use photosynthesis, and animals and humans use plants as a food source. Photosynthesis occurs everywhere, on land and in water. The first photosynthetic organisms probably evolved already more than 3 billion years ago. Then the atmosphere had much more carbon dioxide and much less oxygen than it has now. The carbon dioxide was turned into all kinds of sugars and sugar polymers. Of course, the carbon of the carbon dioxide has also found its way into the oil that was derived from algae and plants.

Sugars even move our muscles. It may not be particularly healthy to eat a lot of ordinary sugar, like the sugar derived from sugar cane or the sugar beet, that is chemically different from glucose. Our bodies turn this type of sugar (and starch) quickly into glucose that also is directly available to us in sweet fruits such as grapes. Glucose, in turn, is used by our bodies to "fabricate" adenosine triphosphate (ATP) molecules that provide the basic energy supply to our muscles. The biochemistry of ATP molecules is important for anyone who is interested in the energy machinery that powers our bodies. It does, however, require a major effort to master this topic which is, therefore, ideally suited for special projects. The interested reader is referred as usual to Wikipedia or other articles on the Internet. There one learns that the chemical process of forming ATP molecules within our bodies is particularly efficient if oxygen is available in addition to glucose. The actual chemical formulae are rather complicated and can only be understood with some advanced knowledge of biochemistry. However, we know about the need for oxygen when we stress our muscles. Then, more ATP is needed, and our bodies need more oxygen to fabricate the ATP from glucose. As a consequence, our breathing is much accelerated when we work hard. If we strain our muscles beyond our capability, the oxygen supply is not sufficient, and the chemical processes that produce ATP go a little haywire. This results in pain that can last for days. Any sports person should understand this process in detail to know how best to manage their body energy.

Humans of modern times need many other forms of energy in addition to ATP molecules. We use a host of machines, ranging from cars to ships and airplanes, and we heat our homes. Agricultural food production is unthinkable without energy. We need energy for using farm machinery and for producing fertilizers. Currently the forms of useful energy that are available for almost all of the above-mentioned needs are some types of gasoline or diesel that power our internal combustion engines. Gasoline diesel are hydrocarbons usually derived from crude oil (petroleum) and derived, in turn, from so-called fossil materials. Crude oil is pumped out of the ground at many places of the earth and is the raw material for gasoline and therefore a precious commodity. As far as I understand it, crude oil has formed from biological materials. For example, algae and certain kinds of plant material may have accumulated in large pools of sea water. They were growing by using the sunshine, the air, and the minerals of the sea. Then over time, mud and sand settled on top, and by heat and pressure, the algae material was transformed into natural gas and petroleum. This process takes millions of years. As a consequence the petroleum reserves of the world cannot easily be replaced by copying the method of nature exactly. The amount of available fossil fuel is naturally finite. As we use them, the ease of accessing petroleum will decrease, because the best fields are used up, and, at some point, we will run out of fossil fuels altogether.

Coal (made mostly out of carbon) is another commodity of this kind and has developed, again in a very slow process, from plants such as trees of large forests that have been covered by mud and dirt over 200–300 million years. Coal is not quite as desirable as oil for a number of reasons. For example, coal also contains a number of other chemicals, such as sulfur, that are undesirable because they cause air pollution. Coal can, however, in principle be transformed into oil because coal is made out of the element carbon and oil contains mainly hydrocarbons. The hydrogen can be supplied from our abundant water (H_2O) resources. Nevertheless, also the coal reserves are finite. As a consequence, one calls coal and oil black gold: precious material with limited availability.

The use of fossil fuels by the burning and combustion of oil and coal also causes many environmental concerns. It results in an exhaust of gases, mainly carbon dioxide (CO_2) but also gases such as sulfur dioxide (SO_2) and other environmentally not so friendly gases. As already discussed, CO_2 is a so-called greenhouse gas that leads, even in rather small quantities, to increases in the warming of the earth's atmosphere. In other words the combustion of fossil fuels has consequences for the properties of the atmosphere of the earth. In addition, the hot exhaust gases represent energy that is wasted. For these reasons, the finite availability of fossil fuels and their undesirable environmental consequences, modern research looks for replacements of these fuels by an energy source that is environmentally friendly and can be replenished in some way. We will see below that this is a very difficult problem, particularly because the new energy source needs to be less expensive or at least as inexpensive as oil and coal. Can we find a way that is somewhat more expensive but does no significant harm to the environment? We would be willing to pay if there would be no negative impact for our planet!

4.1.3 Solar Energy

All energy that is useful to us stems directly either from the sun or from the time of the origin of the solar system when materials formed that can be used to produce energy.

At a recent discussion of energy resources on earth, one of the participants said: "Nuclear energy is not an option, because it is not as clean as that from the sun." It is actually amazing that it was not known until about 1950, and is still not known to everybody that the sun energy is of nuclear origin (see Sect. 5.1). The sun does appear to be "clean" to us, however, because it is so far away and our bodies have become resistant to the higher energy radiation of the sun that we call ultraviolet. People still may get melanoma, a skin cancer, from over-exposure to ultraviolet radiation, and we need to be careful even with ordinary sunlight. However, the sun's radiation is indeed the cleanest energy that we can possibly obtain.

We therefore need to attempt to use the energy of the sun in much more extensive ways than we do already through the natural processes of growing plants. There are many ways of doing this. The big challenge is, that none of them are currently as inexpensive as the use of the available fossil fuels. New technologies are always compared, particularly with regard to the price, with the existing established technologies that have already been shown to be inexpensive enough. Therefore, any new technology is fighting at the start an uphill battle. However, because fossil reserves are finite, and do not represent the most environmentally friendly way to use energy, that battle must be fought and will be an ideal playground for STEM experts. Here are some examples that may work for future energy conversions from sunlight and the broad use of the so obtained energy.

Thermal–Solar

Current developments show a lot of promise for solar energy generation equipment that just uses heating by sunlight. All one needs are mirrors of parabolic shape as shown in Fig. 4.1. Such mirrors reflect the sunlight toward a centerline of the mirror as shown in the figure (search the Internet for parabolic reflectors). Therefore, the energy of all the sunlight that falls on the mirror is concentrated along this line. As a consequence, any object that is placed along this line, is going to be heated significantly. We can place there a thin tube, and we can have a liquid such as water

Fig. 4.1 Sketch of a
parabolic mirror that reflects
all incoming sunlight toward
a line running along its center

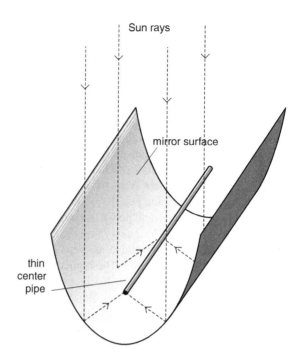

flowing through this tube. The water is going to be heated to such high temperature that it turns into steam. The steam can then be used to generate electrical energy, for example, by driving a steam engine. Alternatively, other types of liquid can be heated up, stored in big containers, and used later to generate steam and drive a steam engine. There are numerous variations of such machinery, and some are very promising for future use, particularly in states and countries that have abundant sunshine.

The heat created by solar mirrors may even be sufficient to split water molecules H_2O into their atomic components hydrogen H (two atoms) and oxygen O in gas form. The hydrogen-oxygen gas mixture is highly explosive and can, in principle, be used to drive combustion engines. Such engines produce virtually no pollution because the product of the burning is just water again. Thus we can obtain explosive gas mixtures from water and sunshine and can turn these mixtures back into energy and water. This is truly renewable and pollution-free energy. Unfortunately things are not quite as easy as this sounds. The temperatures needed to split water into its components are very high, and the oxygen–hydrogen gas mixture is, as mentioned, extremely explosive; both facts pose problems for the use of this process. There are ways around the problems. There exist fuel cells that turn hydrogen and oxygen into heat and electricity without explosion. There are also ways to split water by sunshine without the extremely high temperatures that direct splitting requires. These ways involve chemical catalysis. Catalysis is a branch of chemistry that involves a mediator (a catalyst) to enhance a chemical reaction and represents a very promising area for the creation of hydrogen by sunlight.

Electrical–Solar

As we know from Sect. 3.3.1, there is a way to turn the light of the sun directly into electrical energy by using semiconductor diodes, the so-called solar cells. Diodes made out of a variety of materials including GaAs, CdS, SiC, and pure Si can generate a voltage and current if they are illuminated by light. The efficiency of this energy conversion depends on the material that is used as well as on the geometry of the diode including the metal contacts. Modern designs of crystal-silicon diodes are more than 25% efficient, and even higher efficiencies can be reached by involving crystals of GaAs and similar compound materials. The earth receives more than 170 peta Watts of solar power. Remember, kilo means 10^3, mega 10^6, giga 10^9, tera 10^{12}, and peta 10^{15}. This means that, if we convert less than 1% of the solar power into electrical power, we would still have 300 times of what the USA currently needs. Thus solar energy represents a basically unlimited and renewable energy resource. Diodes are also as clean as can be, at least those made out of silicon or silicon carbide (SiC). The higher efficiency materials may be a different matter. Take, for example, GaAs. Everyone knows that arsenic (As) compounds are poisonous. Thus, if one had GaAs solar cells on the roof-top and the house burned down, that might create problems for the neighborhood. Silicon is a material that is environmentally friendly and surrounds us everywhere; silicon is contained in dirt! Larger silicon crystals are still expensive, particularly since they are used in chip technology. However, solar cells with reasonable efficiency can also be fabricated from polysilicon (made of small crystallites and relatively inexpensive) and even non-crystal glass-type (so-called amorphous) silicon. Silicon carbide (SiC), also known as grinding powder, is inexpensive and has many desirable properties but has not been investigated in any detail yet.

What is it that prevents us from using this very clean solar source of energy? Actually there exist a number of obstacles. The production of solar cells uses energy, because the material needs to be heated and processed. Solar cells also do not last forever they crack and age over the years and need to be replaced. This alone makes the cost-efficiency calculations complicated. Furthermore, the sun does not shine for 24 h. What happens during the night? A smart electrical grid (see below) needs then to supply energy from other sources or other places. Such changes of location for electricity supply are not easy to accomplish, and they are expensive. If we possessed batteries with long life and significantly better storage capability, we could solve this problem. Of course, energy can also be stored in other ways, for example, by pumping water up a hill into a reservoir and running it down to drive an electrical turbine during the night. However, all such solutions create additional expense, and then solar cells are expensive to start with. Yet, I do not see all of this as negative, but on the contrary, see a wonderful area evolving here, ready for innovation by STEM experts.

4.1.4   Renewable Biomass

There are of course many other ways to obtain energy in suitable and renewable form by just using the energy that the sun radiates onto us. A very promising way is to use growing algae, plants, and even bacteria to directly produce carbohydrates or alcohol (e.g., ethanol). Ethanol can be produced efficiently, as is done currently in Brazil, from sugar cane. It can also be produced from corn and beet products. Some current research projects attempt to turn abundant and inexpensive cellulose, instead of valuable food products such as corn and sugar cane, into fuel or other forms of very useful energy. Cellulose is the actual plant material of all plants. Using cellulose would be great and appears possible because cellulose is a polymer composed of glucose. The science and engineering of this type of energy is making great progress as our understanding of biomolecules, including DNA, paves the way for great advances. This is a truly important STEM area, probably more important than any other area for the energy production of the future. The Internet provides a great many understandable descriptions, and we refer the reader to them because this area is changing rapidly.

4.1.5   Energy Supply and the Smart Grid

There are two requirements for a better and renewable supply of useful energy compared to the use of fossil fuels. The first is that we produce the useful energy directly from the sun or from resources that are completely abundant and (relatively) inexpensive on earth. This we have discussed in the above sections. The second requirement is that we distribute the energy in an optimal way and make it available to every person on earth.

The distribution of energy on earth can , at least as far as I understand it, only be achieved by an excellent electrical grid, a smart grid. Think of the existing electrical grid, with all its larger and smaller power lines, just as we have it now. This grid can transport enormous power over large distances because the electrical and magnetic forces that can be derived from electrical currents are very large. We can drive big electric motors, furnaces, air conditioning systems, refrigerators, and many of the needs of a large city by obtaining the energy from remote power plants. If we wish to use not only fossil fuel power plants but also other sources such as direct solar power or wind-driven electrical generators, then our grid needs to be very efficient in distributing electrical energy from different places. Consider, for example, the following situation. Assume that a major city derives its energy from three sources: wind, direct solar, and fossil fuel. Now think of a snowy New Year's Eve with many people having celebrations when it is dark and solar energy is not directly available. Also assume that just by coincidence there is no wind, as rarely occurs in the area of that particular city. Now all the power needs to be supplied by the fossil fuel plant.

However, that plant is not ready for such high power use. Therefore, power needs to be imported via the grid from other places of the country, from places that have wind, that have more fossil fuel plants, or that have water-generated electricity or even nuclear power plants.

Thus what we need is a flexible and smart grid that brings the energy to the right places, exactly where and when it is needed. That smart grid also should have another property: it should not use much energy by itself. Unfortunately ordinary electrical wires do heat up when electrical currents flow through them, and they heat up more when a larger current flows. Actually the heat generation increases in proportion to the wire resistance and to the square of the current. Thus we need low currents and low resistance for the wires of our grid. Remember from Sect. 2.2.2 that the electrical power equals the product of voltage V and the current A in amperes, i.e., V A. Therefore, if we have a high voltage, we can have the same power with lower current A. This is the reason why long distance electrical lines are driven with very high voltages (a more detailed discussion is given in Chap. 5). If the voltage is high, then the current can be relatively low and the heating of the grid power lines is low, while the power V A that is delivered to the city stays the same. What is needed in addition to high voltage, is a low wire resistance. Copper has a very low resistance and is therefore a very preferable material. That has made copper very expensive. What we need are materials with zero resistance. Such materials do exist and are called superconductors (see Chap. 5). Unfortunately these materials work currently only at very low temperatures and are very expensive too.

The grid that we have needs still a lot of improvement and needs to become a lot smarter in the future when different generators of power are used in addition to or instead of fossil fuel plants. This is yet another area for improvement by future STEM experts. We need a smarter and less wasteful (no heat production) grid! As soon as we have it, we may be able to use all energy that is produced by the various methods described above. Then we also can transfer the energy easily to where we need it. The electric cars that are now becoming available will recharge their batteries using power derived from the grid.

4.1.6 Nuclear Energy

Mankind could not survive for long without the special type of nuclear energy that heats the sun. All nuclear energy-generation is based on turning some of the mass of atomic nuclei (composed of protons and neutrons) into energy. The possible conversion of mass into energy was first noted by Einstein. He found the relationship between energy and mass that we have already stated in Eq. (2.16) and that we repeat here because of its great importance:

$$M = \frac{E}{c^2}. \tag{4.1}$$

A more detailed discussion of this famous equation is given in Sect. 5. You may be used to seeing this equation in the form

$$E = mc^2, \tag{4.2}$$

where m is the mass of an object that stands still, the so-called rest mass. For those who do not wish to immerse themselves into the difficulties of relativity theory (as we do in Sect. 5), the difference between the masses of objects that move or stand still does not really matter for the present discussion, although it matters in general. The important point to understand is that the velocity of light c is a very large number in the usual types of units and, therefore, c^2 is an even larger number. As a consequence, a relatively small mass corresponds to a very large energy. If we could turn one kilogram of some substance into energy, we would obtain almost 10^{17} J. To obtain this amount of energy by burning oil, one needs to burn more than ten million barrels of it.

In one of his first presentations about the relationship between mass and energy, Einstein was asked whether he could foresee the use of energy so created. His answer was in essence: "no!" He reasoned that if one could or would build machinery that could transform mass into energy, then one could create so much energy that a single person could use it to blow up a whole city. Einstein's fears were born out by the destruction of Hiroshima and Nagasaki during World War II, and there exist terrorist groups who would readily destroy entire cities were they able to secure nuclear weapons.

The fear of nuclear terrorism and nuclear accidents (as they have occurred in Chernobyl and Fukushima) has prevented the uninhibited large-scale use of peaceful nuclear energy, in spite of the fact that this energy resource would be practically inexhaustible. If produced and administered correctly and wisely (a very big IF indeed) nuclear energy may not even harm the environment as much as the uses of fossil fuels do (remember, they emit greenhouse gases). Because of its general importance, some further explanations are given here as to how nuclear energy can be generated and used peacefully as well as in weapons. Some further details about nuclear energy are also given in Sect. 5. In essence there are two ways of nuclear energy generation.

Nuclear Fission

The first way to obtain nuclear energy is, so-called nuclear fission. Nuclear fission is the process used in all currently existing nuclear plants. This type of mass energy conversion is based on the following fact. Consider the two materials uranium-235 (meaning that this is uranium with a nucleus that contains 92 protons and 143 neutrons) and plutonium-239 (94 protons and 145 neutrons). We know from Sect. 2.3.2 that there are isotopes of atoms with different numbers of neutrons. The number of neutrons is important here, and the processes described below would, for example, not work for uranium-238 which is the most abundant uranium isotope in nature. The actual separation, or "enrichment," of uranium-235 from the more abundant uranium-238 is a difficult process whose details are kept secret by the nuclear powers. The principle of this enrichment process is to use centrifuges (think of the rotating centrifuge of a washing machine) that create significant forces to separate the heavier uranium-238 from the lighter uranium-235.

The uranium-235 and plutonium-239 atoms are special in the following way. If one neutron is added to their nuclei, then these nuclei become unstable and "explode" into smaller pieces including smaller atoms, electrons, protons, more neutrons, and γ rays. There exist atoms in nature, such as californium-252 (Cf-252), that decay spontaneously over time without any external influence and provide neutrons. The French scientist Marie Curie pioneered the discovery and investigation of many of these so-called radioactive materials, materials that decay into others and emit highly energetic byproducts that are usually just labelled as "radiation" or "nuclear radiation."

The important point is that the total weight of the smaller pieces, resulting from the fission of uranium and plutonium, is less than the weight of the original atoms plus the added neutrons. This "lost" mass has been turned into energy. Details of this process can be found on the Internet. The actual energy released is roughly 200 Mega eV per original uranium atom or about $3.2\,10^{-11}$ Joules. Multiply this seemingly small number of Joules by the number of atoms of a small piece of material, which is around 10^{25}, and you obtain $3.2\,10^{14}$ Joules that is enough to heat about eight million barbecues for an hour. Unfortunately, all of this fission-related energy is not just creating heat. As mentioned, the fission of nuclei into smaller pieces is accompanied by all kinds of nuclear radiation, including γ-rays that we have described in Sect. 2.5. γ-rays behave like X-rays and penetrate the human body. They are strong enough to cause damage to the all-important DNA molecules, and they can cause severe illnesses such as cancer even years after radiation exposure. The nuclear radiation also contains highly energized electrons (so-called β-radiation) as well as protons and neutrons and smaller nuclei (α radiation). All of these fission types of radiation, α, β, and γ, are harmful to human life and all life forms. The radiation of radioactive nuclear material can, in addition, persist over many years and even centuries.

A unit that attempts to quantitatively measure the biological effects of radiation is the sievert, denoted by the symbol Sv. This unit measures the radiation that is absorbed by a person, and it weighs the type of radiation by its damaging effects. Neutrons with an energy of 100,000 eV are counted, for example, as very damaging while electrons contribute much less to the possible damage. Tolerable amounts of radiation are in the range of milli sieverts (mSv) meaning one thousandth of a sievert. A chest CAT scan, using X-rays, amounts to around 10 mSv while a dental X-ray amounts to only 0.005 mSv. A recommended limit for volunteers working in severe nuclear emergencies is 500 mSv = 0.5 Sv. It is important for everyone to develop a reasonable attitude toward radiation. We all are exposed to radiation.

Radiation occurs naturally, from the sun, from the materials in the earth, and from the stars of the universe. The average American encounters 0.25 mSv every month of their life. Our bodies have, therefore, developed a natural resistance to radiation. At the same time, our measurement equipment is so sensitive that we can measure not only milli sieverts but also micro- and nano-sieverts. News reports that there are increased radiation levels need therefore be carefully considered and should not induce fear if levels of radiation, induced by natural or man-made events, are close to the average naturally occurring levels.

Whether or not the radiation products of manmade fission can indeed be *safely contained* is a matter of the degree of the actual safety that one requires. The required safety is a matter of endless debates! If we require absolute safety with no possibility of danger for humans at all now, or in the future, then we can not use nuclear energy! No container, or system of containers, is safe enough that it can not be destroyed by terrorists or by enormous earthquakes. It is also not possible to bottle up the waste products of radioactive material that are generated during nuclear energy production and then keep them safe over millions of years. We may bury them in caves, and the buried waste may be safe, but then an ice age may come, and glaciers may open the caves and free the material. If this seems like a silly discussion to you, let me just say that I personally witnessed such discussions and arguments in meetings of leading scientists. Contrary to common thinking, also leading scientists can be silly and follow political agendas. If, on the other hand, we would spend some of the resources of earth, are now wasted in wars, for research, then the author is convinced that we would accomplish relatively safe and environmentally friendly handling and production of nuclear energy. In this connection it is important to remember that no energy production is without danger, and the dangers need be carefully weighed against the advantages. Thus there may still be some chance for the peaceful production of nuclear energy, although the current mood, after the 2011 Japan earthquake and nuclear disaster, does not point in this direction.

Let us return to the actual mechanism of energy production by fission. How can the production of nuclear energy by fission be, on one hand, explosive (the atom bomb) or, on the other, be maintained over long time periods (the peaceful nuclear reactor)? How can one generate the fission of large quantities of uranium or plutonium rapidly to produce a bomb or as peaceful energy source over longer times? As mentioned, neutrons can easily be generated, for example, by Californium, and be brought to hit Uranium-235 or Plutonium-239 atoms and cause them to decay and release energy and additional neutrons. Assume, for example, that with each hit of a heavy uranium or plutonium nucleus two neutrons, are released (actually there may be more). Then the two neutrons can hit two more nuclei that decay and release four neutrons. As this goes on, we obtain 8, 16, 32,..., and for n, such hits 2^n neutrons. You can see that the number of hits is therefore rapidly increasing, of course, only as long as the basic material of uranium or plutonium is all within reach and the reaction can therefore be maintained. For 83 such consecutive hits, $2^{83} \approx 10^{25}$ atoms would be split. If there is enough of the material there to start with, the so-called critical mass of the material, then, much or all of the material will be split before the neutrons go into free space, and, therefore, an energy corresponding to the number of all atoms that were split will be released. This is called a nuclear chain reaction which is the basic mechanism in fission bombs. If the mass of the material that is split is around a few kilograms corresponding to about 10^{25} atoms, then the energy released, basically in an instant, is enough to heat eight million barbecues for an hour as we have calculated above. About this amount of energy was also enough to destroy the cities Hiroshima and Nagasaki. It turns out that the critical mass of plutonium-239 is indeed around 10 kg, and therefore, not much material is needed to create a nuclear bomb. Nuclear bombs are extremely destructive, and their use is very dangerous for mankind.

Peaceful nuclear reactors are designed to create energy slowly without explosive developments. Such reactors drive ships from U-boats to aircraft carriers and supply cities with electrical power. They can work efficiently without fuel exchange for a long time. It is possible to engineer nuclear reactors in such a way that no danger of an uncontrolled nuclear chain reaction, a bomb like explosion, exists. As mentioned above, to run a peaceful nuclear reactor, the rate of splitting of the nuclei must be relatively slow. This means that the mass of nuclear material must never be close to the critical mass. This is ensured in standard reactors by keeping smaller masses of diluted nuclear material, for example, cylindrical rods, some distance away from each other. The nuclear material then heats up because of the slow chain reaction and delivers energy in the form of heat to steam engines. This delivery of the heat to the steam engines also provides the necessary cooling for the nuclear material. If that process of cooling is interrupted, by terrorists or an earthquake, as happened in March 2011 in Japan, then the nuclear material may melt. Uncontrolled melting and overheating may lead to an accumulation of molten material in its container vessel that is usually constructed out of concrete. The heat may become very intensive and, aided by chemical reactions, lead to chemical explosions of the materials used in the reactor, ranging from water to graphite. In the worst case, this can lead to an explosive breach of the container vessel itself. The effects are similar to those of the so-called dirty bomb: the explosions caused by excessive heat and chemical reactions propel radioactive material out into the atmosphere. The radioactive material is then distributed over large areas depending on weather conditions.

The famous movie "The China Syndrome" maintained that the molten nuclear material may burn through from the USA to China. This is more than an "overstatement!" There is not enough material to burn through more than a few meters. The explosions related to such accidents are of the force of ordinary chemical explosions. In a nuclear bomb, the critical mass is present and is held together long enough so that most nuclei split before the parts of the bomb fly apart. This "holding together" of the critical mass enhances the force of the explosion enormously. Nevertheless, even without this enhancement, any leaking of nuclear material to the environment is very dangerous and may spread radioactive material and radiation over large distances, necessitating clean-ups and endangering humans and all life-forms. Explosions due to overheating should not be a possibility in reactors of the future. However, even with such futuristic improvements, the danger related to the radioactive left-overs of the fission processes will certainly be an impediment for a general and broad use of nuclear power for a long time, if not forever.

It is a very interesting project to find out more about radioactive nuclear fission, and generally the interaction of all kind of atoms with the neutrons that are created in the uranium or plutonium chain reaction processes. The neutrons of such chain reactions can, for example, be used to change certain atoms into different ones. They can change, for example, mercury into gold. However, the radiation by-products prevent such gold production from being desirable. Nevertheless, there are great possibilities in such processes of changing atoms by bombarding their nuclei with neutrons. It is even possible to "breed" in that way new nuclear fuel that can undergo fission. In this way, one can create very large resources for energy production.

Nuclear Fusion

The second way to generate energy from nuclear processes is by fusion of nuclei: protons, the nuclei of hydrogen, can be fused together to form helium nuclei (two protons plus two neutrons) with the resulting helium nucleus being somewhat lighter than the 4 hydrogen protons. This results again in a deficiency of mass that is turned into energy with similar effect as just described for fission. More details of this process are described in Sect. 5.

Instead of dealing with fission and all of its problems, why not use the abundant hydrogen in earth's water to create energy by fusing it into helium? This question can be reformulated as to why we do not use here on earth the same process that provides the heat of the sun? The answer is very simple. The fusion of hydrogen into helium, as it occurs in the sun, requires extremely high temperatures, more than about 20 million degrees (in whatever degree units, K, C, or F temperature is measured). The hydrogen for the fusion also needs to be held together under high pressure. These conditions, pressure and high temperature are met in the interior of the sun (and all stars), because the enormous force of gravity compresses the star's core to sufficient density to sustain nuclear fusion. On earth, such conditions are difficult to create in any controlled way. Experiments are ongoing to create such conditions, for example, by sudden intense laser radiation. As these lines are written, no such experiment and corresponding machinery have yet been proven to produce energy on any practical scale. We know only from, so-called thermonuclear bombs that fusion indeed supplies great energy. These bombs use fission bombs to first create the high temperatures and pressures, that then lead to fusion of the hydrogen material that is also part of the bomb (actually it is the hydrogen isotopes deuterium and tritium that are used). However, the slow "burning" that we need for peaceful use has not yet been accomplished for fusion processes here on earth. STEM research in this area goes on, because fusion would give us a truly inexhaustible source of energy.

4.2   The Household

STEM applications are everywhere in the household. Here we have an area that can be partitioned into homeworks or thesis-type projects that all have the goal to readying the student for life in their own home. Einstein suggested that extensive projects of this kind would improve high-school education considerably. This author completely agrees and includes some discussions of electricity in the household as an example of how to use STEM knowledge to master life at home.

4.2.1   Electricity in the Household

There exists a world of electrical equipment in every household: refrigerators, washing machines, microwave ovens, TVs, computers, and lights. Everyone has encountered the helplessness and lack of comfort during power failures! We just depend in almost every way on electrical power. That power comes from the power outlets that are fed by electrical lines that supply the home, or alternatively from batteries and more recently even from solar cells. If some equipment does not work, then the first thing you do is to check the power supply. Is the equipment plugged in, does the outlet have power, is the battery dead? These are the first questions that you need to answer. But how does one answer these questions?

Measuring Electricity and Safety Precautions

One needs to possess and understand an instrument that can measure electrical voltage as well as, for more advanced problems, electrical current and resistance. Such instruments are readily available in many stores that sell electronic equipment. There are several types of such equipment from a variety of companies. The best, for the purpose of teaching, is a universal meter that can measure direct (dc) and alternating (ac) voltages and currents and also can measure electrical resistance. If one has such a meter, it is easy to find out whether or not the power supply works. The voltage of the power outlets can be measured by setting the universal meter to the appropriate voltage range. In the US, that is the range around 110 V of ac voltage.

There are usually three leads in the interior of every power outlet. One is the ground wire that is connected to the "ground," meaning in essence everything around you. It is, of course, not dangerous to touch the ground wire if it is installed properly, and this wire is therefore often bare and not embedded in insulating material. That wire is there to protect you and also is connected to the exterior of any appropriately designed electrical equipment that is plugged into the power outlet. There are circuit-interrupting safety devices that switch the power off if a dangerous voltage would develop (e.g., by some equipment fault) on the outside of the equipment that you might touch. These safety devices are stored in a panel or box called a circuit breaker panel or fuse box.

The second lead inside your power outlet, in the USA usually colored black or red (consult the Internet for color codes), is the hot lead (wire) that actually carries the voltage relative to the ground and relative to the third lead. The third lead, in the USA usually colored white, is the neutral wire that returns the electrical charges to the source of the electricity (e.g., the transformer in front of your home). If you would touch the hot wire or any part connected to it (not at all recommended!), and if some other portion of your body is somehow connected to the neutral wire, then electricity will flow through your body, and shock is certain. Electrical shocks are very dangerous. Everyone knows, of course, that a low voltage such as that of a small battery (around 1 V) does not give you any shock. This is one reason why many household applications, such as flash lights, use low-voltage batteries. The danger for shock increases with the voltage and becomes lethal for the standard voltage of power outlets (110 V in the USA, 220 V in some European countries). It is therefore dangerous to deal with power outlets without the necessary knowledge background of how to deal with electricity. Even light decorations, such as Christmas lights, can be dangerous without that knowledge, because rain water can lead, for example, to unwanted electrical "connections." Shock will also happen if your body somehow forms a bridge from the hot wire to the ground and if you do not have a protective circuit breaker. This is why the location of circuit breakers should be known and regularly tested in every household.

Protective gear made from insulating material, such as gloves or boots, can also help to avoid shock. If you do have a material of very high resistance, such as thick rubber shoes, and if you touch only the hot wire (not the return wire at the same time), then the shock may be diminished and may be too small to notice, as you can understand from the discussion of resistance below. When working on the wiring, the safest way is always to turn off the main source of electricity. That usually can be accomplished by turning off the circuit breakers that control the power outlet and the instrument that you are working on.

Current, Voltage, and Resistance

It is of particular importance for household applications to have a working knowledge about the relations of voltage, current, and resistance (see Eq. (2.54)). If you think of electricity in an analogous manner to the water supply of a garden hose, as we have described it in Sect. 2.2.2, then the voltage corresponds to the water pressure, the resistance to the capacity of the hose to supply water, i.e., mostly its diameter, and the current corresponds to the water flow. If you have an extremely thin hose, then, no matter how large the pressure is, you will not be able to have much water flow. Similarly, if you have a very thin wire, then that wire will not be able to carry much current even if the voltage is relatively high. If you increase the voltage to obtain more current, then the thin wire will burn through similarly a garden hose will explode if the pressure is too high. This means that not much power is available whenever thin wires are used, because the power is given by the product VI of voltage V and current I. One of the consequences of this fact is that you cannot operate powerful equipment by using long thin wire lines. For example, if you would like to have a hot tub way out in your yard and you need a long wire line, then this wire has to be rather thick. There are engineering standards that prescribe the wire size depending on the length of the wire and power requirements. These are also the type of considerations that you need to go through when you try to install garden lights. The more lights you have, the thicker the wires need to be. The actual resistance of the wires and the currents that flow through wires, can also be measured with the universal meter. This, however, requires a lot of practice and could be the topic of nice projects.

Repairing Electrical Equipment

Can you repair household electrical equipment by yourself? There are some repairs you may be able to do. It is relatively easy to replace burned out lamps. If you know you have power, and the lamp does not light even when mounted and fastened correctly, it is likely that it is burned out; just get a new one with the same power consumption (remember power is measured in watts = volts times amperes). It is also often possible to locate a defective switch that turns power on and off. The multimeter usually gives you enough information about whether a switch is defective. Defective light switches can be replaced, with expertise necessary for safety precautions. Switches for other equipment, such as ovens or washing machines, are often concatenated on an electrical circuit board, and, if defective, the whole board needs to be replaced, usually with considerable expense.

Fuses are safety switches that turn off the electricity in cases of some type of overload or danger. The multimeter will tell you whether a fuse has burned out. Other than that, any repair of electrical equipment takes usually an expert. Even an expert can not repair defective computer chips. Defective chips need to be replaced as a whole. Often it is the most effective way to replace a whole circuit board that may contain a large number of chips and other electrical devices. The main circuit board of any household machinery, the board that holds many crucial components that make the machinery tick, is called the mother board. A defective mother board may mean that the equipment as a whole should be replaced. Only an expert for the particular equipment, TV, washing machine, or computer, can tell you the effective way to proceed.

Choosing Projects

The projects that can be executed in high-school STEM education involving electricity in the household are legion and represent really necessary knowledge for everyone in their adult life. STEM teachers can define projects and ask the students to finish them by Internet research: how does a microwave oven work? How does a refrigerator work (see Sect. 2.3.4)? How do you install a low voltage light-system in your yard? What is the voltage of your car battery, how can you measure it, and how long does it take to discharge a car battery if the lights are left on and the motor does not run and recharge the battery? How is the spark generated in the spark plug of the car engine, starting out with a low-voltage battery (see Sect. 2.2.2)? There are essays on all of these on Internet sites, and knowledge in this area is of great use.

4.2.2   Pipes and Pressure

There are many valid and important analogies between wires that carry electrical currents and pipes that transport liquids. Pipes are very important by themselves in every household because one needs water for the bathroom, the kitchen, as well as the yard and garden. The pressure and flow of water in pipes are governed by several rules of physics. We review here a few useful ones.

Pipes that are laid out horizontally and are all connected and filled with water exhibit, when completely turned off (no water flowing), the same pressure everywhere. The water is usually supplied by the city with a certain pressure ranging from around 2 atmospheres to about 4 atmospheres and more depending on the location of your home. There are inexpensive pressure meters that everyone should have and use to determine the water pressure.

If you live on a steep hill, and the water comes from the top, then you have to add for every 10 m height a pressure of 1 atmosphere to the pressure on top of the hill. Inversely, if the water comes from the bottom of the hill on which your house is built, then you have to deduct for each 10 m of height 1 atmosphere of pressure. For example, if the city supplies you with water having 4 atmospheres on the bottom of the hill, and the hill is 30 m high, then the water pressure in your house on top of the hill is only 1 atmosphere. If you need a lot of water, that pressure may not be enough. The planning of your water supply and pressure for your house is thus very important, and you need to know your so-called topology, that is, the elevation of the various places where you need the water. The pressure is often given in pounds per square inch (psi) instead of atmospheres (atm). The conversion factor is 1 atm $\approx$ 14.7 psi.

These considerations of pressure are all for a static situation, meaning that no water actually flows. If water flows, then the thickness (diameter) of the pipe is of great importance for the capacity to deliver water. A pipe that is a few centimeters (1–2 in.) thick can deliver usually enough water for house and garden, if the pressure of supply is about 3 atm or equivalently 44 psi. Very thin pipes will supply less liquid. It is important to remember that the water flow through a pipe is proportional to the pressure difference at its two ends and proportional to the second (or even third) power of its diameter. It is inversely proportional to the length of the pipe. Therefore, if you have a very long pipe, it needs to be rather thick to supply plenty of water. These rules should answer most of the questions that one might have in household water-supply problems.

4.2.3   Other Household STEM-Topics

Household-related student projects should, of course, not be confined to electricity and water pipes. Mechanical problems, electrical motors (Sect. 2.2.2), combustion engines (see Sect. 2.3.3), and their applications in the household should be considered. Einstein wished to make high-school interesting by letting the students perform tasks that would make them master certain areas of great practical importance in their homes. A detailed description of all possibilities would take a separate book, or certainly a long chapter. It is clear, however, that the workings of a car, both electrical and mechanical, will be interesting to many students, and projects around this topic could include a much more detailed description of car engines than the one we have given in Sect. 2.3.3. Of course, biology projects need also be included. One could, for example, select a site in a yard or park that is about 30 by 20 m and have the students list the plants in groups such as trees, shrubs, wildflowers, grasses, vines, mosses, algae, and fungi. Students could be stimulated to master the art how to grow these plants, how to beautify a garden, and how to grow herbs for the household. Such extensive projects and mastery of the subjects by the students need also be rewarded by some kind of STEM awards that can range from diplomas that the school issues to awards from the city, state, or federal level. Such projects may also include the students' family members or friends, who bring a certain expertise in any of the topics mentioned above and who are willing to learn and teach about STEM in the household.

4.3   STEM and Medicine

Everyone occasionally needs a medical doctor, and all doctors rely on medical knowledge and instrumentation to perform their work. The instruments usually are the result of great science and engineering insights, and we describe here a few instruments of great general importance. Is it necessary to understand the science of such equipment?—well, maybe not in detail! However, if you ever encounter magnetic resonance imaging (MRI) to show anatomic details of your body, your knees or especially your brain, then you might be interested in what that severe pounding is that you hear during the imaging procedure. You might also wish to know whether you encounter any dangerous radiation and how such a picture is formed.

4.3.1   Medical Imaging

Basic to medical imaging is the fact that ordinary light does not penetrate our bodies and we therefore cannot look into our body by just shining light at it.

The reason is that the material of our bodies contains energy levels which permit electron transitions with an energy equal to that of visible light, which is therefore absorbed. We have discussed the physics of these energy levels and light absorption in Sect. 2.5.2. Light is not always absorbed by materials. Take, for example, diamonds that are carbon crystals and are so hard that you can scratch glass and metals with them. Nevertheless diamonds are almost completely transparent and do not absorb sunlight. Take as another example jellyfish. They are almost entirely transparent. If our bodies were made that way, doctors would have it easier, they could look into us and see what is wrong. Our bodies do not permit that. They are, however, more transparent for other types of waves. These other waves include sound waves, electromagnetic radiation of very high frequency (X-rays, γ-rays), as well as radio waves in certain rather low-frequency ranges. All of these wave types are in use to let doctors look into our bodies and, of course, other people such as security guards as well.

Ultrasound

Sound waves have a low quantum energy $h\nu$ that is many orders of magnitude lower than the energy that is necessary to break a molecule apart (which is of the order of several electron volts). This is also true for sound waves with a frequency that is higher than our ears are able to hear, that is so-called ultrasound. One therefore considers sound waves of low intensity (low total energy of exposure of the patient) as very safe, and sound waves are even used, and very successfully so, to take images of fetuses in the wombs of pregnant women. The method of taking images is similar to that of taking pictures with visible light. Instead of a light source there is a sound source. In order to transfer the sound efficiently to the body, a little amount of gel is applied to your skin, and the sound generator and detector is put on top of that gel-covered area. The sound is thus transmitted into the body and reflected by internal organs or fetuses. The reflected sound waves are detected and with help of computers transformed into nice images on a screen. To form these images and make them clearer and clearer have been the subject of much research and development. In the past, the images were not very good, and it took a very skilled person to figure out what these images really meant. Currently the images obtained from ultrasound exams are much clearer and still improving, and it is easy for everyone to see a fetus and recognize clearly many features including the sex. The Internet is full of information and photographs of medical equipment that is used for imaging with sound waves.

The Doppler shift of sound waves (see Sect. 2.1.5) can also be used to determine the velocity of motion in a body. The speed of the flow of blood in arteries can be determined that way. This is an important method to determine whether blood clots or plaque impede the flow of blood in arteries.

X-Rays, γ-Rays

Wilhelm Roentgen was expelled from his high school in Germany, because he refused to reveal the identity of a classmate guilty of drawing an unflattering portrait of one of the teachers. Aren't we lucky that times have changed? Nowadays, only the reverse would be possible: teachers might be removed if they did not reveal colleagues who drew unflattering portraits of students. It is unbelievable, but Roentgen never was able to obtain a high-school diploma in his homeland Germany, and had to pursue studies, as Einstein did, in the more permissive Switzerland. Only subsequently did he return to Germany where he had a successful career. As a physics professor in the provincial town of Wuerzburg, he made a very astounding discovery.

Roentgen experimented with vacuum tubes, cylindrical glass tubes of about 20 cm in length. These tubes had electrical contacts, circular plates made out of metal, mounted at each end. The air inside the tubes could be removed by pumps to various degrees, thus the name vacuum tubes. Applying very high voltages to the electrodes caused the remaining gas atoms to be ionized, and the electrons and ions so created gave rise to an electrical current. If a voltage is applied to the metal contacts (electrodes) at the ends of the tube, the electrons acquire an energy equal to the elementary charge e multiplied by the voltage V. Thus the electrons gained energies of several thousand eVs (electron volts).

If these high energy electrons hit the metal contacts at the end of their propagation through the tube, then they can transfer electrons of the metal atoms from lower to higher energy levels. We have described a similar process for the absorption of light by hydrogen and other atoms in Sect. 2.5.2). There is, however a big difference to the case of ordinary light absorption. A metal like copper has energy levels that extend to much lower energy than the ones of hydrogen. The nucleus of copper contains many protons and therefore many positive charges that attract the copper electrons much closer to the nucleus than the single proton of hydrogen does. It therefore takes a much higher energy to excite these copper electrons. These higher energies are available in vacuum tubes because the electrons are accelerated by the very high voltages. Smashing into the contacts, they remove electrons from the lowest energy levels of the metal atoms, thus creating a "hole" that is waiting to be filled by other electrons.

Electrons from much higher energy levels of the metal can subsequently make a transition to the so created holes (missing electrons) of the low energy levels. Such a transition can have typical energies of several thousand (!) electron Volts. Compare that to visible light whose photons have an energy around 2–3 eV. Thus, because $E = h\nu$, a new type of radiation of very high energy and frequency is produced, whenever such vacuum tubes are operated at very high voltage. Roentgen noticed that radiation, and he called these new types of rays X-rays. Roentgen proved that he could cover his tubes by carton material that was totally opaque to visible light, and he still could blacken film that he had put in proximity to the operating vacuum tube. He knew that he had made a great discovery, when he saw the skeleton of his wife's fingers on a piece of film over which her hand was held. He thus became the "father" of diagnostic radiology, the medical specialty that uses imaging by X-rays (and now also by other means) to diagnose disease. Roentgen was awarded the first Nobel prize in physics for this discovery. The STEM student may not find it reassuring that the very first physics Nobel prize was given to a high-school dropout. So what is the point of learning all of this? However, Roentgen did a lot of studies in Switzerland, and his case should not be seen as typical either.

X-rays are still a major tool for medical diagnostic. Many readers probably have already had a chest X-ray taken when they were ill, for example, with pneumonia. A particularly powerful method of using X-rays is the so-called CAT scan or CT scan. CAT stands for computed axial tomography. Tomography means just imaging by sections. What is done here is taking X-rays in multiple directions and feeding all the data into a computer that then can reconstruct, slice by slice, an image of what is inside you. Such tomography can nowadays be performed not only by the use of X-rays but also by the use of γ-rays. The well-known PET scan uses the destruction of electrons by their anti-particles (positrons) to generate radiation of extremely high energy that then is used to produce an image. The acronym PET stands for positron emission tomography. The student of these sections has realized by now that radiation of this power, which also occurs in radioactive decay and nuclear processes, is not without danger for humans. Indeed we can easily imagine that such energetic radiation may change our DNA which indeed it does. CAT scans and PET scans that involve several milli sieverts (mSv) of radiation are therefore only used when serious disease is suspected and when the benefits of the detection outweigh the danger.

Fortunately, a method was developed in more recent decades that uses very low energy electromagnetic waves to also create beautiful images from inside us. How can that be done? This is described below.

4.3.2 Nuclear Medicine

An important area, involving images related to medicine, is the area of nuclear medicine. The key of this discipline is that certain atoms or molecules that are involved in medical problems, such as diseases of inner organs, can be "stained" to become more visible in the images that are taken. The cornerstone of this staining is radioactivity. One needs a radioactive atom-type that quickly decays into parts that are not radioactive and therefore do not harm the human body. Because of the quick decay, this type of atom does not practically exist in nature. Nevertheless, such an atom can be artificially created, and it has the name technetium indicating its artificial nature. Technetium was not known to the chemists of the past and could only be produced by using modern nuclear technology.

Technetium is combined with other chemicals to form the so-called radiopharmaceuticals, radioactive substances that are used for imaging brain, kidneys, and other specific organs, or even cells of the body. The γ-rays that are emitted by the radioactive atoms can be detected and used to produce the medical images.

After the image taking, the radiation diminishes very quickly. You can easily understand the usefulness of this method by noting that certain illnesses such as cancers involve also certain molecules and atoms of our body. If the molecules or atoms of this kind are "stained" by radioactivity, the location of a particular illness can be located by the γ-rays that are emitted. Cancers can also be targeted and destroyed by such radiation.

MRI, Magnetic Resonance Imaging

MRI, magnetic resonance imaging, is based on the resonance of nuclei in strong magnetic fields. With all the fears and prejudice we have about radiation, it is difficult to imagine how a method related to the properties of nuclei can be of medical use and does not involve dangerous radiation. Yet all the atoms of our body have nuclei, and therefore any imaging technique that is sensitive to the properties of nuclei must give us a lot of information. The nucleus that is explored most in the imaging method described below is the proton. The proton is the nucleus of hydrogen and hydrogen is present in the abundant water molecules of our body. Water in turn exists in different densities in the various structures of our body, and these densities are important for the imaging technique described below. This technique uses radio waves to detect the protons, and the radio waves are of very low energy and therefore not dangerous.

The basic scientific effect that is used for this imaging technique is based on the spin of protons and neutrons and consequently of the nuclei of atoms. This spin is an effect analog to the spinning of gyroscopes that we have discussed in Sect. 2.1.4. The spin of protons and neutrons is also linked to a magnetic field. Remember that currents through coils generate magnetic fields and this general idea makes it plausible that spinning objects that are somehow related to a circular flow of charge can have magnetic fields around them. The nuclei of atoms thus have electromagnetic properties and can therefore interact with a radio-frequency electromagnetic field. The actual frequency of interaction is specific to the specific type of atom that is involved. It can also be changed by the application of an additional magnetic field. For medical imaging, that magnetic field is usually generated by a large electromagnet, i.e., a coil through which a large current flows. The coil that is used in magnetic resonance imaging is, in fact, so large that the patient can be moved inside the windings of the coil. While the patient is moved through at a low speed, pulses of high magnetic fields are generated. As a patient you can hear these pulses by pounding noises that arise from the great force that the magnetic field exerts on the coil material. As the body is moved and the magnetic field scans through the patient's body, the various types of atoms of our body resonate at electromagnetic radio frequencies that are predetermined by the strength of the magnetic field. The resonances can be detected by use of suitable radio waves that are typically in the megahertz ($10^6 \, s^{-1}$) range. Again as in the case of the CAT scan, a computer puts all the resonance signals together and obtains a three dimensional image of the atoms and materials inside the patient.

The actual accuracy of resolution of the image, i.e. the quality of the image, depends on how accurately one can locate the magnetic fields in the patients body. Because of this accuracy of location, one needs typically very high magnetic fields, and special coils are necessary to generate them (see superconductivity in Sect. 5). The energies that are involved in this techniques are much smaller than those in X-ray techniques, simply because radio frequencies ν_{radio} are very low compared to the X-ray frequencies and because of the Einstein-Planck law stating that $E = h\nu_{radio}$. Extreme care needs to be taken, of course, that no magnetic material such as a steel paperclip comes close to the patient. The huge magnet would exert extremely large forces on such material that could lead to injury of the patient. Don't put any paperclips into your ear when you have an MRI of your brain!

4.3.3 Microscopes, Viruses, Bacteria and Fungi

STEM has had a great impact on many areas of medicine that we cannot possibly cover. However, it is very clear that making things visible that otherwise are not is one of the most important achievements of STEM for the medical field. Of particular importance for medical discoveries were not only the instruments that enable us to look inside our bodies as described above, but also all kinds of microscopes. Microscopes are instruments that create a magnified image of whatever one wishes to observe. The oldest types of microscopes used glass lenses and light to create such magnified images. We have mentioned previously that the objects that one can make visible with light cannot be smaller than the wavelength of light. The wavelength of visible light is of the order of several micrometers (10^{-6} m). Some objects that are very interesting for medicine, such as viruses, are much smaller than that, and one needs therefore special microscopes that do not just use ordinary light. We have already discussed scanning force microscopes, which can be used to image molecules on surfaces, or the surfaces of viruses or even cells. More often, however, electron microscopes are used to image biological structures such as viruses. Electron microscopes use electrons instead of photons to create the image.

It may sound strange when one first hears about creating images by using electrons, which is what the electron microscope does. We are used to thinking of light in terms of waves and of electrons in terms of particles. However, we know that light can also be seen as being composed of particles, the photons, and electrons can also be seen as being waves, as we have learned from quantum mechanics. It just took a little longer for us humans to handle electrons and create images with them. The relevant wavelength of the electrons is what we have called the de Broglie wavelength, and that depends on the velocity and kinetic energy of the electrons. The higher their energy, the shorter their wavelength. If one accelerates electrons by use of high voltages, then these electrons have a very short wavelength, much shorter than the wavelength of visible light. Therefore one can produce images of much smaller objects by using high velocity electrons. Magnifications of factors of

2,000,000 can be reached with electron microscopes, while only factors of about 2,000 can be obtained with microscopes that use visible light. Images of viruses as obtained by use of electron microscopes can be found on the Internet.

Viruses are large organic molecules of about 100 nano size. They typically contain DNA material that carries the biological information as discussed in Sect. 2.4.3. A protein coats and protects this virus-DNA and some fatty substances (lipids) surround the protein. Viruses can insert themselves into the cells of plants, animals, and humans and multiply and cause damage. Our body has developed a defense system, the immune system, that can usually deal with viruses, particularly if it has encountered the virus before and has adjusted to defeat it. This fact is the basis for the method of vaccination. Vaccination simply means that the immune system is exposed to a molecule that resembles the virus, but cannot multiply as the virus does or is altered so that it does not cause disease. In this way the immune system is primed and knows how to react faster when a real virus is encountered. Mass vaccinations of humans have defeated very serious diseases such as small pox that have infected and killed large sections of the population of cities in previous centuries. The development of vaccines for dangerous diseases is of extreme importance and requires knowledge and research in chemistry, biology, and medicine as well as the development of related technologies. There also exist certain medications that fight viruses effectively and that can be prescribed once a virus is identified.

Bacteria are about 100 times as big as viruses and can, in some instances, actually themselves be infected by viruses. There are many bacteria that are useful and even beneficial for mankind. Any healthy person is colonized by bacteria on their skin and in their body. Bacteria aid your digestion, and there are more bacteria (about 10^{13}) in your intestines than cells in your body (about 10^{12}). There are also bacteria being added to foods such as yogurt, and bacteria are changing sugars into alcohol that can then be used for a variety of purposes, for example, as fuel. Bacteria have evolved with the other forms of life including plants, animals, and humans. There are unfortunately also some bacteria that are harmful, even extremely harmful, to humans. The discovery of medication that helps against many harmful, bacterial infections is one of the nicest examples how major progress in medicine has been stimulated by biology and chemistry. We can give here only a short description of the successful fight against bacteria.

The use of aged bread, with a blue mold that formed on it, had been known in folk medicine since the middle ages as a means of treating infected wounds. Of course, no one knew then what an infection really was and that it could be caused by bacteria. This is a knowledge of modern times. Some bacteria represent the oldest known forms of life and so does the blue mold, the so-called fungus. It is now generally believed that fungi (plural of fungus) and bacteria "clashed" and attacked each other and also developed defenses against each other during their evolution. The stuff in the blue mold that attacks bacteria is now know by the name penicillin! Alexander Fleming gave it this name in 1928 when he showed that the fungus with the Latin name Penicillium notatum was generating a molecule that killed bacteria. That molecule is mainly made of the atoms H, N, O, and C, as are most molecules that somehow relate to life. It is now possible to synthesize such molecules in mass production without the use of molds, and penicillin is still one of the major treatments of choice against many bacterial infections. Some of these infections were deadly before penicillin was known and include wound infections by the dangerous staphylococcus bacteria, infections of the middle ear, diseases that are transmitted by direct contact of body fluids such as syphilis, and many others. The discoveries and use of chemicals that fight bacteria, so called antibiotics, are truly results of science (biology, chemistry) as well as the technology that was developed for mass production—a great STEM success altogether.

Science has also found major medications against fungus infections of our bodies, and medicine can, therefore, effectively fight many microorganisms including viruses, bacteria, and fungi. It is important to know that the various medications work only for specific microorganisms. Penicillin works for a broad range of dangerous bacteria and so do many other antibiotics. These antibiotics do not work for viruses and fungus. Therefore, it would make no sense and be even harmful to take penicillin for the common cold that is caused by a virus. Similarly, the fungus medication for athlete's foot does not help against bacteria or viruses. Therefore, one always has to follow doctors orders, and it is also of great advantage to be informed about what symptoms and disease one fights with what particular medication. The Internet has a lot of information on this topic.

4.4 Games and Probability

This section is about STEM and entertainment. Everyone knows, of course, that entertainment possibilities have been enhanced by STEM-related inventions such as that of TV's, DVD players including play stations, electronic cameras, advanced cell phones that let you use applications (e.g., download music and books), and many other useful tools and gadgets. The author is well aware that these tools and gadgets have often been used to lure students into working on STEM projects and making them believe that STEM is cool. You probably have seen robots jumping around on Internet sites telling you that STEM is pure fun and magically interesting. This is, of course, totally true. However, it is also true that a certain level of mathematical and technical understanding is definitely necessary to fully appreciate all these STEM inventions. We have discussed the technical intricacies of computers, cameras and displays already in great detail. However, we must not only know the technology and be able to switch the instruments on and off and work their menu-driven operations. We also need a rudimentary understanding of the mathematics that underlies the games, in order to influence their outcome or to know that we can not influence it.

I will describe here three well-known games that are based on the mathematics of probability with increasing sophistication. These games were or could have been played without modern technology. However, simulating them with mathematical software definitely helps to understand them better. I hope that, after going through this, you still think that STEM is fun and that you are alerted to the fact, that, professionally speaking, it is a lot of work and requires knowledge of mathematics.

4.4.1 Roulette

Consider a roulette. This is a "toy" that has been used for ages at famous gambling places such as Monte Carlo in France. We do not encourage gambling in any way, particularly not for money. As we will see, that can only lead to losses on the long run. Playing with the roulette for tokens, however, can teach us something about general principles that involve probability and can even be fun, in spite of the fact that the outcome cannot be influenced in this game. The roulette consists in essence of a disc that can rotate and has 37 numbers written on its circumference. Small rectangular grooves are engraved in the disc for each number. The disc is then turned at a quick pace, and a little ball out of an elastic material, resembling the material of billiard balls, is thrown on top of the disc. There are obstacles attached at the side of the disc so that the roulette ball is jumping around randomly between these obstacles and the rotating grooves, before it finally lands in the groove of one particular number. You can find nice illustrations and more detailed descriptions of the roulette on the Internet. It is important that the roulette ball jumps around randomly. "Random" means that there is no preference for the choice of any individual groove and number. This, in turn, means that if you play for a very very long time and throw ball after ball and let it fall into a groove, and if you count how often the ball fell into a particular groove, you will find that all these counts are about the same. This is an experimental implementation of the so-called law of large numbers: for large numbers of trials, we obtain a number of outcomes close to a certain expected value. For the case of the roulette, the expected value Ex_n for any particular number n to occur is related to the number of trials N_{trials} by

$$Ex_n = \frac{N_{trials}}{37}, \tag{4.3}$$

simply because we have 37 numbers $n = 0, 1, 2, \ldots, 36$ and they are picked randomly by the ball falling into a groove.

MATHEMATICA permits you to simulate the roulette in an easy way. Simulation means to model something by computer. Random numbers can be easily simulated by computer and MATHEMATICA has computer, algorithms (procedures) that accomplish just that. All we need to do in order to simulate the roulette, is to obtain a random number between 0 and 36. The way to obtain such a random number R is shown in the MATHEMATICA box.

```
MATHEMATICA
R = RandomInteger[{0,36}]
shift-enter returns you, for example the result 33
```

Using the random number generator of MATHEMATICA you can play the roulette without having a real roulette. Of course, you still need to know the colors of the numbers and other things about the roulette that you can find out from the

Internet. We discuss now some of the mathematical features that you all can easily confirm by using the random number generator from the box.

We denote the number of hits of the roulette number 0 by X_0, of the roulette number 1 by X_1, and so forth up to X_{36} for roulette number 36. For example, if you throw the ball 37,000 times, then it will hit each groove of each number about 1,000 times. Of course it could be that any particular number (e.g., the 3) is hit only 980 times (meaning $X_3 = 980$) and some other numbers more than thousand times. For example, we could have $X_7 = 1,015$, The only thing that is sure is that the total number of throws was 37,000, and so all the results for each number must add up to 37,000. Had we thrown the ball a million times, then the numbers need to add up to a million. If we divide the total hits of a given roulette number by the total number of throws, then we obtain for the just given examples:

$$X_3 = \frac{980}{37,000} = 0.0265 \text{ and } X_7 = \frac{1,015}{37,000} = 0.0274, \qquad (4.4)$$

with both numbers being close to the expected value $\frac{1}{37} = 0.027$. This is the type of result that we obtain if no bias of any kind exists for any roulette number. In the limit of very many throws, that is, very large numbers N_{trials} of throws, we expect according to the law of large numbers

$$X_0 \approx \frac{1}{37} = 0.027, \quad X_1 \approx \frac{1}{37} = 0.027, \text{ and so forth}\dots \qquad (4.5)$$

The fraction of $\frac{1}{37} = 0.027$ is called the probability measure of the roulette outcomes. With it, you measure the chance to hit any particular number. If you sit on a roulette table for an evening and the ball is thrown 370 times, that means it is probable that any roulette number will be hit about $0.027 \cdot 370 = 10$ times. Of course, it is possible that a particular number does not occur all evening. Only for the case of extremely many throws, are we assured by the mathematics of probability that indeed every number is hit about the same number of times. The set of all numbers of the roulette is, mathematically speaking, the so-called sample space of the roulette, meaning that each number may be hit (sampled) as an outcome of each throw. The possible outcomes are called, again mathematically speaking, the elementary events.

The probability measure does, in principle, not have to be $\frac{1}{37}$ for all the roulette numbers. It has been known in the history of the roulette that crooked owners, of the casino in which the roulette was played, were able to introduce a bias for certain numbers. They used, for example, magnetically sensitive balls and an electromagnet under the roulette disc. Whenever this magnet was switched on, then the ball would have a greater likeliness to fall into the grooves under which the magnet was located. If a magnet is, for example, located below the roulette number 7, then we may have a probability measure for X_7, that is, 0.5, and for the neighboring numbers X_6 and X_8, a measure of 0.25 while all other probability measures may be 0. Again the sum of all probability measures for all roulette numbers is 1, as it must be.

Now, however, it is highly likely that only numbers 6, 7, and 8 will occur as events, with probability measures 0.25, 0.5, and 0.25, respectively. Thus the casino owner could use this trick in the following way: if all players made bets on numbers other than 6, 7, 8 and the casino owner switches on the magnet, then all players will lose, and the owner will win big. The owner of the roulette actually always wins anyway, even if they play entirely honestly, as we will explain below. Before that explanation we just like to recapitulate the important points and definitions of the mathematics of probability.

- Every game has a "sample space." The sample space is the set of all possible outcomes. For the roulette the possible outcomes are the numbers $0, 1, \ldots, 36$. If you throw just a coin the possible outcomes are heads or tails. The sample space consists therefore of head (H) and tail (T).
- The probability measure tells you how likely each outcome is. If you multiply the number of trials (throws of the roulette ball or throws of the coin) by the probability measure, then you obtain, for the limit of many throws, the expected number of such events. For an unbiased roulette the probability measure for the event of the occurrence of any number is $\frac{1}{37}$; for a fair coin (that falls on each side equally likely) the probability measure for both head and tail is 0.5. If you throw a fair coin 10,000 times, it likely that you will have about $10,000 \cdot 0.5 = 5,000$ heads and tails.
- The sum of the probability measures of all the possible outcomes must be 1. You can see this from the following reasoning: if we had a probability measure of 0.5 for head and 0.4 for tail, then we obtain from 10,000 throws about 5,000 heads and 4,000 tails. However, now there are thousand outcomes missing in our prediction because $5,000 + 4,000 = 9,000$. Where are the missing 1,000? This is the reason why the sum over all probability measures must be 1! This is always so, no matter how complicated the game. If somehow you are told that it is not one, they are cheating you.
- The sample space and the probability measures form together what one calls a probability space. The probability space is all you need to calculate your chances to win in a game.

In some instances the probability measure is not given as a number between 0 and 1 but is given in percent (meaning per 100). Thus we may hear from a news media poll that the likelihood for a candidate to win the presidential race is 40%, meaning the probability measure for the event that this candidate wins is 0.4. Here the probability space is just given by all presidential candidates. We could have 3 candidates and correspondingly 3 probability measures, , for example, 0.4, 0.1, and 0.5, respectively. Written in percent, that would correspond to a 40%, 10%, and 50% chance, respectively, for the candidates to win the presidency. The total of all percentages is and must be 100%. If you see percentages that do not add up to 100% then you know you have deficient information, and the pollers or media have not given you the whole story necessary to predict the outcome.

By now you might ask, is that really fun? All we have discussed is related to probability measures and percentages. So let's talk about how to really play the roulette and whether and how one can win! To find out about this, we first need to discuss the roulette rules in more detail. The roulette numbers between 1 and 36 are colored red and black, and the 0 is colored green. If you put your tokens on any number, then if the number comes you obtain 36 times the number of tokens that you had bet on that number. However, the probability measure of any number is $\frac{1}{37}$. Let's say, therefore, you have 10,000 tokens and you bet, one at a time, on the number 10. Then it is likely that the 10 will occur about 270 times and each time it does occur, you will obtain 36 tokens, meaning you win about 9,720 tokens. This means you have lost 280 tokens, and the longer you play, the more you lose. The same is true if you bet on red or black. Then, if a red or black number occurs, you obtain twice the number of tokens you have put down on that color. If you bet one token at a time and you bet 10,000 times, you win and lose with equal likelihood except if the 0 comes that is green, then you get nothing because your bet was on red or black. The 0 likely will occur 270 times, and that is the number of tokens you lose. And this is the way it is with all your possible bets on the roulette: betting on even or odd numbers, betting on numbers below and above 18, and whatever is possible. On the long run you lose. So when can you win? You can win only by playing a relatively short time! It is possible, that one and the same number will occur several times during a short period. If that number is your favorite number on which you have bet, then you win, and if you stop then playing, you leave as a winner. Of course it is also possible that your favorite number or color does not occur all evening. Then you lose and it is better to stop also. So the only way you can win is if you play only for some short duration. The best approach, of course, is not to play at all, at least not for real money.

4.4.2   Deal or No Deal

The roulette does not permit you to influence the outcome in any way: it has a fixed probability space; both sample space and probability measure are not changed during the game. The more interesting games have a probability space that changes as the game goes on. Take, for example, a game similar to the famous TV game "deal or no deal." Let's say that our simplified game starts with 20 cases. 19 cases are empty, and one contains one million dollars. The cases are numbered 1–20, and the million is being placed into one case by a random choice. Again, MATHEMATICA can do that easily. You just generate a random number between 1 and 20 as shown in the box. In this case we have obtained the number 13. That number has to be kept secret, and the player must not know it. The player picks now a box with her or his favorite number, say 15.

```
MATHEMATICA
R = RandomInteger[{1,20}]
shift-enter returns you, for example the result 13
```

She or he does not know whether case 15 really contains the million or not. The chance that the million is in the case is just one in 20 or $\frac{1}{20}$, meaning the sample space consists of all 20 cases and the probability measure is $\frac{1}{20}$. Because the million may be in the case that you picked and you would win the money if it is in your case, that case has a certain money value. Accordingly, the game host (actually the so-called banker in the real TV game) offers you now as much money as she or he thinks it is worth and wishes to buy that case from you. The host then asks the player the crucial question "deal or no deal," and the player has to accept or decline the deal.

For example, the game host knows that the player has picked the million with a chance of $\frac{1}{20}$ to win. Because $\frac{1,000,000}{20} = 50,000$ dollars, the value of the case of the player may be considered close to 50,000 dollars. Let's say, that the game host offers half of that sum, 25,000 dollars, to the player. If the player takes the deal, she or he walks away with 25,000 dollars. If the player declines the deal, then she or he is forced to open several more cases, say six more cases. If the million is in one of these six cases and therefore not in the one originally chosen, then the player loses everything. Otherwise they continue playing. After the six additional cases are open, the sample space is different and consists of only 14 cases. The new probability measure for each of the remaining cases (including the one that the player picked at the start) is correspondingly $\frac{1}{14}$. Now the player receives another offer from the host who wishes to buy the case. That offer is higher, because the chance that the player's case contains the million has increased from $\frac{1}{20}$ to the higher probability measure of $\frac{1}{14}$. So the host may offer about one half of the value estimated by the given chance to win which is $\frac{1,000,000}{14} \approx 71,428$, resulting in the offer of about 35,714 dollars. Again, the player has the choice of making the deal or otherwise has to open six additional cases and lose everything if the million is in one of them. This whole process continues until the originally chosen case and one other case are left over. Now the host permits the player to exchange the originally chosen box against the left over box. After that choice, the last box not in the players possession is opened. If it contains the million, then the player lost; otherwise, she or he has become a millionaire.

Obviously, in this simplified version of the game, the chances are against you each time you decline the deal because then you have to open cases and if they contain the million, you lose. Therefore, you should take the money offered as soon as the offer is high enough so that you really like to have it and would terribly regret loosing it. The real "deal or no deal" game is a little more difficult and tempting because there are several cases with big money in them. Note that this game has a different probability space after each of the "deal or no deal" decisions. The players can therefore influence the outcome of the game by their decisions. We will return to further intricacies after the description of the next game.

4.4.3   Changing Probability Spaces

The third game that we discuss here is the following: You have three doors that you can open and that are numbered 1, 2, and 3. Of course you could also play this with cases or boxes as in "deal or no deal" instead of doors. Behind one door is a treasure. You can pick one door and if the treasure is behind that door it is yours. Thus, as you start, you have a sample space of three doors numbered 1, 2, and 3. The probability measure that indicates your chances to actually pick the treasure is $\frac{1}{3}$ for each door. After you have picked your door, say number 1, the game host reveals that the treasure is not behind a certain other door. Say the host reveals that the treasure is not behind number 3. Now the host gives you the choice to keep the door 1 that you picked at the beginning or to exchange it to door number 2 (exactly as the game host does at the end of deal or no deal). What should you do? Here we have a typical situation of choices that arise in such games. It may be that the number 1 is your lucky number and therefore you like to keep it. However, this would not be a smart choice because of the following: After the game host has *revealed* that the treasure is not behind door number three, the probability space has changed from three doors to two doors, and the probability measure that indicates your chance of winning is $\frac{1}{2}$ if you pick door number 2. That chance is greater than the chance of your original choice which was $\frac{1}{3}$. Thus your chances of winning increase if you change the door and choose 2 instead of 1. The main point is that, because of the revelation of the host that the treasure is not behind the third door, both the sample space and the probability measure have changed. If the host would not have revealed anything and just had asked you whether you would like to change the door, then, of course, the probability measure has actually stayed the same, and you do not need to make a different choice but can stick with your favorite number. Naturally, the probability of the outcome of the game is not influenced by what your favorite number is, but you can choose it anyway.

The logic of all of this becomes even clearer if we consider the following variation of the "deal or no deal" game. Assume that you start as outlined above with 20 cases, one containing a million dollars. You choose number 15 as your case. Now assume that the host *reveals* that all other cases except for number 13 are empty, and then asks you whether you keep the case number 15 or you change to number 13. Of course, you change. You have picked number 15 with a sample space of 20 cases and a probability measure, i.e., a chance to win, of $\frac{1}{20}$. Now you have a sample space of two cases. If you change to number 13, you have a probability measure, a chance to win, of $\frac{1}{2}$ that is 10 times as high as your original chance of $\frac{1}{20}$. The actual "deal or no deal" host does not reveal anything. The player makes random choices that reveal the content of cases, and several cases contain major money amounts. Therefore the answer to the question, how much of an advantage a switching of the last two cases will give, is more complicated. The discussion of this question should be a great class project.

4.4.4 More Complex Games: Poker

Comparing the two games, deal or no deal and the roulette, we see the following: The roulette has always the same sample space and probability measure, 37 numbers, and a probability measure of $\frac{1}{37}$ for each number to occur. Deal or no deal is different. Each time we open cases, we are dealing with a new probability space. That makes the game more interesting and gives us a chance to influence the winnings with our choice of making the deal or rejecting it. The game with the three doors introduces one more twist. The game host changes the probability space by *revealing* something and lets you influence the game by changing your choice of the door. Probability theory is the branch of mathematics that tells us about all of this and more. A very important part of probability theory is how to find and assign probability measures in complicated situations such as presented by the next game.

The player's choices and corresponding changes of sample space and measure bring us to the very well known and complicated game of poker. In one variation of the game, the host (dealer) hands out cards, two with face down for each player. Subsequently the dealer starts putting down five cards face up, one after the other. Each time the game host reveals a new card face up, the probability space is changed, because the open cards together with the ones of the players (face down) determine the winning chances of the players. In addition, each player can look at their own cards but not at the cards of other players. This means that each player sees a different probability space and winning chance. They also may make their choices of continuing the game (holding) or quitting (folding). An additional twist of the poker game is the following: The players can raise their bet even if they have bad luck and their cards (the face down cards that can be seen by them but not the other players) are not winning. If someone bets a lot of money on a bad hand of cards, this is called a bluff. To be a master in poker requires great expertise and a great understanding of the probability spaces and measures that are involved, at least on an intuitive level! Even with the greatest understanding and mastery, winning depends on a lucky hand or, mathematically speaking, on the random chance of having great card combinations. Don't play poker for money! If you must, then just play it for tokens and fun!

To give you an idea about how many possibilities are involved in a poker game, we discuss a well-known question of probability theory. The question is, in how many different ways can I choose a number of items, say k items out of a total of n items? Related to poker, the question is in how many ways can one pick five cards (or some other number of cards, say 7) out of the complete deck of 52 cards. In other words, we are looking for all possible outcomes of having 5 (or 7) cards out of the deck of 52. This would then be our sample space and, if we knew how many such elementary combinations (events) of 5 (or 7) cards can occur, then we can find the appropriate probability measure for our winning chances. To make things simple we assume that the order of how we obtain the cards does not matter. Mathematics (or just laying out the cards in all possible ways) tells us that the number of ways to select k distinct items from n items is

$$\frac{n!}{(n-k)!k!}. \tag{4.6}$$

Here $n! = n(n-1)(n-2)\ldots.1$ is the "factorial" of the number n as we have learned already in Sect. 1.3.1. To confirm this result suppose you choose two marbles at random from a group of three differently colored marbles. If the order matters, then you can rearrange the three marbles in $3! = 6$ different ways, hence the $n!$ in the numerator of Eq. (4.6). We have assumed, however, that the order does not matter. Therefore we still need to divide by the number of ways to rearrange the items that were not selected, which is $(n-k)!$ or in the case of the marbles just $(3-2)! = 1$. We also need to divide by the number of possible rearrangements of the chosen pieces, again because order does not matter. This number is $k!$, or in the case of the two marbles $2! = 2$. This leaves us with Eq. (4.6).

A detailed understanding of Eq. (4.6) takes some work with a lot of examples, and the interested reader should consult the Internet for the solution of these combination problems. The mathematical symbol for the number of possible choices of k out of n is $\binom{n}{k}$ (in words n choose k). We therefore have from Eq. (4.6)

$$\binom{n}{k} = \frac{n!}{(n-k)!k!}. \tag{4.7}$$

Thus we obtain for the number of possibilities of choosing five cards out of a deck of 52 cards:

$$\binom{52}{5} = \frac{52!}{(52-5)!5!} = \frac{52 \cdot 51 \cdot 50 \cdot 49 \cdot 48 \cdot 47!}{47! \cdot 5 \cdot 4 \cdot 3 \cdot 2 \cdot 1} = 2{,}598{,}960, \tag{4.8}$$

and the chance to obtain a royal flush, the five cards 10 to ace of one suite, is one in 2,598,960. The chance to get a royal flush of any suite is just four times as large. If you just shoot for any flush, then you choose 5 of the 13 available cards of a suit which gives you $\binom{13}{5} = 1{,}287$ possible flushes. Thus we have 5,148 possible flushes for the four suits and therefore a chance of $\frac{5{,}148}{2{,}598{,}960} \approx \frac{1}{505}$ (in words a chance of one in 505) to get any flush. In this way, the whole probability space of poker can be and has been mapped out. The chance to get a pair is about 42% (a little less than $\frac{1}{2}$).

4.4.5   Computer Games and Game Projects

Recently developed computer games are, fortunately, less dangerous for your wallet than stud-poker and also permitted below the age of 21. They still can be expensive, because they involve and require sophisticated machinery, mostly computers, to be played. I do not think that these computer games need be described here. They are widely known and based on the progress of computer visualization, the rendering of objects and persons by computers, and computer vision. Computer vision means that the computer recognizes objects and persons by use of a camera and handles that information. For example, while you look at a list of restaurants, the computer may see you smile when you look at a pizza and therefore suggests more pizza restaurants. This whole area of human computer interactions is very attractive for future STEM-related applications. Sometimes people ask what more can we do than we are already doing? The answer is that developments and innovation will not rest until the computer can do exactly what humans do in their games and their work. This means that future computers will be able to render 3-dimensional images as well as register and deal with three-dimensional situations. Some of the novel games are already close. They detect the player and transform the motions of the player into the actions of an image that moves like the player does in the situations that the computer game creates. These developments approach some of the capabilities of a holodeck. A holodeck is a facility that simulates reality by use of computers in a completely believable way and is well known from the TV series Star Trek. Holodeck dreams stimulated the work of researchers and will continue to do so. The author remembers a research laboratory at the Beckman Institute of the University of Illinois that was called "the cave" and did accomplish 3-dimensional rendering even in the early days of computer simulation. More is to come for sure, in both vision and visualization. As these lines are written, 3-dimensional "printers" already exist. These printers can create 3-dimensional objects according to your design. They can build statues and objects of art as well as objects of engineering such as car models. STEM work and fun are bound to take quantum leaps by using these novel ideas to pursue new opportunities of the future.

Nice educational projects of inventing and constructing new games can be assigned and performed in the following way: Take a game such as the simplified "deal or no deal" that was described above. Now vary the game by deciding on different ways of putting different amounts of money into the cases. Vary the game further, by letting the game host *reveal* something that changes the winning chances during the game. For example, the player may be permitted to buy, during the game, another box with the money that the dealer has offered. There are large numbers of possible variations that change the probability space and the winning chances depending on decisions of the players. All that you need, in order to design such a new game, is a random number generator and time to play. Have fun!

Chapter 5
Some More Advanced STEM Problems

I know that I know nothing. Socrates (Σωκράτησ) of Greece

In this chapter, we discuss a few advanced topics. The first, Einstein's space-time is advanced in the sense that it challenges our ways of thinking about time and space. We are used to thinking of time the way Newton did as something that "flows" without being influenced by anything that we can control. As it turns out, however, time cannot be seen as independent of the velocity of the system in which it is measured. It also is not independent of gravitational forces. Why do we not notice that in our daily life? The reason is that time changes only by very little for the changes in velocity that we actually can achieve. Even very accurate wristwatches will not show a different time after traveling in an airplane at high speed and at high altitude, where the gravitational forces of the earth are already reduced. Einstein derived the dependence of time on velocity and gravitation from certain laws of physics, not by measuring time directly with a clock. There existed no clock accurate enough to measure these dependencies. Nowadays we do have atomic clocks that are accurate enough to register the difference in time when moving at different speed and also to register the difference when the gravitational force changes. In fact, at the time these lines are written, these atomic clocks are accurate enough to measure the difference in time caused by moving one foot up from the earth's surface. One foot higher the gravitational force is a little lower (see Eq. (2.37)) and that changes the time registered by the clock. The clock at higher altitude works at a faster rate, as if time would flow faster compared to clocks at lower altitude, and that difference can be measured! It turns out that these changes in the flow of time are of great importance for the clocks that are used for the GPS (global positioning system). Therefore, every engineer working on GPS must know about Einstein's theory. This theory can be presented, as you will see below, by using very elementary mathematics that should be mastered by every mathematically inclined 8th grader.

K. Hess, *Working Knowledge: STEM Essentials for the 21st Century*,
DOI 10.1007/978-1-4614-3275-3_5, © Springer Science+Business Media New York 2013

The difficulty with Einstein's so-called special theory of relativity lies not in the mathematics. The theory is rather difficult, because of our prejudices about time and space, and because it is not easy to abandon prejudice in favor of clear thinking. Thinking hurts! :-)

The other advanced topics presented in this chapter require, in addition to clear thinking, some advanced mathematics like differentiation and integration. This type of mathematics, called the calculus, should be mastered by interested juniors and seniors in high school and is certainly taught in detail in the calculus courses of colleges. Using differentiation and integration, we present two more advanced science topics that also have many engineering applications: (1) the equation for electromagnetic waves that is the basis for wireless communications and (2) the wave equation of quantum mechanics that is the basis for all of chemistry and much of physics. The final two sections deal with engineering and technology problems that require science advancements for their solution.

5.1 Einstein's Space-Time

About everyone has already heard of Einstein's concept of space-time and that we are living in a "four-dimensional space-time" world. Four-dimensional means that we need four numbers to characterize the world that surrounds us. This is, at least in principle, not difficult to understand. We know already from the discussions of coordinates, such as the Cartesian coordinates, that in order to find any point in the space that surrounds us we need three numbers denoted by x, y, z or by x_1, x_2, x_3. We also often need to include a time coordinate t that represents a clock time or, for a GPS system, the GPS time. We have explained this in Sect. 1.3.4. The GPS is tracking moving objects such as cars or airplanes that are found at different places for different times. Therefore our GPS is based on the four coordinates (x_1, x_2, x_3, t) where t stands for the GPS time. We know from discussing the concept of force, that mathematical symbols like (x_1, x_2) may represent two-dimensional points or radius vectors and (x_1, x_2, x_3) three-dimensional points or vectors. Therefore, no one will have a problem with calling (x_1, x_2, x_3, t) a four-dimensional point or radius vector, and naturally we speak about a four-dimensional world, a world in which the four numbers (dimensions) of space-time are very important.

How important are they really? We certainly can agree that the GPS is something important and therefore the four-dimensional GPS coordinates are important. However, the significance of space-time in our life goes far beyond the GPS. We know that coordinate systems were created by Des Cartes and his disciples with problems of geometry in mind. The distances of geometry are measured with real instruments such as meter long rods, and times are measured with real clocks; the corresponding coordinates are important for all discussions of geometry.

Beyond this, it turns out that we cannot even say a logical sentence without the use of space-time. Plato and the Greek philosophers had a rule about logical statements, or any statement for that matter. If we have any statement and wish to discuss whether it is true or false, then the rule is that a statement is either true or false and there exists no third possibility. If the statement was denoted by the symbol "A" then the ancients said: "A" is true or not true, there exists no third possibility. Now take a simple sentence such as "The moon shines." Is that sentence true or false? Well, if we ask someone in New York during bright daylight, the answer will probably be "false." If we ask someone in Tokyo, where it is night while it is day in New York, the answer might be "true." However, there may be clouds just at that time in Tokyo, and so the answer may also be "false." Thus, if we wish to use statements that follow the ancient rules of logic, we need to be more precise and give coordinates. For example, we can say, "We have seen the moon shine in Tokyo at the GPS coordinates x, y, z, t". Clearly, coordinates have a great importance, if we cannot even utter a generally true or false sentence without them. And we need four coordinates! When we speak, we make statements about four dimensions; we speak in terms of Einstein's space-time.

Is that what Einstein found out? Well, yes, he surely knew this. However, what he actually found out was even more astonishing. Einstein asked himself a question that we put here into a more modern form by using what we know (and Einstein did not know) about the global positioning system. Assume that we are here in our solar system and galaxy, the "milky way," and that we have developed a GPS system for the milky way and for the whole universe that we can see with our telescopes. Assume further that there is a galaxy named G123 far far away from us that moves very very fast with a velocity v relative to the milky way. Assume also that this galaxy G123 has intelligent inhabitants and they also have a GPS system. Now say that we meet these life forms from G123 and we compare our GPS coordinates with theirs because we wish to talk with them about the universe. How would these GPS coordinates compare?

We would find stunning differences between the two GPS systems. The atomic clocks of the two systems that are supposed to agree to fourteen decimals would show significant differences. The same meter measures would appear to be shorter in G123 when viewed from the milky way, and the distances would therefore be different in the two GPS coordinate systems. This means that the location of some stars would have unexplainably different GPS coordinates in the two GPS systems. Furthermore, what appears to the G123 inhabitants as pure time appears to the inhabitants of the milky way as a mixture of space and time. All the different data in the two GPS systems would look like one big mess, and it seems impossible to even relate the two GPS systems to each other. Actually, however, there is no mess, and Einstein could make perfect sense of all data even without knowing anyone from G123 and without knowing what a GPS system was. To explain how Einstein made sense of it all and how we can relate the two GPS systems to each other, we just need to remember and consider three special and very important laws of physics. These laws are so "self-evident" that they usually are not even mentioned. We list them here and also tell you from the start the two laws that are absolutely true and the one that is only an approximation. At the time of Einstein all three laws were thought to be absolutely true. Only Einstein disagreed. The three laws are:

1. The first law is related to the fact that the two galaxies are moving with high speed relative to each other. For simplicity, we say that our galaxy, the milky way, is standing still and G123 is moving with velocity v. Of course we cannot really say that. It would also be possible that G123 stands still and the milky way is moving with velocity $-v$. We further assume, just for simplicity, that we need only one space coordinate that we call x in the coordinate system of the milky way and x' in the coordinate system of G123. This means simply that we consider only one direction and all the objects that we consider move only in this one direction. Einstein did everything for three directions but the principle is exactly the same. The time coordinate in the milky way is labeled by t and that in G123 by t'. Now assume that one of the G123 life forms is driving a car with velocity v'_{car} while G123 is moving with velocity v compared to the milky way. Then it is plausible that, if we could see the car from earth through a telescope, it would move with a velocity v_{tot} that is given by

$$v_{tot} = v + v'_{car}. \tag{5.1}$$

In fact, one calls this the theorem of the addition of velocities that has been used throughout history. Of course, we need to let the velocities be positive or negative, depending on the direction in which the galaxy and the car are actually going. If they are going in the same direction, the velocities add, if they are going in different directions the velocities subtract.

Everyone who has watched a bowling team knows about that. The person with the bowling ball runs a few steps in forward direction and moves the hand with the bowlingball in forward direction which gives the ball, when it is thrown, a velocity that is the sum of the running velocity plus the velocity of the throwing hand. So the velocities add. Naturally, if the person who is bowling drops the ball by mistake, while the hand moves in the back direction, then the velocities would subtract. We know this effect of adding velocities also from soccer and football. The player runs and kicks the ball while running which adds the velocity of the running to that of the kicking foot. Eq. (5.1) is thus very well established. It does not only apply to ball games but essentially to every situation with two or more velocities involved. The equation has been tested over and over, almost as well as Euclidean geometry, and has been found always accurate to many decimals. The famous physicist Newton took it as one of the self-evident truth of science that Eq. (5.1) was valid. Einstein did not think so. He had good reasons to believe that Eq. (5.1) was not exactly true. However, we know from many experiments that Eq. (5.1) is indeed valid to many decimals. Below we will see how Einstein resolved this apparent contradiction.

2. The second law of nature that is relevant for our considerations here is the so-called "principle of relativity." This principle says: all the laws of nature have the same form for a given system of objects, whether or not that system

moves with a constant velocity relative to some given coordinate system. We can understand this law by remembering that earth not only rotates around its axis but also with high velocity around the sun. Yet, we have the feeling that, when we stand still, all the landscape around us stands still also. We know that we can move at 800 km/h in a plane, and we feel nothing of that high velocity when sitting in the plane. Similarly, if we check any laws of nature such as chemistry or electricity laws in a plane or on a ship, we find that these laws are all the same no matter how fast the plane or ship moves. Of course, if the plane accelerates, then we notice that, because we feel a force like gravity pressing us toward the back of the seat. The principle of relativity that we discuss here is valid only for a constant velocity of movement, not for accelerations.

Einstein believed that the principle of relativity was an absolutely true law of nature. This principle has often been distorted in the popular press to sentences like "all is relative," and silly jokes are commonly made. A famous joke is "time seems to be endlessly long when you sit on glowing coals, but it appears very short when you are on vacation; so everything is relative." The principle of relativity says something totally different, it says that all the laws of nature have the same form *independent of the given (constant) velocity of motion*. This means, for example, that the laws of falling stones and planets are the same in all galaxies no matter how fast they are moving relative to each other.

3. The third law that we need to remember is a law that follows from Maxwell's wave equation that we will discuss in detail in Sect. 5.3. This law says that the velocity of light in vacuum has a constant value $c = 300,000$ km/s. Einstein believed that this law was proven by so many electricity-related experiments and by the whole accuracy of Maxwell's theory as well as Faraday's experiments that one could not have any doubt about its absolute validity.

The problem with these three laws is that they contain a contradiction. The contradiction is very simple and should be clear to the reader. The third law tells us that the velocity of light is the same in the vacuum of the milky way and in the vacuum of G123. We know that the velocity of light in the milky way is $c = 300,000$ km/s. 0 However, the first law tells us that the velocity of light in G123 as seen from the milky way must be

$$v + c, \tag{5.2}$$

because G123 moves with velocity v relative to the milky way. Therefore, Einstein concluded that the first law, as expressed by Eq. (5.1), can only be approximately true and must be replaced by a more accurate law. When searching for this more accurate law, it became clear to Einstein that the space-time coordinates of the GPS system of the milky way and those of the GPS system of G123 must be different and connected to each other in a rather special way. Why is this so?

Consider the two coordinate systems x, t of the milky way and x', t' of G123 as shown in Fig. 5.1. We can freely choose the two coordinate systems in such a way that they have the same origin (zero of the coordinates) at some given instance

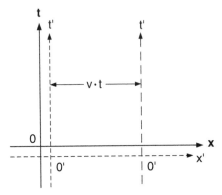

Fig. 5.1 Space-time coordinates (x, t) of our milky way, and (x', t') of galaxy G123. Galaxy G123 moves with velocity v relative to the milky way. If one chooses the two coordinate systems such that they coincide at a certain instance of time, then at time t after that instance in the milky way, the zero point of the G123 galaxy coordinates has moved to $x = vt$ because the whole galaxy G123 has moved by that distance

of time. We encourage the readers to convince themselves of this fact. Then after a time t has elapsed in the milky way, the primed system must have moved by a distance vt as seen from the milky way, because the whole galaxy G123 has moved, and because coordinate systems are connected to real measures of distances and times. This is of course also consistent with Eq. (5.1). One can see this fact in the following way: consider running and kicking football players and someone who observes them. Now associate a coordinate system (x, t) with the observer and a coordinate system (x', t') that moves with the football players with velocity v. Then the ball gets kicked with the velocity v_{ki} of the kicking foot in the coordinate system of the football players. However, seen in the coordinate system and from the eyes of the observer, the velocity is $v + v_{ki}$. For the same reason, we must have the following relation between the two coordinate systems of the milky way and G123, respectively:

$$x' = x - vt \tag{5.3}$$

and

$$t' = t. \tag{5.4}$$

This would then be the way that two persons from the respective galaxies would connect the two GPS systems. This approach, however, does not work because Eq. (5.1) contradicts the third law about the velocity of light in vacuum! It follows therefore that if we have to modify Eq. (5.1), we also have to modify Eqs. (5.3) and (5.4). This is done in the next section, and it is elementary and uses only 8th-grade algebra skills. However, it is conceptually not easy. . . so be alert.

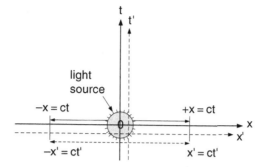

Fig. 5.2 A light source sends a signal from the zero of two coordinate systems that move with velocity v relative to each other. The coordinate systems coincide (are identical) at the instance of time when the light is first sent. After a time t (in the first coordinate system), the light has progressed to the point $x = ct$ as well as to $x = -ct$ on the negative x-axis. The same must be true in the (x', t') coordinate system, because light propagates in vacuum with velocity c. The consequences are discussed in the text

5.1.1   Einstein's Special Theory of Relativity

We first search for an equation that relates the two coordinate systems to each other and at the same time obeys our third law which says that the speed of light is the same in vacuum no matter from which system we look at it. This can be mathematically formulated by considering the following "thought experiment" that is illustrated in Fig. 5.2.

We consider an instance during which the two coordinate systems of the milky way and G123 coincide. We assume that a light source, at the common origin of the two coordinate systems, sends light signals to both sides (positive and negative x-axes) at that instance during which the two coordinate systems coincide. Light propagates in the two systems with the same velocity to the left as well as to the right. Therefore we can say that, in the milky way, the signal to the right arrives at a point x exactly at the time t if

$$x = ct \text{ or equivalently } x - ct = 0. \tag{5.5}$$

For the signal propagating to the left we have

$$-x = ct \text{ or equivalently } x + ct = 0, \tag{5.6}$$

as also indicated in the figure. From this we conclude that we also must have

$$(x + ct)(x - ct) = 0 \text{ or performing the multiplication } x^2 - c^2t^2 = 0. \tag{5.7}$$

Because the inhabitants of G123 also think that their galaxy is standing still and the velocity of light in vacuum is the same everywhere, we must also have

$$(x' + ct')(x' - ct') = 0 \text{ or performing the multiplication } x'^2 - c^2t'^2 = 0 \quad (5.8)$$

From Eqs. (5.7) and (5.8) we have

$$x^2 - c^2t^2 = x'^2 - c^2t'^2. \tag{5.9}$$

Equation (5.9) must hold if the third law, the law for the velocity of light, is correct. This equation is actually more general than its derivation with the simple thought experiment suggests. We know that at the times t and t' when we actually measure the light signal, the coordinates of the G123 galaxy must have changed as seen from the milky way and vice versa because the coordinate systems are moving relative to each other. Nevertheless Eq. (5.9) must be valid if we wish to fulfill our third law that the velocity of light is the same in vacuum everywhere. Equation (5.9) must also be valid for any and all velocities v that any other galaxy might have, and it does not even contain any reference to these velocities v. We therefore will regard Eq. (5.9) as basic to Einstein's theory.

We have thus obtained the desired result: we have derived a relationship between the two coordinate systems or GPS systems of the milky way and G123, respectively, that satisfies our third law. The question is then how we can make this result consistent with Eqs. (5.3) and (5.4) that are representing the first law? The answer to this question was, of course, Einstein's great discovery. We cannot describe here Einstein's original derivation, because his way of doing things was even more complicated than what comes here. That is why, when you see the answer below, you may ask yourself the question how can anyone discover that? Well, the way that one makes such great discoveries cannot be easily taught. It also happens only every 1,000 years or so. Here we can only try to make the result clear by showing it mathematically. And we can do that with relative ease with only high school mathematics, because we already know the answer from Einstein.

To make things even more transparent, we introduce a special unit of length. We have previously used "meters" or "kilometers" as units of length. Now we introduce a much bigger unit: the "bigmeter." This unit is not often used by others. It is our secret weapon to make the algebra easier. We define one bigmeter as being equal to 300,000 kilometers or 300,000,000 meters. Remember this is exactly the distance that light travels in vacuum during one second. Therefore, we have in this unit system a special value for the velocity of light: $c = 1$ bigmeter/s. The reason we do it this way is because then we do not need to write down c in the equations. c is just equal to 1, and anything multiplied by 1 stays the same. Of course there are also other consequences of this new unit system. The velocity of a jet plane which is about 800 km/h becomes now 0.000000742 bigmeters/s. Therefore all the velocities that we normally deal with on earth, such as the velocity of a car, a ship, or a plane, are all represented by very small numbers. The velocity v of G123 might be a little larger, like $v = 0.001$ bigmeters/s. With all of this said, Eqs. (5.3) and (5.4) from above stay the same because they do not contain c, and we have again

$$x' = x - vt \tag{5.10}$$

and

$$t' = t. \tag{5.11}$$

We just have to remember that, in our special unit system, the values of v and all other velocities, except for the velocity of light itself, are small numbers. With $c = 1$ Eq. (5.9) becomes

$$x^2 - t^2 = x'^2 - t'^2. \tag{5.12}$$

We can see now at one glance that there we have a contradiction, because Eq. (5.10) contains the velocity v and Eq. (5.12) does not. The two equations can agree only if $v = 0$. However, this is false because G123 moves with high velocity compared to the milky way. So how can we reconcile all three equations? Einstein's great genius discovered that Eq. (5.11) for the times t and t' must be changed also, meaning that time flows differently in the milky way and in G123. Einstein's thought was if something is subtracted from the space coordinate in Eq. (5.10), why should the time t' not change in a similar way? Now, this is a big thought! Clocks must then work differently if they travel with high speed. This is indeed exactly what happens. We check therefore, whether the following equations work:

$$x' = x - vt \tag{5.13}$$

and

$$t' = t - vx. \tag{5.14}$$

Well, the unfortunate thing is that they do not work as you can see easily by inserting x' and t' into Eq. (5.12). So we try one more thing. We multiply the right hand side of both Eqs. (5.13) and (5.14) by some constant β and try again to find out if things work out, and what this constant β might be. Thus we try

$$x' = \beta(x - vt) \tag{5.15}$$

and

$$t' = \beta(t - vx). \tag{5.16}$$

Science often involves such ideas and random trials. A great scientist needs a great intuition, and Einstein certainly had one. If we now insert x' and t' into Eq. (5.12) we obtain

$$x^2 - t^2 = \beta^2[(x - vt)^2 - (t - vx)^2] = \beta^2(x^2 + v^2t^2 - t^2 - v^2x^2). \tag{5.17}$$

As can easily be checked out by multiplication (use of the distributive law) we have

$$\beta(x^2 + v^2t^2 - t^2 - v^2x^2) = \beta(1 - v^2)(x^2 - t^2). \tag{5.18}$$

Equation (5.17) is therefore always fulfilled if we choose

$$\beta^2(1 - v^2) = 1, \tag{5.19}$$

which gives us for β:

$$\beta = \frac{1}{\sqrt{1 - v^2}}. \tag{5.20}$$

Using Eqs. (5.15) and (5.16) this gives us the desired connection between the two GPS systems which is

$$x' = \frac{1}{\sqrt{1 - v^2}}(x - vt) \tag{5.21}$$

and

$$t' = \frac{1}{\sqrt{1 - v^2}}(t - vx). \tag{5.22}$$

It is now easy to understand that Eqs. (5.21) and (5.22) fulfill all our requirements. They clearly fulfill the third law because we have used this law to derive them. Furthermore, assume for the moment that we are not looking at another galaxy but just at some moving object on earth. For example, we can look at a car or a ship or even at a jet plane. Let the velocity of these cars, ships, or planes be denoted by v, and remember that we are using this very strange unit system involving bigmeters. We know from above that v is therefore very small, $v = 0.000000742$ for a plane and still much smaller for a ship, a car, or a running and kicking football player. Then we can, for most practical purposes, neglect in Eqs. (5.21) and (5.22) all the terms that contain $v^2 = 5.5 \cdot 10^{-13}$ and also the term vx (which is of similar magnitude). We therefore end up with Eqs. (5.3) and (5.4) which are the equations that follow from the first law. Thus, the first law is valid to a pretty high degree of accuracy for the small velocities that we humans usually deal with on earth.

The real miracle is that with use of the above equations, also the second law is fulfilled, the principle of relativity. If we take any true law of nature and write it down for a coordinate system (x, t) and then we transform it for a moving object by using Eqs. (5.21) and (5.22), then we obtain the law in exactly the same form but now with coordinates x', t'. To show this is beyond what can be shown here with elementary mathematics. One needs to know more about differential equations, because most laws of nature are given as differential equations. We discuss differentiation and the wave equations in the following sections, and, as a special project, you can try to show that Maxwell's wave equation follows the principle of relativity. This equation also shows that the velocity of electromagnetic waves in vacuum equals c.

What does all of this tell us, and do we really need to know it? If we confine ourselves to small velocities and not too high accuracy in decimals, then we do not need this elaborate theory. However, the GPS system works with atomic clocks that can be accurate to 14 (fourteen!) decimals, and these clocks are mounted on satellites that move very fast. Therefore, as discussed below in more detail, the engineers of the GPS system needed to use the theory of Einstein. Very high velocities, in fact very close to the velocity of light, can also be achieved in the particle accelerators that have been built such as Fermi laboratory close to Chicago and the "Large Hadron Collider" in Switzerland (see Sect. 5.4.2). Galaxies can move with very high speed also, and the theory of Einstein is necessary for their understanding.

Here is why the GPS engineers needed Einstein's theory. Use Eq. (5.22) and take the difference of two times t_1', t_2' and t_1, t_2, respectively. Assume for simplicity that $x = 0$ which just means that we are at the 0 point of the (x, t) coordinate system and we are looking at the moving system observing a clock in that system. We see the time difference in the moving system:

$$t_1' - t_2' = \frac{1}{\sqrt{1 - v^2}}(t_1 - t_2).$$ (5.23)

This means that the time that is seen to elapse in the moving system is shorter than the time difference that the clock shows in the system of the observer who is at rest. Nowadays, atomic clocks are so accurate that this time difference can be easily, and with high accuracy, measured and indeed has been measured in full agreement with Einstein's theory. Consider a GPS system with atomic clocks in a satellite that moves with a typical velocity of 0.00001 bigmeters/s. Then we have

$$0.99999999990(t_1' - t_2') = (t_1 - t_2).$$ (5.24)

Although this correction is extremely small, it accumulates and becomes much larger than can be tolerated for the accuracy of time in GPS systems. After a day or so, the difference of the two systems becomes of the order of microseconds, which is 1,000 ns, while the accuracy of the atomic clocks in GPS systems is required to be about one nanosecond. GPS has made the theory of relativity important for our daily life, and the theory of relativity, conceived in 1905, has been fully confirmed by an instrument of our daily life, the GPS system. We have to note, however, that the clock speed is also influenced by gravity as Einstein found out in his general theory of relativity. This effect is described in the next section.

Big differences start to occur when v comes close to the speed of light. For example, if $v = 0.99$ then we have

$$0.1(t_1' - t_2') = (t_1 - t_2),$$ (5.25)

and the moving clock is 10 times slower! Such velocities of moving systems can indeed be observed with certain particles that are emitted from the sun, and these particles that usually decay very fast (see radioactivity) "live" indeed much longer when they travel with high speed. The effect described here means the following: We observe twins, both 10 years old, and send one out on an ultrafast spaceship while keeping the other on earth. The spaceship travels for 50 earth years with velocity $v = 0.99$ and then returns. The twin that stayed on earth is now 60 years old while a 15 year-old teenager climbs out of the spaceship. This effect has been called in the past the "twin paradox" because it was paradox for people to believe that this can happen. In addition, the effect involves changes of the velocities because one of the twins returns. Such changes are only included in Einstein's general theory of relativity and not in the treatment above. Nowadays the paradox no longer exists. The GPS system has amply confirmed this effect, and one can even take atomic clocks on airplanes and see the correct difference in time when the plane returns.

Because of the importance of Eqs. (5.21) and (5.22), we rewrite them now so that they are valid for any unit system. Remember we chose bigmeters to make $c = 1$ bigmeter/s and then dropped all the factors containing c's, because they just resulted in a multiplication by 1. Now we restore all the c's, and to do this is actually a medium difficult homework problem. We obtain

$$x' = \frac{1}{\sqrt{1 - \frac{v^2}{c^2}}}(x - vt) \tag{5.26}$$

and

$$t' = \frac{1}{\sqrt{1 - \frac{v^2}{c^2}}}\left(t - \frac{v}{c^2}x\right). \tag{5.27}$$

Historically, these equations had been derived by Lorentz before Einstein and are called after him the "Lorentz transformations." Lorentz derived these equations from very complex thoughts about Maxwell's equations. He realized from these equations that moving objects appear shortened to the resting observer. This shortening effect is called the Lorentz contraction and also has been experimentally confirmed. It can be derived from the above equations that a measuring rod of length L in the resting system has a length L' in the moving system that is given by

$$L' = L\sqrt{1 - \frac{v^2}{c^2}}. \tag{5.28}$$

Lorentz found this result. However, he did not have the bravery to really think that time and clocks would actually change as well when moving. Einstein derived and explained this daring conclusion. That is why we associate mostly Einstein with the theory of relativity.

Einstein did, in addition, more than just being daring. He also found that the mass of a body depends on the velocity of the body in the given system of coordinates. Thus mass is not a constant as was assumed over thousands of years. How can that be? How was this fact going unnoticed for so long? The reason is that the mass changes are very very small for the ordinary velocities that we encounter on earth. Einstein found the following famous relationship between energy E and mass that we give here without proof:

$$E = \frac{mc^2}{\sqrt{1 - \frac{v^2}{c^2}}}. \tag{5.29}$$

Here m is the so-called rest mass (mass for velocity $v = 0$) of the object that is under consideration, and E is its energy. This equation of Einstein is most often printed incompletely as $E = mc^2$ by the popular media. As can be easily checked, the mass changes are negligibly small for velocities of cars or airplanes, because the constant c, the velocity of light, is so large. Notice, however, that as the velocity v approaches the velocity of light c, the energy E becomes larger and larger and approaches infinity. This fact is the reason that Einstein maintained that massive bodies can never reach the velocity of light. The velocity of light in vacuum is, accordingly, the highest possible velocity that can be reached. One can also formulate this fact in the following way. We can define the total mass M of a moving body by the equation:

$$M = \frac{m}{\sqrt{1 - \frac{v^2}{c^2}}}. \tag{5.30}$$

Then, we see that M goes to infinity as v approaches the velocity of light c. Remember that c is the velocity of light in vacuum. Light does slow down when it moves within materials such as glass, and this lower velocity is not the limit of speed.

Equation (5.29) tells us also that if a mass can be somehow turned into energy, then that energy will be very very large because the factor c^2 is very very large. Inversely, Einstein's theory tells us that any object with a total energy of E will behave as if it had a total mass M given by

$$M = \frac{E}{c^2}. \tag{5.31}$$

Mass can indeed be turned into energy. There exist atoms, such as uranium 235, that permit the change of mass into energy as we have already learned in Sect. 4.1.6 when we described the fission of uranium into smaller atom pieces. Invariably a noticeable change of mass involves processes in the nuclei of atoms. The uranium nucleus can decay into smaller pieces that fly apart with enormously high velocity v. Because the velocity of these pieces is so high, their total rest mass m is noticeably smaller than the mass of the original uranium nucleus. In other words, mass has changed into kinetic energy in the process of the fission of the uranium atom, and this generally happens in processes of radioactive decay. Einstein compared atoms with misers. They usually do not change and keep all their energy, just as misers do with their money. Then, however, if a miser dies (the nucleus decays), a large amount of money is taken away by the tax authorities (is transformed into kinetic energy of the smaller and lighter nuclei), and the heirs are left with smaller portions of the fortune.

One can also gain energy by the inverse process of fission the so called fusion of nuclei, for example, by the fusion of hydrogen atoms into helium. This process supplies the energy of our sun and of all stars. Helium has a nucleus of two protons and two neutrons while four hydrogen atoms have just four protons. Neutrons, however, can change into protons by emission of electrons and another particle (called antineutrino).The actual process of fusion of hydrogen into helium occurs in multiple steps and involves also the hydrogen isotopes deuterium and tritium. Deuterium is a hydrogen atom with one additional neutron, while tritium has two additional neutrons in its nucleus. The interested reader can find many illuminating descriptions on the Internet. Here we give only a simplified version with the essence of this nuclear fusion process. The mass of four hydrogen atoms is $6.6927 \cdot 10^{-27}$ kg. The mass of one helium atom is $6.6443 \cdot 10^{-27}$ kg. Thus the helium mass is slightly smaller than that of four hydrogen atoms. The difference in mass, that is the mass that vanishes in the fusion process, is $4.838 \cdot 10^{-29}$ kg. But mass cannot just vanish. It is turned into energy. The energy that one obtains according to Einstein is derived by multiplication of the missing mass with $c^2 = 9 \cdot 10^{16}$ (measured in meters square per seconds square). Therefore we have

$$E = 4.838 \cdot 10^{-29} \, \text{kg} \, 9 \cdot 10^{16} \frac{\text{m}^2}{\text{s}^2} = 4.3543 \cdot 10^{-12} \, \text{kg} \, \frac{\text{m}^2}{\text{s}^2} = 4.3543 \cdot 10^{-12} \, \text{J}.$$

$$(5.32)$$

This looks like a very small number and therefore like a very small energy. But wait, this was obtained for just four hydrogen atoms and one helium atom. A few liters of water contain more than 10^{25} hydrogen atoms. That number of atoms yield then $4.3543 \cdot 10^{13}$ J of energy! With that amount of energy you can heat about one million (!) barbecues for one hour. This is the reason why the sun and all stars can deliver enormous amounts of energy over billions of years. Their energy is derived from nuclear fusion processes. If the sun would burn just as a coal fire does, it would have burned at most for a few thousand years. It is interesting that this mechanism of energy creation has only been understood since about 1939. Nevertheless, only very few scientists were wondering why the sun was still shining. This is not untypical. We often take the things that surround us for granted and assume that someone will know why things work that way, but this is not always so.

5.1.2   Einstein's General Theory of Relativity

Einstein's general theory of relativity is based on mathematical methods that are not taught in high school. Remember that we have based all our derivations on the fact that the systems that are considered move with constant velocity v relative to each other. However, we know already from previous sections that many physical processes include accelerations which represent changes in the velocity. Einstein noticed that accelerations have locally the same effect as the gravitational fields

Fig. 5.3 The light rays of two stars are being bent by the gravitational forces of the sun. This can only be observed during an eclipse that is also shown in the illustration. The *dash–dotted lines* are drawn to the points where the stars (*empty circles*) would be if the rays would not be bent. The actual stars (*full circles*) are located at different positions (follow *full lines*). The sum of the angles $\alpha'_1 + \alpha'_2 + \alpha_3$ adds up to 180° (π radians). The sum $\alpha_1 + \alpha_2 + \alpha_3$ does not. Note that the drawing is not to scale because of the cosmic distances that are involved here

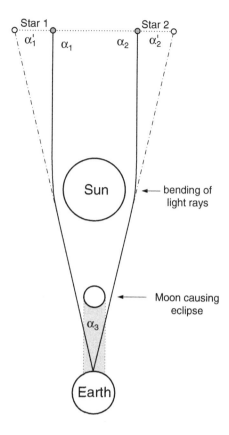

of the earth, the sun, or any massive body. We know about this from everyday experience. When a car accelerates, we are pushed back toward our seat, just as if a gravitational force would act on us. Our mass that resists acceleration is exactly equal to the mass that determines our weight. Einstein used this equality of acceleration-mass and weight-mass as basic to his general theory of relativity. Indeed, if we would fly in a space-ship and if we could not see the outside universe, any acceleration of the capsule would be for us just the same as the pull of gravity that we feel on earth and that determines our weight. This notion enabled Einstein to obtain a general theory of gravitation and relative motions that need not occur with constant velocity. We mention here only a few of the consequences of Einstein's general theory; its mathematics is too involved to be presented on an elementary level.

Energy and mass are related to each other by Einstein's formula of Eq. (5.29). This suggested to Einstein that gravitational forces act on both mass and energy and that light (be it particle or wave) will also be attracted by massive bodies. In addition to attracting light, the gravitational forces of massive bodies also distort space itself in their proximity. These two effects lead to one of the most famous consequences of Einstein's general theory that is illustrated in Fig. 5.3.

The rays of two faraway stars propagate toward the earth and are bent by the gravitational forces of the sun if the rays indeed pass close to the sun. In this case, the two stars appear separated by a larger different distance as compared to an observation of the same stars at times when the sun is not close by. This difference, indicated in Fig. 5.3, can indeed be measured. However, this measurement requires special circumstances: if the sun is close to the path of the rays of the starlight toward earth, then the bright sunlight overpowers the star light and renders the stars invisible. The scientists that performed the actual measurements, led by Arthur Eddington, had therefore to wait for an eclipse of the sun that is also illustrated in the figure. Eddington and his colleagues performed a very accurate measurement in 1919 and confirmed Einstein's theory of the bending of light by gravitational forces beyond any shadow of a doubt.

The bending of light rays by gravity goes to the heart of what we know and can say about geometry: light rays are the straight lines of geometry. If light can be bent, then there is a difficulty with Euclid's geometry. Lasers give us the straightest lines that we know and that we can use. Therefore, one usually insists to define light rays as the straight lines of Euclid's geometry. Now, however, the "straight" rays only represent the shortest connection from one point to another, from the stars to earth, and they are not really straight. We then have the following problem that can also be seen from Fig. 5.3. The angles $\alpha_1, \alpha_2, \alpha_3$ of the "triangle" between the actual location of the stars and the point of observation on earth (full lines of Fig. 5.3) do not add up to two right angles (or 180°). Only the sum $\alpha'_1, \alpha'_2, \alpha_3$ does add up to two angles. Therefore, Euclid's fifth axiom is not fulfilled, and his axioms involving straight lines need also be modified. There are no straight lines, only lines of shortest connections. The real geometry of the universe is therefore not Euclidean but only approximately so.

In some cases the geometry of the universe is not even approximately Euclidean. At the center of most or even all galaxies are, as far as I understand it, very heavy objects with masses of millions of suns. These objects have such a powerful gravitational pull that light cannot escape from them. It is pulled back toward them. Therefore these objects are called "black holes." There exist many creative stories about black holes on the Internet and in movies. Black holes are attributed with lots of interesting, albeit often highly speculative, properties. The author becomes particularly suspicious when the deviations from Euclidean geometry are presented as opening possibilities for time travel to the past. This would enable us to go back in time and correct our mistakes. Einstein himself never believed in such time travel. Direct experiments with black holes that could decide such questions cannot be performed, because the gravitational forces would easily destroy any exploratory vehicle. All the knowledge about black holes is bound to stay very indirect.

Clocks do, however, show different times in gravitational fields than they do without gravitation, and this has been completely confirmed by experiments. Clocks are fastest when gravitational forces are absent and slow down in high gravitational fields. This effect is large enough to be of importance for the atomic clocks of the GPS system. Einstein's equations which are relevant for these time dilatations in gravitational fields have all been confirmed by atomic clock measurements.

They are a little more complicated to derive, however, than the time dilatation of moving objects that we discussed in connection with Eq. (5.25). Here we just give plausibility arguments for the gravitational time dilatation by the following example.

Consider an object with mass M that we move from the surface of the earth to a height l against the gravitational pull of the earth. The force that we need to overcome is according to Newton gM with $g = 9.81 \text{ m/s}^2$. The energy that we thus need to bring this object up to height l is glM. We can see this fact from our discussions of Eq. (2.5). If we wish to talk about photons instead of massive objects, then we need to replace the mass M by $\frac{h\nu}{c^2}$, where $h\nu$ is the energy of the photon as discussed in Sect. 2.5. Therefore, a photon moving up to height l loses the energy $gl\frac{h\nu}{c^2}$. The new frequency ν' of the photon at height l can, therefore, be calculated from the new energy $h\nu'$:

$$h\nu' = h\nu \left(1 - \frac{gl}{c^2}\right). \tag{5.33}$$

Clock time and frequency have an inverse relationship. The speed of a clock would therefore change by the inverse factor

$$\frac{1}{\left(1 - \frac{gl}{c^2}\right)} \approx \left(1 + \frac{gl}{c^2}\right). \tag{5.34}$$

For a height of $l = 12,000 \text{ km}$ we have then a factor of $(1 + 1.3 \cdot 10^{-9})$ by which the speed of a clock would increase. The change of clock speed needs also be included when atomic clocks are used in the satellites of the GPS system. This effect increases the speed of clocks in satellites, while their rapid motion slows the clocks down as we know from Eq. (5.25). There is an important and noticeable net effect that has been included by the engineers of the GPS system.

5.2   Some Advanced Mathematical Concepts

The laws of nature can be formulated in very elegant ways by using advanced mathematical techniques. These techniques have developed alongside the increased understanding of the basic laws of physical and chemical sciences, and they are now of general use in science and engineering. We discuss in this section an area of mathematics that was introduced in the seventeenth century by the famous mathematicians and scientists Isaac Newton and Gottfried Leibniz. This mathematical area is called the calculus of differentiation and integration or just "the calculus." Calculus can and is being presented at colleges and universities in terms of purely logical and algebraic analyses. Here we restrict ourselves to a geometric interpretation that is related to curves. This geometric interpretation of the calculus can be presented by using only mathematical concepts that can be taught in high school and still covers most of what is needed for college science and engineering.

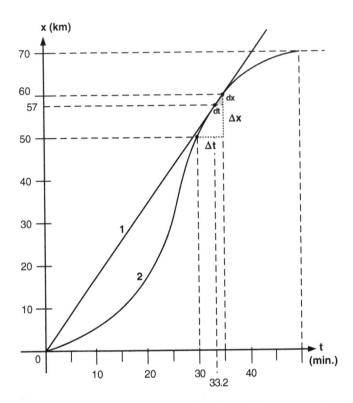

Fig. 5.4 Graphical representation of the position x (*vertical axis*) of a car moving with different velocities at different times t (*horizontal axis*). The straight line 1 is for a car with constant velocity. The curve 2 is plotted for a car that changes its velocity starting from 0 and ending with a stop (0 velocity) after 70 km. The straight line 1 is also the tangent to curve 2 at the 35 min and 60 km point; meaning line 1 touches curve 2 at that point. In principle line 1 may touch curve 2 in more than one point and in this illustration line 1 and curve 2 are indeed very close over some range

5.2.1   *Differentiation*

We have learned in Sect. 1.3 about straight lines and about curves such as the sine or cosine functions. Physics often deals with such graphs. For example, we can monitor the position of a car as a function of time that is measured with a clock. For the following explanations, it is convenient to use a Cartesian coordinate system with the horizontal axis representing time and the vertical axis representing the position x of the car. Note that in the standard plots the horizontal axis is often denoted as the x-axis. Here, however, our purpose is to derive the velocity of the car in a mathematical way, and for this purpose we need the special coordinate system shown in Fig. 5.4. Such a graph could have been created with help of our GPS system that measures, as we know, both position x and time t.

The shape of the graph tells the expert a lot of facts. The straight line in the middle of the graph is the graph of a car moving with constant velocity. For the given line, we can see that the car has moved 60 km in 35 min. If an object moves with constant velocity, then the definition of the velocity is

$$\text{velocity} = \frac{\text{distance}}{\text{time}}. \tag{5.35}$$

Therefore the car, described by the straight line, moves with a velocity $v = \frac{60}{35} = 1.7143$ km (kilometers) per minute or, as one obtains by multiplying with 60, about 103 km/h. The following equation describes the position of a car at time t if it starts at the origin of the coordinate system and moves with constant velocity v:

$$x = v \cdot t. \tag{5.36}$$

Thus we associate exactly one position x of the car with a given time t. Therefore we deal with a function. As we know, a function is a mapping of a domain (in the present case the times t) to a codomain or range (presently the locations x). For the linear part of the graph, the mapping is obtained by just multiplying the given time t with the velocity v.

If the path of the car and the corresponding curve of Fig. 5.4 does not form a straight line, the car does not move with constant velocity, and the velocity may be different for each different position x. How does one then find the velocity at any given point x? As we will see immediately, this is not so easy. However, we have plotted the curve of Fig. 5.4 in a way that suggests the method of calculation for the case of varying velocity. At precisely 35 min and at the 60 km point, the curve is just "touched" by the straight line. At this touching point, the car proceeding on the curved graph must have exactly the same velocity that the car characterized by the straight line has. One says in such a case that the straight line is the "tangent" of the curve and the velocity of the car is equal to the "slope" of the tangent which is given by Eq. (5.35). This definition of the velocity as the slope of the tangent is indeed true for any point of the curve.

How do we calculate the slope of the tangent of a curve? We can, of course, plot that tangent at every point and measure the distance and corresponding time of a car that would proceed along the straight tangent line. There is, however, another method that may at first glance appear quite cumbersome but at the end represents the royal road to determine slopes of tangents. In Fig. 5.4, we have included three dots on the curved line. One dot is located at the 35 min, 60 km point where the tangent just touches the curve. The other two dots are to the left of this dot at the 33.2 min 56.9 km and 30 min 50 km marks, respectively, and are located below the straight line labeled 1. Assume now that we try to determine the velocity by dividing the distance Δx between the first and the third dot by the corresponding time difference Δt and take this value as an approximate value of the slope of the tangent and therefore of the velocity v. We then have

$$v \approx \frac{\Delta x}{\Delta t} \approx \frac{60 \, \text{km} - 50 \, \text{km}}{35 \, \text{min} - 30 \, \text{min}} = 2 \, \text{km/min} = 120 \, \text{km/hour}. \tag{5.37}$$

We see that the velocity so obtained is a little higher than the actual velocity of 103 km/hour that we have calculated above. The reason is that we need to get the points much closer to each other to approach the real slope of the tangent. To check this out, we take the first and the second dot on the curve and have

$$v \approx \frac{dx}{dt} \approx \frac{60\,\text{km} - 56.9\,\text{km}}{35\,\text{min} - 33.2\,\text{min}} = 1.7222\,\text{km/min} = 103.333\,\text{km/h} \qquad (5.38)$$

which is very close to the actual 103 km/h. To get the exact value, we would have to use points in even closer proximity to the touching point of the tangent.

The precise mathematics of finding the slope of the tangents involves the concept of the "limit" of a sequence of numbers. Take, as an example, the sequence of numbers that we obtain when we calculate with smaller and smaller differences Δt. We say that our sequence of numbers Δt approaches the limit 0, if and only if Δt becomes smaller than any positive number ϵ that we can think of. Note that Δt can never actually be equal to 0, because we are not permitted to divide by 0. In the limit of Δt approaching 0 arbitrary closely, we obtain the exact result for the slope of the tangent or the "derivative" that we denote by

$$v = \frac{dx}{dt}. \qquad (5.39)$$

The fraction $\frac{dx}{dt}$ is pronounced in words dx over dt and is called by mathematicians the "derivative" of the curve or its corresponding mathematical function at the time t. The process of finding the derivative is called "differentiation." The concept of the limit is important for all advanced mathematical analysis and has developed over a long period of time, with major contributions by Augustin Louis Cauchy and Karl Weierstrass. For our purpose, it is sufficient to think of the quantities dx, dt just as very small numbers, and we need not be concerned about the mathematical intricacies of the concept of a limit. STEM lovers are alerted, though, that mathematical analysis including limits is a very beautiful area full of surprises.

The procedure to measure smaller and smaller distances Δx for smaller and smaller time periods Δt is, of course, very cumbersome. However, there exist simple rules to obtain the slope of the tangents for most curves, if they are given as functions such as $\sin(t)$, $\cos(t)$, t^2, and t^3. MATHEMATICA performs differentiation with ease for practically all curves that we can imagine. Here are a few examples. First we investigate the curve $x = at^2$ and find the derivative at any point t. As shown in the box, MATHEMATICA delivers this derivative on a silver platter and finds as the result the function $2at$.

MATHEMATICA

The derivative of any function $x = f(t)$ such as $x = at^2$ is obtained in the framework of MATHEMATICA by using the symbol Dt in the following way: The command $Dt[f(t), t]$ shift-enter, gives you $\frac{dx}{dt}$ at any point t.
$Dt[at^2, t]$ shift-enter gives the result $2at$ which is exactly what it should be
$Dt[bt^3, t]$ shift-enter gives the result $3bt^2$

Had we used a different function such as bt^3, then MATHEMATICA would have found the derivative $3bt^2$. There is a general rule behind such derivatives, and one can obtain them without the help of MATHEMATICA. One just needs to remember the following.

- The derivative of any constant function such as $f(t) = c$ is 0.
- The derivative of any function of the form ct^s where t is the element of the domain of the function and c, s are constants, is given by $cst^{(s-1)}$. Thus we can write the following:

$$\text{If } x = ct^s \text{ then } \frac{dx}{dt} = cst^{(s-1)}. \tag{5.40}$$

It is relatively easy to derive these rules from our Eq. (1.59) in the following way. Let's say we have $x = t^3$ and let's assume that ϵ is a very small quantity. Then we have

$$dx = t^3 - (t - \epsilon)^3 \approx t^3 - t^3 + 3t^2\epsilon = 3t^2\epsilon. \tag{5.41}$$

Because $dt = \epsilon$ we obtain in the limit of ϵ going to 0 (but never actually being 0)

$$\frac{dx}{dt} = 3t^2. \tag{5.42}$$

We are now close to our goal to calculate the velocity of the car at any point of any curved graph. We have just learned how to find the velocity when the graphs contain only powers of time t and constants. The good news is that all well-behaved functions of time (or any other variable for that matter) can be expressed in terms of sums that involve only powers of t (or any other variable). "Well-behaved" means in essence that the function looks like a curve without interruptions, which is called a continuous curve (a curve with the property of continuity). Modern calculus can also deal with functions and curves that are not well behaved (have a lack of continuity), but this takes some effort and special techniques that are taught in advanced mathematics courses. However, if we have any given well-behaved function of time, and we call this function $f(t)$, then the mathematician Karl Weierstrass has shown that one can write

$$f(t) = \sum_{n=0}^{m} a_n \cdot t^n. \tag{5.43}$$

Such a sum, with only powers of a variable involved, is called a polynomial. The a_n are constants that we need to find and m is an integer depending on the shape of the curve. Note that $t^0 = 1$ for any t (in words, t to the power 0 equals 1). How do we find m and the constants a_n? This is, in general, not so easy. However, MATHEMATICA does this job quickly and reliably. All we need to know are the values of the function for every point of a certain domain. Let's say we know the following values of a function: $f(1)=1$, $f(2)=5$, $f(3)=19$, $f(4)=49$, $f(5)=101$, all expressed in the units of our curve, of course. Then, all we have to do, to obtain the polynomial that we are looking for, is to write down this precise sequence of values as shown in the MATHEMATICA box. MATHEMATICA returns the polynomial as also shown in the box.

MATHEMATICA

The command:

InterpolatingPolynomial[$\{1, 5, 19, 49, 101\}, t$] shift-enter

gives the result

$1 + (t - 1)(4 + (t - 2)(t + 2))$. Performing the multiplications one obtains $t^3 - t^2 + 1$, and from this we can easily derive the differential corresponding to the velocity v at any time t

$v = 3t^2 - 2t$

In terms of Eq. (5.43) we therefore have $a_0 = 1, a_1 = 0, a_2 = -1, a_3 = 1$, and all other constant coefficients are zero.

The derivative of any polynomial is the sum of the derivatives of all the terms of the polynomial:

$$\frac{dx}{dt} = \frac{df(t)}{dt} = \sum_{n=1}^{m} n \cdot a_n \cdot t^{n-1}. \tag{5.44}$$

Note that the summation starts now with $n = 1$, because the derivative of the constant term ($t^0 = 1$) equals 0.

We end this section with a remark on repeated differentiation. We have plotted in Fig. 5.4 the distance versus the time (as horizontal axis) and obtained from the tangent the velocity at any given point of distance. We can repeat the method and plot the velocity versus time. Then the slope of the tangent $\frac{dv}{dt}$ gives us the change of the velocity with time that is called the acceleration. Because the velocity itself is a derivative $v = \frac{dx}{dt}$ the acceleration is a second derivative that is written as

$$\text{acceleration} = \frac{d^2x}{dt^2}. \tag{5.45}$$

Here the superscript 2 as indicated in the numerator and denominator simply means that the second derivative is taken.

We have covered in the above section a whole year of teaching related to the calculus of differentiation. Yet, believe it or not, this short section should enable you to handle many problems of engineering and physics, in principle almost all problems of differentiation that one can think of, even at the college level of engineering. One needs, of course, considerable practice to do problems in this area, and one needs also often some further factual knowledge. However, once this section is understood, applications of it should be learned with relative ease. Of course, the help that one can have from MATHEMATICA is very important for the actual solution of problems, and many engineers and scientists use this or similar mathematical software packages.

Fig. 5.5 An area of a piece of land plotted by using a coordinate system (x, y). The land has straight line boundaries except for the top that has a curved shape. The total area is covered by squares as well as possible. However, we can see that if the squares are not infinitely small, there will always be left over areas (*shaded*) that have not been counted

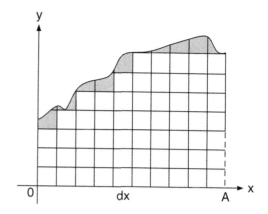

5.2.2   Integration

Integration deals with finding the area under a curve. We can, for example, think of an area, say a piece of land that has the shape shown in Fig. 5.5. The land has straight lines as its lower, left, and right boundaries, but the upper end is curved. The curve can also be described by a function as discussed above. The question that we asked in Sect. 1.3 was how we could calculate the area of any complicated landscape. Our answer was that we should just cover the land with very little unit squares and then count the number of squares which tells us how many square feet, square meters, or square millimeters the area is. Integration amounts thus to the summation of squares. If we wish to have a very precise result, we need to use a lot of squares. If we wish to be exact, we actually need infinitely small squares which means an infinitely large sum of squares. The mathematicians use the name integral for such a large sum and denote it by the symbol:

$$\int_0^A (\text{Polynomial}) \, dx. \tag{5.46}$$

The point 0, the zero of the coordinate system below the $\int$ sign, indicates were the land starts and the point A where it ends. The $\int$ symbolizes a big S and hints that a very big sum is involved. The dx means that we involve in the process of integration the small distances dx, and the word "polynomial" that is written after the integral symbol is the polynomial, or in general any function, that describes the curve.

There exists a very interesting and surprising mathematical theorem that helps us to determine the big sum $\int$ and thus the exact area under a curve. This theorem was developed under more and more general assumptions over hundreds of years. Newton and Leibniz started to prove the theorem, Cauchy extended their work, and Henri Lebesgue finally stated the theorem, now called the fundamental theorem of calculus, in its most general form. They found the following. Given any function

$F(x)$ that has a derivative that assumes values $\frac{dF(x)}{dx}$ on the domain of the function between 0 and A (including the end points), we have

$$\int_0^A \frac{dF(x)}{dx}\, dx = F(A) - F(0). \tag{5.47}$$

Therefore, this theorem links differentiation, i.e., the finding of slopes of curves, with integration, i.e., the finding of areas.

Assume now that the function $f(x)$ is equal to the derivative of $F(x)$, and we therefore have

$$f(x) = \frac{dF(x)}{dx}. \tag{5.48}$$

Then $F(x)$ is called the antiderivative of $f(x)$, or (equivalently) $F(x)$ is called the indefinite integral of $f(x)$. One writes

$$F(x) = \int f(x)\, dx. \tag{5.49}$$

Thus the fundamental theorem of calculus tells us that we can calculate the area under a curve corresponding to a function $f(x)$, if we can find the antiderivative or indefinite integral $F(x)$.

It is easy to find the antiderivative of a polynomial. Assume that the polynomial describing the curve of Fig. 5.5 is given by

$$y = f(x) = \sum_{n=0}^m b_n \cdot x^n. \tag{5.50}$$

Then the antiderivative is the following sum:

$$\sum_{n=0}^m \frac{b_n}{(n+1)} \cdot x^{(n+1)} + k. \tag{5.51}$$

Thus, we have obtained this sum by a process that is the exact inverse to the process of differentiation. We have increased the power of x by one and divided by that increased number, while for differentiation, we had decreased the power of x (or t) by one and multiplied by the original number n. We have also added an arbitrary constant k because the derivative of a constant is 0. Therefore, we have to add, in general, a constant to obtain the antiderivative. Clearly, if we differentiate Eq. (5.51), then we obtain the original sum of Eq. (5.50) which in turn means that Eq. (5.51) is the antiderivative of Eq. (5.50). Therefore the fundamental theorem tells us that

$$\int_0^A f(x)dx = \int_0^A \frac{d}{dx}\left[\sum_{n=1}^m \frac{b_n}{(n+1)} \cdot x^{(n+1)} + k \right] dx = \sum_{n=1}^m \frac{b_n}{(n+1)} \cdot A^{(n+1)}.$$

$$\tag{5.52}$$

Note that this equation is only correct as long as the integration starts at 0, the origin of the coordinate system. Had we considered the area of a curve starting at a point B (instead of the origin), then we would have obtained

$$\int_B^A f(x)dx = \sum_{n=1}^{m} \frac{b_n}{(n+1)} \cdot A^{(n+1)} - \sum_{n=1}^{m} \frac{b_n}{(n+1)} \cdot B^{(n+1)}. \tag{5.53}$$

Of course, MATHEMATICA and similar software packages know how to integrate very well and return you antiderivatives and areas under the curves also for functions that are not given as polynomials. An example is shown in the MATHEMATICA box, which integrates the curve $x^3 + \sin(x)$ between the points 0.5 and 1.7.

Mathematica
Integrate[x^3 + *Sin*[x], {$x, 0.5, 1.7$}] shift-enter
which gives the result 3.07883 in the appropriate units of the graph (e.g. square meters)

5.2.3 Applications of Differentiation and Integration

A lot of material about integration is posted on the Internet, and many useful exercises can be conceived by the teacher. Many laws of nature are best explored by using the concepts of differentiation and integration. As an example we show how two of Newton's important results can be formulated with the calculus of differentiation and integration. We know already from Eq. (5.39) that the velocity is given by

$$v = \frac{dx}{dt}; \tag{5.54}$$

similarly one obtains for the acceleration a

$$a = \frac{dv}{dt}. \tag{5.55}$$

Therefore, if we have a constant acceleration like the gravitational acceleration g of the earth, then we can find the velocity of any object at any time t_1 by using the fact that $g dt = dv$ and integrating both sides:

$$v = \int_0^v dv = \int_0^{t_1} g \, dt = g t_1. \tag{5.56}$$

Here we have assumed that the object starts to be accelerated at $t = 0$. To obtain the height h, that a stone falls in the time period t_1, we integrate again and have

$$h = \int_0^{t_1} v dt = \int_0^{t_1} gt \, dt = \frac{gt_1^2}{2}, \tag{5.57}$$

which proves Galileo's law of Eq. (2.24).

One can calculate the kinetic energy of a falling stone by using the definition of energy. Energy is the sum of forces times distances and becomes now an integral. Using Newton's law, which states that force equals mass times acceleration, one obtains

$$E_{kin} = \int F dx = M \int a \, dx = M \int \frac{dv}{dt} dx = M \int \frac{dx}{dt} \frac{dv}{dt} dt = M \int v \, dv = M \frac{v^2}{2}. \tag{5.58}$$

This proves Eq. (2.18) that we gave for the kinetic energy. The last few equations look easy but require a lot of integration practice by the students to be understood in all their details.

5.2.4 Monte Carlo Integration Using Computers

We finish the discussion of calculus topics by showing how one can integrate using modern computers. The particular method that we discuss here is called Monte Carlo integration. This method is used for all kinds of statistical predictions, ranging from the stock market to nuclear bombs, and from the weather pattern to retirement accounts. The method does not use advanced concepts and can, at least in principle, be understood by an 8th grader.

Consider a painting of a tropical landscape with an area of exactly one unit square. This means the area equals one foot times one foot, one yard times one yard, or one meter times one meter. Imagine that the painting contains a palm tree with lots of fuzzy palm leaves. How can one determine the area of all these palm leaves? Determining the area is by definition an integration problem. As we have suggested previously, we could inscribe very tiny squares into all palm leaves and then add up all the squares. This is, of course, a lot of work. A simpler way is provided by the Monte Carlo method. This method is based on probability theory. The name derives from the famous gambling place Monte Carlo. Gambling with the roulette, for example, can be described by the mathematics of probability as we have seen in the Sect. 4.4.

Imagine that you throw darts at the painting and you throw them totally randomly. Random means that there is no preference to any point of the painting. All points can be hit with equal chance. Imagine also that you have thrown lots and lots of darts so that the painting is totally covered with them. Now associate with each dart a little square. It is then plausible that the number of darts that hit the palm leaves will relate to the number of darts that hit the picture, just as the ratio of the area of the palm leaves relates to the total area of the picture. Thus we have

$$\frac{\text{area of palm leaves}}{\text{area of picture}} \approx \frac{\text{number of darts in leaves}}{\text{total number of darts}}. \tag{5.59}$$

Throwing darts randomly has given us therefore the approximate area of a complicated surface. If we would have considered the area under a curve, then we would have in essence performed an integration. This is what one calls Monte Carlo integration, and it represents a great method of integration by use of computers.

To perform a Monte Carlo integration with a computer, we use random numbers instead of darts. We have seen in Sect. 4.4 how to generate integer random numbers with MATHEMATICA. Now we need pairs of real random numbers between 0 and 1 in order to obtain random coordinate pairs for a point (x, y) in a unit square. Consider the curve corresponding to the function $y = x^2$ as well as 20 randomly chosen points all in a unit square. The random points are generated in the way shown in the MATHEMATICA box.

MATHEMATICA
The command
$x = $ RandomReal[{0, 1}] and shift-enter,
gives a random number between 0 and 1 e.g. $x = 0.57071$.
$y = $ RandomReal[{0, 1}] and shift-enter,
gives another random number between 0 and 1 e.g $y = 0.397652$.
From this one obtains, for example, the random point ($x = 0.57071$, $y = 0.397652$) representing a dart-hit.

It is, of course, a lot of work to plot a curve in a unit square, to include all the generated random dots (x, y) and then count those that are below the curve in order to obtain the area below the curve. A computer can do that automatically. The next MATHEMATICA box points the way toward that goal. The key to understand this is the following. Considering the curve $y = x^2$ we know that for the points (dart hits) that lie below the curve, we have $y < x^2$, where $<$ stands for "smaller." Thus we have

$$r = y - x^2 < 0. \tag{5.60}$$

MATHEMATICA permits you to go through expressions N times and print out the result by writing the command: Do[expression, {N}]. The following MATHEMATICA box does just that for $N = 20$ and prints out all the 20 results some of them are positive others negative.

MATHEMATICA
Do[r = RandomReal[{0,1}] − RandomReal[{0,1}]^2;Print[r], {20}] shift-
enter gives you 20 results some positive some negative

After executing this last MATHEMATICA box, all you need to do is count the
number of negative results and divide this number by 20 or any number N that you
have used. This gives you the approximate value of the integral. The approximation
becomes better the higher the number N. Thus we have

$$\int_0^1 x^2 \, \mathrm{d}x = \frac{1}{3} \approx \frac{\text{number of negative results}}{N}. \tag{5.61}$$

This type of calculation is great for projects. You can convince yourself that the
result comes closer to the exact result of $\frac{1}{3}$ as you increase the number N. With
$N = 20$ the result may still be 30% off, while with $N = 100$ you will be very
close to the exact result. You can try all this for different functions, and you can also
extend the MATHEMATICA calculation in such a way that it adds the number of
negative results automatically. In this way you can become a master of Monte Carlo
integration.

5.3   Wave Equations in Electromagnetics and Quantum Mechanics

This section and the next are for the connoisseurs of quantum mechanics, the theory
of wavicles. Readers who were happy with the phenomenological treatment of
Sect. 2.5 may wish to proceed to Sect. 5.4.4.

The methods of differentiation and integration are, as we have seen, useful to
describe the motion of falling stones and even planets. Their importance, however,
goes beyond that. These methods can also be used to describe waves, for example,
the electromagnetic waves of wireless communication and the strange particle wave
mixtures, the wavicles of quantum mechanics. Waves can be described by analogies
to water waves or mathematically by the sine function, and we have done that
in Sect. 2.1.5. However, to describe waves completely and for all scientific and
engineering purposes, one needs a so-called wave equation. Wave equations are the
subject of this section.

5.3.1   Maxwell's Wave Equation

We start with waves that we have discussed already in some detail, electromagnetic
waves. These are fully described by the rules that Maxwell and Faraday have
developed (see Sect. 2.2.3). From these rules, one can derive a wave equation

that is a great tool to calculate most phenomena connected to electromagnetic waves. The mathematics of this equation and its solutions is advanced and involves differentiation and integration.

Waves depend on both time and space coordinates because waves propagates in space and, as we know from water waves, rise and fall with time at any given point in space. To describe a wave we need therefore the time coordinate t and at least one space coordinate, say x. Therefore the functions that describe waves are functions of at least two variables such as $f(t, x)$. Discussing differentiation, we have used up to now only functions of one variable. Fortunately, we can use all the ideas and concepts of differentiation of one variable when we deal with waves and thus with two or more variables. The reason for this fact is that, for all our purposes, we need to know only the differential with respect to one variable for fixed values of the other variables. Therefore, we can treat the other variables like fixed constants. To indicate that the other variable(s) is (are) indeed treated like a fixed constant, we write the differential in the form

$$\frac{\partial}{\partial x} f(t, x). \tag{5.62}$$

As an example consider the function $f(t, x) = 2t + 3x^2$. Then we have:

$$\frac{\partial}{\partial x} f(t, x) = \frac{\partial}{\partial x} (2t + 3x^2 = 6x), \tag{5.63}$$

which is exactly the result that we obtain from our differentiation rules by regarding t as a constant (like $t = 2\,\mathrm{s}$). Note that it does, of course, not make any sense to add times and distances. This would be like adding apples and bananas. However, in any correctly stated problem, the factors will have some physical meaning so that at the end, we add equal things. For example, the factor 2 in Eq. (5.63) could mean, for example, frequency (like 2 Hz, which means 2 oscillations per second) so that the product of this factor and time t becomes a mere number. Similarly, the factor 3 could have a physical meaning of cm^{-1} so that $3x$ also represents just a number and we are justified to add.

If we consider only the time coordinate for the differentiation, then we write

$$\frac{\partial}{\partial t} f(t, x). \tag{5.64}$$

The geometrical interpretation of this type of differentiation, called partial differentiation, is again related to the tangent of the curve. In the case of Eq. (5.62) we obtain the spatial slope of the tangent of the curve $f(t, x)$ at a given fixed point t in time, while from Eq. (5.64) we obtain the slope of the curve as it depends on time for a given fixed spatial point x. MATHEMATICA can, of course, also perform this

type of differentiation. The partial differentiation in MATHEMATICA is performed
by using the MATHEMATICA command $D[\ldots]$ in the following way:

$$\frac{\partial}{\partial x} f(t, x) \rightarrow D[f, x]. \tag{5.65}$$

An example of this is given in the MATHEMATICA box.

MATHEMATICA

$D[2t + 3x^2, x]$ shift enter gives the output

6x

or alternatively

$D[2t + 3x^2, t]$ shift enter gives the output

2

One also can define a function of one or more variables. As an example
we define the function for the wave of Eq. (2.40):

$f[x\_, t\_] := A Sin[kx + \omega t]$

Note the underscore for the variables in the function and the $:=$ defining
sign, both showing only in the definition of the function not below. Then
with this function defined we can write

$D[f[x, t], x]$ to obtain after shift enter the output

$Ak Cos[kx + \omega t]$, that corresponds to $\frac{\partial f}{\partial x}$.

We can even take a second partial derivative by writing:

$D[D[f[x, t], x], x]$ to obtain after shift enter the output

$-Ak^2 Sin[kx + \omega t]$

that corresponds to the so called second partial derivative $\frac{\partial^2 f}{\partial x^2}$

The wave equation contains partial derivatives of second order, meaning that the
differentiation is performed twice as we have seen when talking about acceleration
as defined it in Eq. (5.45). If we use the Greek letter ϕ to denote either the electric
or the magnetic field, then Maxwell's wave equation reads:

$$\frac{\partial^2 \phi}{\partial x^2} = \frac{1}{c^2} \frac{\partial^2 \phi}{\partial t^2}. \tag{5.66}$$

Here c is the velocity of light and of any other electromagnetic wave in vacuum. If
we desire to write the equation in terms of the frequency ν and wavelength λ or in
terms of $\omega = 2\pi\nu$ and $k = \frac{2\pi}{\lambda}$ then we have

$$\frac{\partial^2 \phi}{\partial x^2} = \frac{k^2}{\omega^2} \frac{\partial^2 \phi}{\partial t^2}. \tag{5.67}$$

With a little proficiency in differentiation, or with the help of MATHEMATICA
and the double differentiation as shown in the box (now also for time t), it is easy to
show that a wave as given by Eq. (2.40) fulfills indeed the wave equation. The wave

equation does, however, a lot more than just explaining a simple wave. Maxwell's wave equation can be used to calculate the electromagnetic field as it is emitted from resonance circuits, or as it is emitted from instruments like a cell phone in certain environments. Thus the wave equation is very important for problems in electronics engineering. The wave equation had also a major influence in science, far beyond the science of electromagnetic waves. Einstein noted that all solutions of Maxwell's wave equation propagate in vacuum with the same velocity c no matter where and how they are created. This follows immediately from Eq. (5.66) because c is the only constant of this equation. For example, the light emitted from a fast-moving galaxy or that from a slow-moving galaxy has the same velocity according to Eq. (5.66). This is very surprising. As we have discussed above, if you are on a ship and you throw a ball in the direction in which the ship moves, then someone observing the ship sees the ball moving with the velocity of the ship plus the velocity with which the ball was thrown. Not so with light! If you have a flashlight on the ship, the light out of the flashlight still moves with velocity c, no matter how fast the ship moves. You can see from our section on Einstein's theory of relativity the interesting results that Einstein deduced from this fact.

5.3.2 The Wave Equation and Quantum Mechanics

Quantum mechanics also uses a wave equation. The Austrian physicist Erwin Schrödinger built on the work of Louis de Broglie and Niels Bohr and looked for a wave equation that would describe the standing wave patterns of electrons around atoms and molecules. He started with the wave equation that we discussed in connection with electromagnetic fields given by Eq. (5.67). This equation is valid for all types of waves, moving waves or standing waves. A standing wave can be written as the product of a function that depends on the space coordinate multiplied by a function that just depends on time. Therefore Schrödinger tried the function ϕ:

$$\phi = \psi(x)sin(\omega t). \tag{5.68}$$

Here we are only considering the x-direction as the space coordinate. Schrödinger actually calculated everything in 3 dimensions and considered the function $\psi(x, y, z)$. For us, one dimension (the x-direction) is complicated enough. Schrödinger inserted Eq. (5.68) into Eq. (5.67) and obtained

$$\frac{\partial^2 \psi}{\partial x^2} + k^2 \psi = 0. \tag{5.69}$$

The solution of this equation is indeed a standing wave, meaning that we have

$$\psi(x) = A \sin(kx), \tag{5.70}$$

where A is a constant. This follows from the differentiation shown in the Mathematica box from which we obtain

$$\frac{\partial^2 A \, \sin(kx)}{\partial x^2} = -A \, k^2 \sin(kx). \tag{5.71}$$

Therefore Eq. (5.69) is fulfilled by the function $\psi = A \sin(kx)$. The factor A that gives us the amplitude of the standing wave can only be determined after we agree upon what the meaning of $\psi(x)$ really is. In other words, one needs to know the nature of the wave of that wavicle (e.g., an electron) that we consider. This is discussed in the next section.

MATHEMATICA
We define the Mathematica function
$f[x_-] := A Sin[kx]$ and differentiate the function twice by
$D[D[f[x], x], x]$ and obtain after shift enter the output:
$-Ak^2 Sin[kx]$

At this point, Schrödinger used the work of de Broglie and his wavelength λ_{dB} for electrons. He replaced k of Eq. (5.69) by $k = k_{dB} = \frac{2\pi}{\lambda_{dB}}$. The resulting equation

$$\frac{\partial^2 \psi}{\partial x^2} + k_{dB}^2 \, \psi = 0 \tag{5.72}$$

is now called Schrödinger's wave equation.

Interpretation of Schrödinger's Waves

Schrödinger thought at first of these waves as "matter waves," i.e., some oscillations of some substance related to the nature of the electron and its mass. However, all such attempts of explanation ran into difficulties. The way out of these difficulties was found by Max Born who, in a stroke of genius, suggested that one should regard this ψ-wave just as a wave of the probability of finding the electron. If one does not really know the nature that underlies certain phenomena, then it is natural to take recourse to probabilities. Einstein did not like the idea of bringing probabilities, out of the blue, into the fundamental equations of physics. Of course, we do introduce probabilities for obtaining heads or tails when we throw a coin. In this case, however, we have a reason for it. We say that the coin jumps around and hits bumps of the floor in such an uncontrolled way and that it makes more sense to talk about probabilities than solving all the equations for all the bumps and hits.

Born's idea prevailed in the end, and ψ is now generally related to the probability of finding the wavicle (electron, proton, etc.) in question. The probability is a number between 0 and 1 and therefore a positive number. Because the amplitude $\psi(x)$ of a wave can also be negative, it cannot directly be equal to the probability of finding a wavicle. Born proposed, therefore, that the square of $\psi(x)$ should be equal to that probability. Actually, the situation is even more difficult. It turns out that in order to achieve a generally valid wave equation, it is necessary to admit imaginary and complex (real plus imaginary) numbers for ψ. Because we know that $i = \sqrt{-1}$ is the basic imaginary unit and therefore $i^2 = -1$, we need to use the so-called absolute value of numbers to always obtain positive probabilities. The absolute value is written by using two parallel vertical lines. For example, the absolute value of i^2 is $|i^2| = +1$. $|\psi|^2$ is , therefore, regarded as the probability of finding an electron or any wavicle. We mention this only for completeness and will not go further into this more general type of treatment that is the subject of extensive lectures at universities. Note that imaginary numbers are also used when dealing with electrical circuits and Sect. 5.5.1 may make this use in quantum mechanics more plausible to the reader. Below we deal only with cases where ψ is represented by real numbers. Then we have

$$\text{Probability} = \psi^2(x) = A^2 \sin^2(kx). \tag{5.73}$$

The value of the amplitude A must be properly chosen to assure that the probability of finding an electron is a number precisely between 0 and 1 and that it is certain (probability 1) that the electron is found at any place at all. This can always be accomplished by requiring that the integral of ψ^2 over all space equals 1.

Vibrating Strings and the Solution of the Schrödinger Equation

Solutions of the Schrödinger equation for electrons in atoms are complicated, because the de Broglie wavelength, as given by Eq. (2.75), contains the velocity v of the electron which is not known. Electrons in atoms interact with the positively charged nucleus, and this interaction needs to be included if we wish to obtain the electron energy and velocity. An explicit solution of the Schrödinger equation can still be found for the hydrogen atom even in three dimensions and is usually presented in university-level quantum mechanics courses. For all other atoms and for molecules, one needs to consider not only the attraction of the nucleus but also the repulsion of all the other electrons when calculating electron energies. Then, a solution of the Schrödinger equation is only possible by computer.

A straightforward solution of the Schrödinger equation can be found for electrons that are just confined to a very small volume but do not interact with a nucleus or other electrons. This solution does illustrate the main features of Schrödinger's quantum mechanics. Consider the case of a single electron confined to a very small range $0 \leq x \leq L$. Inside this range $[0, L]$ the electron is permitted to roam freely, but it cannot get out of this interval. Then, inside the interval $[0, L]$, the electron has the total energy $E = \frac{Mv^2}{2}$ and we have

$$E = \frac{Mv^2}{2} = \frac{h^2 M^2 v^2}{2h^2 M} = \frac{h^2}{2M \lambda_{dB}^2}. \tag{5.74}$$

Fig. 5.6 Two energy levels
of an electron confined to a
region of length L
corresponding to the integer
numbers 1, 2. Also shown are
the two wave patterns
corresponding to these two
energy levels. Infinitely many
of such levels with shorter
wavelength follow at higher
energies corresponding to
$n = 3, 4, 5 \ldots$

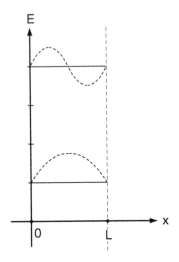

Remember now that $\psi(x)$ describes the standing wave pattern that is given by
$A \sin(k_{dB}x)$. Because we consider only one dimension, such a pattern corresponds
exactly to that of a vibrating string. We have shown one wavelength of a vibrating
string in Fig. 2.13. Actually one can have all types of standing waves within the
interval $0 \le x \le L$ as long as

$$n\lambda_{dB} = 2L, \qquad (5.75)$$

where $n = 1, 2, 3, 4, 5, \ldots$. You can see this by trying to plot a wave in the
interval $0 \le x \le L$, with the requirement that the amplitude of the wave is
0 at the boundaries $0, L$ as well as outside this interval. To fulfill this boundary
condition, Eq. (5.75) must be true. This is illustrated in Fig. 5.6 for $L = \frac{\lambda_{dB}}{2}$ and
$L = \lambda_{dB}$. Using Eq. (5.75) together with Eq. (5.74) one obtains for the possible
electron energies:

$$E = \frac{n^2h^2}{8ML^2}. \qquad (5.76)$$

The lowest energy is, of course, obtained for $n = 1$ and the next higher for
$n = 2$. The corresponding energies are also shown in Fig. 5.6. The assumption
of a wavelength and the solution for a wave that vanishes at the boundaries tells
us therefore that the electrons occupy "states" or "levels" with discrete energies.
This represents the explanation of the discrete energies numbered $1, 2, 3, \ldots$ that
we discussed in the chemistry Sect. 2.4.2 and in Sect. 2.5.3 on the spectra of atoms.
This result follows from the requirement that the wave amplitude vanishes at
the boundaries of the interval $[0, L]$, and this is why Schrödinger chose the title
"Quantization as a boundary value problem" for his celebrated paper on quantum
mechanics.

The probability of finding an electron in the interval $0 \leq x \leq L$ is given by $A^2 \sin^2(\frac{n\pi}{L})$. A is obtained from the condition that the probability of finding the electron anywhere in the interval equals 1. We leave it to the interested reader to find out from the Internet how A is calculated by integration and give only the result, which is $A = \sqrt{\frac{2}{L}}$.

We thus have derived the standing wave model of an electron confined to a short interval of length L with the following important results. The electron can only assume certain discrete energies related to the de Broglie wavelengths that correspond to the standing waves of a string of length L. The energies of the atomic energy levels that we discussed in Sect. 2.5.3 can also be explained in this way. The energy numbers of that section correspond directly to the numbers n of Eq. (5.76). The work of Schrödinger gave thus a beautiful explanation for the discrete energies that had been known from the spectral lines. Note that these results are typical for almost all problems of quantum mechanics, but our results of Eq. (5.76) are not quantitatively valid for atoms, because the attraction of the electron by the proton was not considered in our calculation. As a consequence, the energies of our simplified model are proportional to n^2, while those of the hydrogen atom are proportional to $\frac{1}{n^2}$. There are, however, cases for which our simple model is entirely correct. Nanostructure researchers have created narrow wells, so-called quantum wells, by using modern crystal growth techniques. These wells can actually be fabricated by creating very thin layers of certain materials sandwiched between other materials, and electrons in such wells do indeed exhibit the energies given by Eq. (5.76). Nanoscience and engineering considered and produced a large number of such structures, including artificial atoms (so-called quantum dots) that are not only interesting from the viewpoint of science but also promise new and interesting applications.

More Recent Developments of Quantum Mechanics: Spin and Gauge Fields

Quantum mechanics, as described by the Schrödinger wave equation, has developed considerably further after its "invention" in 1926. Paul Adrien Maurice Dirac generalized Schrödinger's work to include relativity theory. His work elucidated a property of electrons called spin. That property relates, as indicated by the name, to a rotation of the electron wavicle around some axis. It is now known that every energy level can accommodate exactly two electrons with opposite spin. This principle is derived from the "Pauli principle," named after the Austrian physicist Pauli. It supplies the reason why the s-type standing wave pattern that we discussed in Sect. 2.4 can accommodate two electrons. This accommodation of two s-type electrons happens for the helium atom, and helium is, therefore, a noble gas and chemically not reactive. The 3 p-type standing wave patterns can also accommodate two electrons each, and the next heavier noble gas, neon has two 1s, two 2s, and six 2p electrons, 10 in total. This is, according to the Pauli principle, the possible occupation of standing wave patterns with energies 1 and 2. Therefore, neon does neither like to accept nor to donate any electron and is chemically not reactive.

The unit of electron spin is the same as that of an action: eV seconds (eVs). Planck's constant h is also given in eV seconds and the electron spin equals

$$\text{electron spin} = \pm\frac{\hbar}{2}. \tag{5.77}$$

Here, $\hbar = \frac{h}{2\pi}$ is Planck's constant divided by 2π. Remember from the treatment of waves that a factor of 2π often occurs in physical theory. The additional factor of $\frac{1}{2}$ in Eq. (5.77) is important and occurs not only for the spin of the electron but also for the spins of the proton, neutron, and other elementary particles. All of these particles or wavicles with such half numbered spin obey the Pauli principle and are called fermions, named after Enrico Fermi. If electrons would not be fermions, all of chemistry would be different. The Pauli principle is one of the very basic principles that explains the periodic system of elements. The precise energy levels of atoms and molecules are a lot more difficult to calculate, when spin, the Pauli principle, and relativistic effects are included. Very accurate calculations can be performed by use of computers.

The fact that the spin can assume only two values when measured is typical for phenomena related to quantization. These two values are written as positive and negative (the $\pm$ sign in Eq. (5.77)). The choice of positive and negative is arbitrary. We might as well have defined the two spin values as clockwise and counterclockwise indicating the relation to rotation. Physicists often use the expressions spin up and down. Quantum mechanics does not predict which value of the spin will be measured in any given measurement. All that quantum mechanics predicts is again the probability of measuring a positive or negative (or up/down) spin.

Another major advance of quantum mechanics was the realization that electrons do not just "sense" the classical electric and magnetic fields as described by Maxwell and Faraday. Electrons sense another type of field, a so-called gauge field that is usually denoted by the vector **A** and called vector potential. This gauge field is the actual basic field from which the electric and magnetic fields can be calculated. Surprisingly, this theoretically deduced field can influence the path of electrons even at locations at which both the electrical fields as well as the magnetic fields are zero. Experiments, suggested by the work of David Bohm, have confirmed this fact and indicate that we need to revise our notions about the vacuum as space with nothing there (see also next section). Gauge fields need be introduced to explain the finest details of atomic spectra. The theory that includes all these details and describes the interaction of electrons and photons with greatest precision is called quantum electrodynamics. This name indicates the fact that the gauge field can be mathematically described and also visualized as if it would originate from an infinite number of tiny separate oscillating entities. Only on a macroscopic scale, the scale that humans deal with, do all these oscillators then correspond to and result in the effects of the electromagnetic fields of the classical equations of Maxwell.

Quantized fields have, in general, also a spin property. For example, we know that light is described by polarized waves. Polarization means that the actual direction of the electric field plays a role. Many readers are familiar with polarized sun glasses.

Such glasses permit the transmission of light only when the electric field of the light points in certain directions. You can look at the reflection of the sunlight from water and rotate your polarized glasses, and you will see more or less of the reflected light depending on the rotation. The actual direction of the electromagnetic field of light can also rotate by itself and that brings us back to the spin property of fields. If we regard the field as a quantized object, then we can attribute to it a spin, just as we have done it for the electron, proton, and neutron. However, the spins related to the quantized fields of light and other electromagnetic waves do not have the factor $\frac{1}{2}$ as described above for the electron. These spins are described by integer numbers, and all such wavicles with integer spin, for example, the photon, and the gluon are called Bose particles or bosons, named after Satyendra Nath Bose. Bose developed with Einstein a theory that tells us how many such wavicles with integer spin will occupy a given energy level. Bosons do not follow the Pauli principle, and more than two bosons can occupy a given energy level. In fact, bosons follow the rule of thumb "the more the merrier."

5.4   Secrets of Matter and Fields: From Smashing Protons to Astronomy

This section discusses further developments of quantum theory. These most recent developments are taking a fresh look at the concepts of vacuum and matter. Neither vacuum nor matter seems to correspond, in the final analysis, to the obvious explanations: vacuum means nothing is there, and matter is something that fills the vacuum. The modern view is that a lot is going on in the vacuum and a lot is actually there. Matter on the other hand is actually mostly empty containing just a few wavicles buzzing around. To explain these new views, we need to back up and review the standard concepts of vacuum, fields, and matter.

I have mentioned how amazed I was when I saw electromagnets for the first time and when I could take one in each of my hands and feel how strongly they attracted each other. It was a complete puzzle to me that there was this powerful attraction, and nothing could be seen in between the magnets. How could that be? That was magic for all my feelings. There must be something in between the magnets that carries that powerful force! What puzzled me as boy and still puzzles me today was, in fact, the subject of long and intensive scientific debates. Newton and his disciples assumed that there existed some "action or influence at a distance" with no trace of this action in between the objects that attract each other (such as the earth and the moon). This action at a distance was not to the liking of several well-known scientists. As discussed in Sect. 2.2.2, Faraday and Maxwell did not think of the interaction of magnets as caused by action at a distance. They introduced the concept of a "field" in the case of the magnets the magnetic field.

A field is, in principle, nothing but a number or a vector (or even more complicated mathematical entity such as a matrix) that is given at every point of space-time. A vector is needed if the actions of the field do not only depend on the space-time coordinates but also on a direction. Magnetic and electric fields can act in certain directions and are, therefore, mathematically speaking vectors. Faraday made the magnetic field (its strength and direction) visible, by putting iron dust around the magnet as we have shown in Fig. 2.21. The dust arranges itself in lines that show the field direction and accumulates more in the areas where the field is stronger. Since Faraday's work, and particularly due to the additional work of Maxwell, the assumption of action at a distance was made by fewer and fewer scientists and was replaced by the action of a field. The electromagnetic field was shown to propagate with very high speed, the speed of light, and was assumed to act only on the immediate vicinity, but not magically at a distance. The puzzle still remained that the fields that represented such strong forces, appeared to be there without anything else being there. That did not seem to make sense and, therefore, Maxwell and disciples introduced a hypothetical "medium" that would carry the field. They thought of this medium in the way we think about some ordinary material or liquid. For example, one can take a liquid and heat it up at different locations and times. The liquid exhibits therefore a space- and time-dependent temperature and can be described by a "field" of numbers corresponding to these temperatures. The problem is, of course, that the medium that carries the electromagnetic field cannot be detected as easily as some ordinary liquid can. However, Maxwell did think that some "material" was actually there, and this unknown material that carried the electric and magnetic fields was called the "ether" or "aether."

5.4.1 Modern Views of Vacuum and Matter

What was this ether made out of? No one knew, and all the attempted explanations of what it was ran into some kind of problems. Because of these problems, Einstein and others wanted to do away with the ether altogether and wished to find a way to explain things without making any assumptions about any "ether" that was the carrier of the phenomena that we observe. Einstein just talked about the vacuum. In his later years, however, Einstein did get away from the idea that one would just need to use the concept of "vacuum," meaning that there was nothing there. In recent times, scientists have endowed the vacuum with more and more properties that they needed to explain their experiments. There is really a lot going on in this vacuum that carries a large number of fields and, in addition, so-called quantum fluctuations as well as dark energy. Dark energy is related to the expansion of the universe as discussed in the next section. Dark matter is something different than dark energy and resides also in the vacuum. Dark matter derives its name from the fact that it neither emits nor absorbs light or any other electromagnetic radiation. It can be detected, however, from its gravitational effects that are clearly observed by the instruments of astronomy. What is this "vacuum" that is endowed with such mysterious properties? What is it that appears like nothingness and can carry enormous forces in the form of a variety of fields including magnetic, electric, gravitational, and the very strong fields that act in the nuclei of atoms? There are currently two theories in development that attempt to explain all of this. One is "quantum chromodynamics," and the other is "string theory."

Quantum chromodynamics is a theory similar to Maxwell's theory and its modern extension, quantum electrodynamics, that was briefly described in Sect. 5.3. However, quantum chromodynamics accounts for the much stronger forces that are important in the nuclei of the atoms. The main difference to quantum electrodynamics is that quantum chromodynamics uses different symmetry laws when dealing with the wave functions and the quantum fields that are involved. We have learned about symmetry on several occasions in this book. For example, a symmetry valid for numbers is the commutative law of addition discussed in Sect. 1.1. You can add numbers in any sequence you wish, and the outcome is the same. Such a rule or law that permits changes in a procedure (such as the change in the sequence of addition) but does not change anything for the outcome is called a symmetry or a symmetry law. Rotations have a different symmetry than numbers as we also have learned in Sect. 1.1, and such different symmetries play a big role in quantum chromodynamics. Considerations of laws of symmetry are involved when deriving the equations for the fields of quantum chromodynamics and the wavicles of this theory, the quarks and gluons that are discussed below.

String theory derives its name from using mathematics similar to that of vibrating strings in order to derive a theory of all existing wavicles. We have discussed in Sects. 2.1.5 and 5.3 the importance of standing waves and vibrating strings for quantum mechanics. In spite of this nice name and the string analogies that are used, string theory is mathematically very involved and exists in a variety of forms, one of them is the so-called M-theory. In M-theory, the vacuum is a space of 11 dimensions! As far as I understand it, we currently do not have any explanation of both vacuum and matter by M-theory that can be presented in terms of understandable analogies to a student in high school. You can see this void of analogies and understandable explanations by searching the Internet. The deeper one digs into the scientific descriptions of matter and the vacuum, the more "Latin" one finds that can only be understood by the "high priests" of string theory. We therefore do not attempt here to explain string theory any further and instead add a few paragraphs about the amazing facts and results that have been found experimentally and that are reasonably well understood by quantum theory as we have discussed it.

5.4.2 Smashing Protons: Quarks and Gluons

Up to this point, we have learned about several types of particles or wavicles. These were electrons, protons, neutrons, and photons. The first three have a so-called rest mass. This means that these particles have, even if they would stand still, a measurable mass, meaning that they resist acceleration and are attracted by gravitational forces. Photons always move with the velocity of light and are attributed zero rest mass. As far as I understand it, no one has ever seen any of these elementary particles taking a rest, and the whole concept of rest mass may be up for some rethinking. For example, scientists, working with the particle smashers that are described next, are searching for the so-called Higgs particle that may be closely related to a deeper understanding of how wavicles become massive. Key to this understanding are experiments that smash protons and neutrons against each other. These experiments have made it possible to unravel the "inner workings" and "composition" of neutrons and protons. The result of these investigations was the recognition that protons and neutrons can be further divided into even more elementary particles: quarks and gluons.

It was known early on, from experiments related to radioactive decay, that free neutrons could and would turn, typically after 15 min, into an electron, a proton, and another particle, a so-called antineutrino. Thus the neutron could split into pieces and was, therefore, not elementary in the sense that it could not be divided. The question whether protons can be divided into smaller pieces has stimulated a lot of basic interest that was very much enhanced by the interest in nuclear energy and weapons. Particle accelerators were built that could accelerate protons to enormous energies and to velocities very close to the speed of light. Large accelerators exist, such as the Fermi accelerator close to Chicago and more recently the CERN accelerator in Switzerland that is financed by many countries from all over the world. This accelerator, called the Large Hadron Collider (LHC), uses electromagnetic fields to accelerate protons and nuclei of lead atoms. The word hadron designates nuclear particles that are much heavier than electrons. Beams of particles move in large circles (27 km or 17 miles in circumference), and they gain higher and higher energy with each circular motion by the use of electromagnetic fields for their acceleration. The goal of the LHC is to energize the particles to many tera-electron-volts (1 TeV = 10^{12} eV). At these energies, the mass of the protons must be increased by more than 1,000 times according to Eq. (5.29), as can be deduced from that equation as a nice homework problem. A special twist of the CERN experiments is that two beams of protons or lead atoms are set in motion in opposite directions and at some point, when they are both at the highest possible energy, brought to collide against each other. The debris of these collisions is monitored by elaborate large 3-dimensional detection machines.

Feynman compared this experiment to smashing two clocks against each other and then deducing from the debris how the clocks were engineered. He made an important point with this analogy. It is very difficult to deduce the inner workings of protons and nuclei from the debris of such collisions. A definite result was obtained, however, after many years of investigation. To understand the enormity of this result, let's back up and recapitulate a few facts known from many other experiments. Atoms consist of a nucleus (protons and neutrons) and electrons buzzing around it in the standing wave patterns of Sect. 2.5.2. The size of the hydrogen atom (meaning the average "buzzing" distance of the electron from the proton) is about 0.1 nm. The size of the nucleus, in the case of hydrogen the proton, is much smaller than that. We do not wish to give here a definite number because these sizes depend on the actual way one tries to measure them. Rutherford bombarded gold foils with small nuclei that he obtained from radioactive decay. He found that these small nuclei were passing almost freely through the gold foil and hinted toward very few collisions with hard cores, the gold nuclei. From such experiments and the frequency of collisions he deduced that there were indeed gold nuclei (the hard cores) and that they were about 100,000 times smaller than the gold atoms. The great discovery of the big accelerator experiments is that the protons are not hard cores as suggested by Rutherford but again appear, at least in these collision of extremely high energy, as mostly transparent except for a few hard cores that are still much smaller than the protons! These hard cores are now, after much research, associated with new elementary particles that have been given the name quarks by their discoverer Murray Gell–Mann.

Quarks have never been observed in separation from a proton. In fact it is generally believed that they cannot be separated. They are held together in the protons by so-called gluons, the quantum particles of the strong nuclear forces. The existence of quarks and gluons is altogether deduced from complex evaluations of the smashing experiments that are based on symmetry considerations as mentioned above. The fact that complicated symmetry laws are invoked to deduce the existence of quarks gives some justification to their strange name that Gell-Mann gave them in memory of a poem that crossed his mind. There is really no analog in our experience that would permit us to give the quarks some other name that would supply us with some hints what quarks are about. That is a bit different, of course, with the gluons that derive their name from holding things together within the protons. Quarks also carry either a charge of $+\frac{2}{3}e$ or $-\frac{1}{3}e$, where e is the absolute value of the charge of the electron, also called elementary charge. The proton contains two quarks with charge $+\frac{2}{3}e$ and one with charge $-\frac{1}{3}e$ and carries therefore a positive charge of $+\frac{4}{3}e - \frac{1}{3}e = e$.

5.4.3 The Music of the Grid

The smashing experiments have also told us much about still another basic particle, the neutrino. Neutrinos are very elusive and do not interact much with atoms. Billions and billions of neutrinos pass through our bodies in every second. They do not interact with us, in spite of having large energies. Their existence is mainly deduced from conservation laws such as the conservation of energy: some energy is missing in an experiment; thus there must be additional parts in the smashed "clock." This may sound a bit unreliable. However, after many years of research and reasoning, there is no doubt that such neutrinos exist. Another well-established fact, deduced from the smashing experiments, is that each particle has its antiparticle. When particles and antiparticles collide they destroy each other leaving their energy in the form of radiation. An electron and its antiparticle, the positively charged positron, can "destroy" each other in a collision resulting in γ-rays. As we know from Sect. 4.3 these rays have practical applications in PET scans. Thus, the work on these particles is not entirely abstract, and scientists become increasingly familiar with them as time goes on. It is, however, without any doubt mind-boggling that what we see with our eyes is not hard and massive, not static and standing still, but is subject to the high-speed motion (or "buzzing") of ever smaller wavicles, electrons around protons, quarks within the protons, and everything held together by photons and gluons, respectively. Are we just holographic ghosts in some kind of fabric called the vacuum?

Frank Wilczek gave us more to think about in his book *The Lightness of Being*. He compared the Planck–Einstein equation

$$E = h\nu, \tag{5.78}$$

with Einstein's equation for the mass M of any object with a total energy E:

$$M = \frac{E}{c^2}. \tag{5.79}$$

This equation was given previously in Eq. (2.16) and discussed in Sect. 5.1. These two equations suggest a connection of mass and frequency and therefore of mass and waves. The frequency that we obtain from the two equations is

$$\nu = \frac{Mc^2}{h}. \tag{5.80}$$

The question is with what type of analogy can we understand "wavicles" with such frequencies. We know that the vibration of the body of instruments causes their sustained tones. For example, the strings of violins form standing waves and communicate their vibration to the wooden violin body that also vibrates then in a very complicated three-dimensional pattern. All of these vibrations, when they persist, can be mathematically described as standing wave patterns. As far as I understand it, this analogy is a very complete one, and we can think of the electrons of an atom as forming such standing wave patterns and even go further and see the inner machinery of a proton or neutron also in terms of such standing wave patterns formed by the quarks and gluons. Equation (5.80) even lets us assign a frequency ν and a de Broglie wavelength λ to single electrons. For an electron that frequency would be $1.24 \cdot 10^{20} s^{-1}$ which means that we would have more than 10^{20} (a number with 20 zeros) vibrations per second. What is it that vibrates, and what is the medium in which it vibrates? No one really knows, and we can only hope that we will come closer to this secret of nature and find better analogies as we go on with basic research. For now we just imagine particles, wavicles, that move in what we call the vacuum. Frank Wilczek called it the "grid" and associated the frequency of Eq. (5.80) with the "music" of the grid.

Pause here a second and consider what we have discussed. The matter that we know, and that appears to us as tangible as a piece of gold does consist mostly of empty space with electrons swirling around in it. There are gold nuclei made of protons and neutrons that in turn are mostly empty space with quarks swirling around in them. And everything is held together by forces including gravity, electric fields, and gluons that make up—or are at least somehow existing in—Wilczek's grid or Maxwell's ether. All of these facts have been found by scientists "looking" at smaller and smaller sizes as well as larger and larger energies.

Scientific research is often most effective and fruitful if it can be pushed in certain directions further and further with virtually no bounds. We have just discussed the push that scientists have made and are making toward smaller and smaller entities, toward atoms, protons, quarks, and electrons. We have also seen the great results of science and engineering that followed temperature to ever lower degrees by refrigeration or to ever higher degrees as necessary for the fusion processes in the sun. One can also push the scientific investigation to larger and larger objects and objects that are far away. This brings us to astronomical science that is described next.

5.4.4 Telescopes Exploring the Universe

Astronomy is one of the oldest forms of science and has been developed during different ages by several different cultures, including Egyptians, Greeks, and Aztecs. Astronomy provided the rulers of these cultures with certain powers, because they could predict certain occurrences in the night sky, for example, how the moon and planets would move and appear in various shapes and forms, and even when a comet might show up. This knowledge also gave them information that was vital to navigate ships and find their way on travels over the oceans. Early astronomy was also often connected to superstitions that we now describe by the name "astrology" and that attempt to predict our fate by the configuration of stars and planets. Astrology has, of course, nothing to do with science. Most stars of the universe are extremely far away, too far to have any influence on us. Many stars may not even exist anymore, they may have exploded (see novae and supernovae on the Internet), and we see only the light that they have radiated sometime in the past. The actual influences that stars can possibly have on humans are limited to the effects of their radiation and their gravitational pull. Very significant radiation is only received from the sun. The reason is that the intensity of the radiation from stars (that includes a wide spectrum from radio waves to light and to γ radiation) decreases with the square of the distance from us. This law is called the "inverse square law." The same is true for gravitational forces.

The stars that we see with our eyes, and with smaller telescopes such as those of Galileo and the first modern astronomers, must be relatively close. These are the stars of the so-called milky way, and no other stars were known when Einstein was born and even when he developed his theory of relativity. We know now that the milky way is only one of billions and billions of so-called galaxies. These, and other important facts about the existence and possible origins of all matter in the universe, have been discovered by use of the most powerful modern telescopes. The most recent telescopes have access to large portions of the universe and monitor planets, stars, and clusters of hundreds of billions of stars, the galaxies.

Modern telescopes are very different from Galileo's and use slightly curved polished mirrors instead of lenses. The larger the mirrors, the more photons (light) can be collected and the more distant objects can be observed and investigated.

The light of the stars of the night sky, which one currently can observe with the best telescopes, may have, and indeed has for most cases of observation, travelled a long way and for long periods of time, years, millions of years, and even billions of years. This is a very amazing fact. If light can travel billions of years, then the space that it has passed must be clearer than glass or crystals. Light travels $3 \cdot 10^8$ m or $3 \cdot 10^5$ km or about $1.9 \cdot 10^5$ miles in a second. In Sect. 5.1.1, we called that distance a "bigmeter." Light traveling for one year travels then about 32 million bigmeters, and light traveling for a billion years travels 32 times 10^{15} bigmeters, an unimaginable distance! Thus space must be almost totally transparent, at least for the parts that let us see very distant stars. Compare that with very clear water in the ocean. Maybe you can see to about 100 m deep, if the water is extremely clear. An optical fiber that is used for Internet communications, is clearer than the purest water and transmits light over many miles, but not over one bigmeter!

Thus, when we observe distant stars, we see only the light that these stars have emitted in the past. As mentioned, the star itself may have exploded a long time ago and may not even exist anymore. Its light, however, is still propagating through space and arriving at earth. Even the light of our moon takes about a second to reach us, and the light of our sun takes about 8 min to arrive here. We do not know where the distant stars actually may be "now", because we see only light that was sent out millions or billions of years ago; meanwhile the stars have moved to different places in space-time. The actual movement of stars that we see every day, the rising of stars in the east and the setting in the west, is not any indication of the movement of the stars themselves but of the rotation and motion of the earth. Most of the stars are so far away that their motion can only be detected during long times of observation or not at all. If a star is a billion light-years away, its position, as seen from earth, does barely change in a year, even if it moves with a velocity of a hundredthousand miles or kilometers per hour. These facts are really mind-boggling and deserve to be thought through in detail.

The construction of the modern telescopes with their large mirrors, currently as large as 10 m in diameter and 30 m in the planning stage, represents a very difficult engineering problem. The mirrors need to be extremely precisely shaped in order to produce a clear image. The curvature of the mirrors needs to be exact within the wavelength of the observed light, which requires exactness on a micrometer scale. It is a big technological problem to form the curvature of a mirror of 10 m diameter with a precision that is better than the width of a hair. You can imagine that gravity will distort big mirrors easily by one micrometer (one millions of a meter), simply because of the mirror's weight. Therefore very special engineering designs need to be made to avoid all the possible distortions and corresponding problems. Many of the big existing mirrors are, therefore, composed of smaller mirrors that are arranged together and can be moved mechanically to get into perfect position. Figure 5.7 shows a sketch of one of the twin Keck telescopes, located at the top of

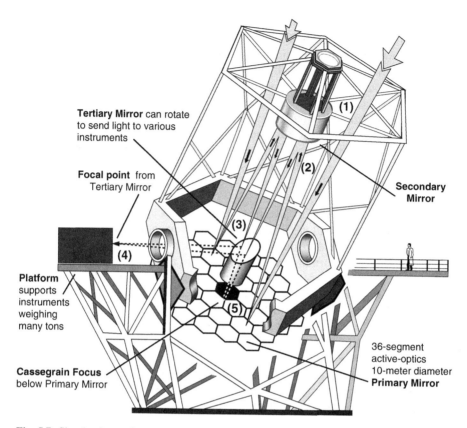

Fig. 5.7 Sketch of one of the twin Keck mirror telescopes on Mauna Kea (Information courtesy Keck Observatory)

Mauna Kea. The Keck telescopes are the most powerful telescopes existing on earth as these lines are written. We, therefore, give a more detailed description of how they (and all modern telescopes) work.

The main (primary) mirror is labeled (5) in the illustration and contains 36 hexagon-shaped segments. The light from the stars and galaxies, indicated by the arrows and the corresponding light beams in the illustration (e.g., with label (1)), is collected by this 10 meter wide system of mirrors and then reflected and focused onto a secondary much smaller mirror (label (2) in the figure). The light path, subsequent to the secondary mirror, can be channeled in two possible ways.

The first possible path of the light is toward a focus that is behind the dark opening in the middle of the primary mirror. This is called the Cassegrain focus. At this focus, the image of the observed planet, star, or galaxy is formed. That image is usually received and processed by electronic cameras that are connected to computers. The name Cassegrain refers to a mirror design by Laurent Cassegrain that features a parabolically shaped (convex) large primary mirror and hyperbolically shaped (concave) secondary mirror. Consult the Internet for detailed explanations and the definitions of convex and concave. The electronic cameras of the Cassegrain focus form images by using (usually) very long exposure times. Long exposure is necessary to collect as much light as possible. To work with long exposure, one needs to compensate for the rotation of the earth by turning and pointing the telescope always toward the observed star or galaxy. This complicated process of long exposure and turning of the telescope leads to the beautiful images of galaxies, stars, and planets that everyone has seen in news around the world, and that you can find on the internet. If you would observe the stars at the Cassegrain focus with your eye, the image would be much less impressive and actually disappointing. The times of direct eye observation are long gone. Cameras, computers, and printers form and process the images.

The second possible way to investigate the star light is to use a tertiary mirror (3) that can rotate and send the light to various instruments located on a platform. These instruments can analyze the light, for example, by use of gratings as discussed in Sect. 2.5. Instead of forming nice images of stars and galaxies, the instruments determine the finest details of the starlight spectrum. These spectral details, in turn, provide much information on the atomic composition of the universe as we know from Sect. 2.5 and as is discussed below.

It is important to remember that modern telescopes have these two essential capabilities (to form images and analyze the spectrum of the star light), and they have revealed in this way much information about the universe. As mentioned, only a little more than half a century ago, the universe was thought to be in essence our galaxy, the milky way; the collection of several hundred billion stars that also contains our sun. The naked eye and smaller telescopes cannot even detect a fraction of the stars of the milky way, not to mention the stars of other galaxies. No planets other than those of our solar system were known at the time this author was in high school. We know now that the milky way is only one galaxy out of possibly a trillion or more galaxies and that fuzzy objects, called "nebulae" in the past, are other galaxies. Most well-known galaxies have the spiral form that the milky way also has. We know now that stars other than our sun also have planets, and many planets have been observed and even directly photographed by the large telescopes. These planets are called exoplanets.

Photographs of the universe that are taken by these telescopes can be found on the Internet, and you can study the location of planets and stars of the night sky on your computer or even on your cell phone by using certain applications. Studies that make use of Internet applications can be topics of very interesting STEM astronomy class projects.

Astronomers observing the enormous extent of our universe have naturally pondered the question: where does all of this come from? Part of modern astronomy is based on the hypothesis of a big bang, an almost instant expansion of the universe from smallest size to enormous extensions. The initial expansion is commonly assumed to be that of a homogeneous entity of incredible energy and speed. Descriptions of this initial expansion do invoke assumptions that cannot be proven by science in any direct fashion. Naturally, we can neither establish that, at the event of the big bang (if it occurred), currently known laws of physics were valid, nor that they were invalid. We do know, however, that the universe is still expanding today. This is deduced from the well-established observation that the stars, which are farthest away from us, appear to move away with the highest velocity (see redshift of spectra below). The following analogy makes this effect plausible. Assume that we would be two-dimensional beings living on the spherical surface of a balloon-like structure. If the balloon is blown up, then all the points on the surface move away from each other and the further away a point is, the faster it moves away. Scientists believe that this is a very good analogy to the expansion of the universe that is observed by the redshift of spectra in all directions that we can look. However, when we discuss such models, we have to be aware that we are leaving here science at its most solid and proven, because we cannot recreate anything close to a big bang in any experiment. When we look at the night sky, we are just inspecting the past and receive some clues of what might have been going on in the past, not the whole story.

As emphasized, the telescopes do not only provide us with beautiful images of planets, stars, and galaxies. They also provide us with information about the chemical composition of the stars. This is accomplished by measuring the spectra of the starlight. Investigating the spectra means that the frequency of the light is measured. We know from Sect. 2.5 that atoms emit and absorb light in various ranges of frequency, so-called spectral lines, that are characteristic for them. The telescopes detect and measure these spectral lines and receive, therefore, information about the material compositions of the universe. These findings are very interesting and tell us, among other things, that the universe appears to follow the same or very similar laws everywhere. We see everywhere the same atoms ranging from hydrogen to helium and heavier atoms. We see everywhere the same effects of gravity, planets orbiting suns, suns orbiting each other, and galaxies moving in spirals around a center.

Other important clues about the universe can be derived from the finest details of the spectra of the stars. We know from Sect. 2.1.5 about the Doppler effect. Doppler has taught us that the frequencies of waves are lowered when we move away from the sources of the wave and increased if we move toward the sources. The higher the speed of movement, the larger is this Doppler shift. For the visible light, such a frequency lowering is observed as a shift toward the red, while a frequency increase means a shift toward the blue. This is indeed what the telescopes detect.

The measurements of the Doppler red or blueshift can be done extremely accurately. One can measure how fast a star moves away from us or toward us with ease. Beyond this, one can measure much finer details of the movement of the stars. An important example is the following. We know that stars attract planets and the planets orbit in ellipses around the stars. The planets, in turn, also attract the stars, and they also cause the stars to move a little, not very much, because the stars are usually very massive. However, even this small movement can be measured by the Doppler effect, and one can gain in this way a lot of information about the orbit and mass of exoplanets that surround stars outside our solar system.

Another important aspect of the spectral lines of stars has already been pointed to above. The more distant a star is, the larger is the redshift of its spectral lines. The most distant stars have the highest redshift and thus appear to move away with the highest speed. This fact is consistent with an expansion of the whole universe. Indeed such expansion is thought to be the cause of this type of redshift characteristic for all distant stars. Recent research has even shown that the expansion of the universe is accelerating.

Big telescopes do not only tell us about the matter of the universe that emits and absorbs light. One also can detect massive objects that do not emit light. Black holes (see Sect. 5.1.2) are in the center of all or most galaxies and may have a mass of millions of suns. They attract close-by stars that then orbit around them with very high speed. Using the methods of Newton and our calculations of falling stones and planets, one can estimate the mass of the black holes from the time that it takes these stars (that are seen, e.g., at the center of the milky way) to complete their orbit. Accumulations of so-called dark matter, a previously unknown type of matter that does not emit or absorb light, have also been detected by their gravitational effects, although we do not know much else about the nature of dark matter.

You may ask, what benefits do we get from all of this knowledge? The increase of our knowledge is a very important benefit by itself. However, there is always also a necessary increase in engineering skills to achieve such goals, and the increased engineering skills have, when used in positive ways, a great positive feedback effect on our lives. As these lines are written, bigger telescopes with 30 m mirrors are in design. These telescopes will penetrate further into the distant universe and will give us more information about other planets, stars, and galaxies. The bigger telescopes are not only increasing our knowledge of what is but also contribute to our capabilities in technology and engineering.

Take as an example the "twinkling" of stars that is known from the song "twinkle, twinkle little star." This twinkling arises because of fluctuations in the atmosphere. The air of the atmosphere above us moves due to strong winds, and the temperature of this air fluctuates. These fluctuations lead to fluctuations of the transparency of the atmosphere (to fluctuations of its optical properties) and therefore to fluctuations of the starlight detected by the telescopes. This effect is very detrimental for astronomy, because the fluctuations blur the images of the stars. Instead of a beautiful spiral-formed galaxy you might see or photograph just a blurred blob, even if you have a great telescope available. Modern technology gets around this problem in two ways and is therefore able to produce the most stunning images of the universe that you can see on Internet sites. One way is very expensive: send your telescope into space carried by a satellite. This has actually been done with the Hubble telescope, and other telescopes are in the planing. There is also an elegant technological way to remove the twinkle of the stars for telescopes here on earth. The earth is surrounded, at about 65 miles height, by a thin layer that contains sodium atoms. If a laser is pointed toward the sky just next to the telescope, then the laser hits the sodium layer and lights it up creating a bright yellow spot. Remember that sodium emits yellow light, for example, when you throw salt into a flame. That yellow dot excited by the laser simulates an artificial star, and it twinkles exactly like the real stars do. This twinkling can now be used to calculate (by computer) how the earth atmosphere fluctuates and that information can be used to undo (again by computer) the twinkling of the image that is taken with the camera. In other words, the whole optical system of the telescope is "adapting" to the atmospheric fluctuation and removes the twinkling of the stars. This is what one calls adaptive optics. That type of optics has, of course, also other applications, and adaptive optics represents a nice example of how science and engineering are interwoven and interactive.

5.5 Examples of Advanced Engineering Problems

We have discussed the progress that was made from the first invention of Edison's phonograph to the DVD and beyond and have shown how science and engineering are intertwined and enable each other. We have seen this symbiosis of science and engineering over and over, from our discussions of the jet engine to astronomy with adaptive optics. In all of these cases, the science of the problem was mostly understood and was then only slightly expanded by the research and development of the project, while big progress was made on the side of engineering and technology. Here, we discuss two examples. The first relates to the solution of more elaborate electrical circuit equations. This section should give you an idea how well understood the area of electrical circuits is and that all the necessary mathematics involved is related to the solution of linear equations. The second example is of different nature and deals with excessive heat generation by electrical currents that presents problems in many applications ranging from computer chips to the power lines of bigger cities. In this problem area, engineering necessities hint toward the need of major progress in science and technology. Without that science and technology progress, no progress can be made for the engineering related to heat generation that prevents us from increasing the capabilities of our computer chips and wastes energy by heating the wires of power lines. Solutions to these problems require new materials that are superconducting at room temperature.

5.5.1 Electrical Circuits

The laws and definitions that we discussed in Sect. 2.2.2 are sufficient to deal with arbitrarily complex circuits. These laws are valid for electrical currents that do not change with time, as the so-called ac currents do that are discussed further below.

We discuss first a circuit that is elementary and typical for devices that one can find in any household. In Fig. 2.18, we have a voltage that is equal to the product of current I and resistance R which is IR. As mentioned in connection with this figure (Sect. 2.2.2), one can generalize this rule for any closed circuit with any number of galvanic elements (batteries) with voltages V_m where $m = 1, 2, \ldots, M$, and any

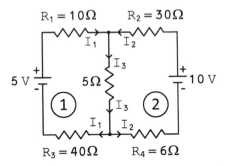

Fig. 5.8 Circuit with five resistors and two batteries. The currents are chosen such that the node equations are fulfilled automatically. The directions of the current are chosen arbitrarily, meaning that if the actual current flows in the opposite direction, the calculation will yield a negative sign. The numbers (1, 2) shown in *circles* just enumerate the left-side and right-side loops, respectively

number of resistors R_n and currents I_n where $n = 1, 2, \ldots, N$. M and N can be any integer number, for example, $M = 3$ and $N = 10$. We thus have

$$\sum_{m=1}^{M} V_m = \sum_{n=1}^{N} R_n I_n. \tag{5.81}$$

Of course, currents and voltages that are oriented in different directions must have different signs to validate this loop equation.

A circuit with five resistors and two galvanic elements is shown in Fig. 5.8. To calculate the currents we use first Eq. (2.53), the equation for currents at nodes, to obtain

$$I_3 = I_1 + I_2. \tag{5.82}$$

From Eq. (5.81) we obtain for the left loop:

$$5 = 10 I_1 + 5 I_3 + 40 I_1 \tag{5.83}$$

and for the second loop:

$$10 = 30 I_2 + 5 I_3 + 6 I_2. \tag{5.84}$$

These are three equations with three unknowns that we can solve by hand as described in Sect. 1.2 or by using MATHEMATICA. We first rewrite the equations to bring them into the appropriate form:

$$I_1 + I_2 \quad - I_3 = 0,$$
$$50 I_1 + 5 I_3 = 5,$$
$$36 I_2 + 5 I_3 = 10. \tag{5.85}$$

These equations can be solved as shown in the MATHEMATICA box.

> MATHEMATICA
> $m = \{\{1, 1, -1\}, \{50, 0, 5\}, \{0, 36, 5\}$
> $v = \{0, 5, 10\}$
> LinearSolve[m, v] and shift-enter gives the exact solution:
> $I_1 = \frac{31}{446}$, $I_2 = \frac{105}{446}$ and $I_3 = \frac{68}{223}$ Amperes.

The solution of the above circuit problem is thus, in its essence, reduced to the problem of solving linear equations. We know that linear equations can be solved efficiently by computers, and one can solve practically any number of them. Our best computers will have no problem solving the circuit equations of a whole city or even the whole country.

A problem that we have not discussed yet is that the power supplies of households, cities, and states work with ac currents and voltages of around 60 Hz frequency and not with dc currents and voltages. This choice of ac instead of dc is a very interesting topic by itself and was discussed (and fought about) by two great engineers: Thomas A. Edison and Nikola Tesla. Tesla won the fight, and our homes are supplied with ac electricity. Current I and voltage V of our power lines have therefore a time dependence. This time dependence follows a sin- or cos- law, just as we know it from the time dependence of waves. One typically deals then with equations such as

$$V = V_0 \sin(\omega t) \tag{5.86}$$

for the voltage and

$$I = I_0 \sin(\omega t - \phi) \tag{5.87}$$

for the current. Here V_0 and I_0 are the amplitudes that determine the largest and smallest values of voltage and current, respectively. ω equals $2\pi \nu$ where ν is the frequency of 60 Hz in the USA and 50 Hz in Europe. It is important to note the so-called phase ϕ that is included in the equation for the current. It is this phase that makes all the difference in the treatment of dc and ac currents. Mathematically such a phase simply means that the wave of the current may lag behind that of the voltage or vice versa.

Remember, for example, a capacitor. If we connect a capacitor to a voltage source, current will flow as charge accumulates on the capacitor plates. The voltage across the capacitor will increase until it charges up to the voltage of the voltage source, then the current will stop flowing. During this charging process, the current "leads" the voltage, and this can be accounted for by the phase angle ϕ. Circuits can, in general, have resistors with a resistance R, capacitors with a capacitance C, and also inductors with an inductance L. All these circuit elements determine the voltages, currents, and phases. It turns out that there is a simple way to calculate these ac quantities. Simple, of course, only after one is used to dealing with it. What one needs to do is to formulate the same circuit equations as we have done for the dc case above. However, now we have to admit also imaginary numbers in the equations. We treat capacitors exactly like resistors but with an imaginary resistance of $\frac{1}{i\omega C}$ where $i = \sqrt{-1}$ is the imaginary unit. Inductors are also treated like resistors but now with a resistance of $i\omega L$. This treatment gives us the amplitudes V_0 and I_0 in a form that involves imaginary numbers from which the real amplitudes and the phase ϕ can be easily obtained. Remember that advanced quantum mechanics as discussed above also deals with imaginary numbers. The reason is that a phase like ϕ is also important for wavicles.

The solution of such a problem goes already beyond what one reasonably can teach in a regular high school class. This is the type of problem that a college student or an electrical engineer should understand in detail. However, one may be able to teach the solution of such problems within a special project, like a project for those who think they are gifted in electrical engineering, or a project for the household as discussed in Sect. 4.2. There exists, of course, software for electrical engineers that solves all kinds of circuit problems including the case of currents and voltages. A very well-known software of this type is SPICE, and the interested readers can look up details on the Internet.

5.5.2 Heat Generation in Computer Chips

There are many advanced engineering projects, meaning that they involve complicated engineering approaches, such as the circuit equations that we have just discussed. These projects need to be dealt with by experts who have studied engineering and technology topics in detail. There also exist important engineering problems that cannot be solved by what we currently know and that require new developments in science. One such problem is the heat generation of computer chips. Whenever we wish to process information fast, we need computers. Whenever we handle a lot of information, as is the case of streaming videos, a lot of heat is generated. In fact, the heat that is generated by hundreds of millions of transistors (as they switch) is too large to be easily disposed of. One can cool the chips with fans, and one can even cool them by use of liquids. Nevertheless, the heat generation of chips presents a limit to how many transistors can be put on a chip and how fast they can switch.

Science fiction authors often predict that we will have tiny chips everywhere, in our clothing, implanted in our skin and our sensors (eyes, nose), and also to support our brain function. One only needs to watch movies or TV series to get inspired about what is thinkable and possible, at least in principle. However, if we look and see what actually is developing, then we notice that powerful chips require a lot of battery power and generate a lot of heat. You can feel the heat generated by your laptop when you process a lot of high-definition photos. The heat that is generated that way is very difficult to overcome or avoid. The reason is that heat generation follows from a very basic rule of physics. This rule states that any electron that is accelerated by some force in free space will radiate electromagnetic waves. A variation of this rule, equally valid and important, is that an electron that is accelerated in a crystal (or any other solid or liquid) will excite vibrations of the atoms and thus loose energy. Vibrations of the atoms of a solid represent heat, as we have learned in Sect. 2.3. Electrons are certainly accelerated when electrical currents are generated by the applied voltages that switch the transistors of chips. Therefore these electrons create vibrations and thus heat. The faster the transistors switch, and the more transistors we have, the more heat is generated. What is the solution to this problem? Does nature do it better? Why does our brain not heat up as much as chips, and we still can do things fast? The answer is that our brain can work on several things in parallel and while the transistors of the brain, the neurons, switch rather slowly, the parallel brain architecture helps. One can, of course, do the same with computer chip technology and process arithmetic and other commands in parallel. This is currently done in so-called parallel "cores" that form a central processing unit. How far this technology can be pushed is still a question, because it still creates heat and one still wishes to have ultrafast switching and more transistors. Can one ever get around this limitation of chip technology? No one knows. Attempts are being made by using nanotechnology and different principles (architectures) to produce computers. These new principles try to imitate nature and the parallelism of our brain, or even try to use entirely new mechanisms provided by our knowledge of quantum mechanics. We will hear about one such possibility based on "superconductivity" as discussed below.

As mentioned, the heat problem is not unique to chip technology. Light bulbs with glowing wires generate a lot of heat and you burn your fingers if you touch them. Light bulbs are therefore very inefficient and need to be replaced by something better, by some devices that do not generate as much heat such as light-emitting diodes. This is not a simple problem, however, because of the currently higher cost of such diodes and because we are used to the warm light of glowing wire bulbs. There exists a good chance, however, that this problem will be solved with the existing knowledge about light-emitting diodes.

5.5.3 Heat Generation in Power Lines

Another difficult problem that may require new science for its solution is the heat generation of power lines. The heat energy H_{heat} that is generated by an electric current I in a wire having a resistance R and being used during a time period t is given by

$$H_{heat} = I^2 R t. \tag{5.88}$$

The power lines that supply your city present a big problem with heat generation, because that heat energy is lost before the electricity can be used in the city. This is a very important factor when energy is supplied by electricity over large distances. We therefore explain what is involved here in more detail. What we wish to have is the smallest possible heat generation in the wires of the power lines that run to the city. Therefore we wish the heat generation H_{heat} of Eq. (5.88) to be as small as possible while the power lines still need to supply the required energy H_{city} to the city. This energy is given by the following equation:

$$H_{city} = IVt, \qquad (5.89)$$

which simply represents the power (current I multiplied by voltage V that is usually measured in watts), multiplied by the time period of use. V is here the voltage of the long-distance power line, which may differ from the voltage in your home (110 V in the USA). The above equation is equivalent to

$$It = \frac{H_{city}}{V}. \qquad (5.90)$$

Inserting Eq. (5.90) into Eq. (5.88), we obtain

$$H_{heat} = \frac{IRH_{city}}{V}. \qquad (5.91)$$

As mentioned, H_{city}, the energy that the city needs, is given. Therefore $H_{city} = IVt$ must be unchanged by any choices of voltage V, current I, or resistance R of the long-distance power lines. Therefore, in order to minimize the heat generation with given power supply, we need to minimize $\frac{IR}{V}$. This means that the voltage V must be as high as possible, because a high voltage V results in a lower current I with unchanged power H_{city}. The resistance R needs to be also as small as possible. The power supply lines of cities are, therefore, operated at very high voltages, as high as safety permits, and use wires made out of metals with low resistance. Such a metal is copper, and this is the reason why copper is very much sought for and is also very expensive. Search the Internet for "electric power transmission," and you can learn a lot more about this. With all these considerations fulfilled, there is still a lot of power wasted due to heat generation when cities are supplied. How much is wasted depends on various conditions of the supply and on how smart the network is to redirect power to the right places. For this reason, a smart grid is of great importance and a great topic for present and future STEM experts. Many problems would be solved, however, if we had a wire that has no resistance at all, i.e., $R = 0$. Then no heat would be generated. Is that possible? In fact, it is, at least in principle. However, the current possibilities are not yet practical, and science needs to make some progress before we can hope solving this problem. This is what we discuss next.

5.5.4 Superconductivity: No Heat Generation!

Do accelerated electrons always loose energy either by emitting electromagnetic waves or by exciting vibrations of atoms? The answer is actually no! The rule of physics that says that they do is not a general law! We know that electrons roam around the nuclei of atoms being accelerated by the electric field that corresponds to the positive charge of the protons. Nevertheless, the hydrogen atom and all the other atoms are stable and do not loose energy. Electrons around atoms behave differently than electrons that are accelerated or decelerated in free space and that are located far away from atomic nuclei. The electrons and positrons can obviously, as far as I understand it, perform a dance around each other in such a way that no energy is lost by radiation of any kind. As we have seen in Sect. 5.3, quantum mechanics describes this dance by the Schrödinger equation for standing waves. Is it possible that electrons in solids do not excite atomic vibrations if accelerated? Yes, that is possible too. The Dutch physicist Heike Kamerlingh Onnes investigated the electrical resistance of metals such as lead at very low temperatures. He found that below a certain critical temperature, these metals have a resistance $R = 0$ as accurately as he could measure. He called this phenomenon "superconductivity," meaning that some materials were conducting electrical currents far above their normal capability when cooled down.

A theory of superconductivity was worked out by the US physicists John Bardeen, Leon Cooper, and J. Robert Schrieffer (the BCS theory). This theory is a so-called many-body theory and involves in its bare essentials two electrons interacting by crystal lattice distortions. We know such complicated many-body situations already from astronomy. For example, the exact path that the moon takes is influenced by the gravity of both the sun and the earth. Even the greatest mathematicians, including Euler, had difficulties to find approximate solutions to this problem. Nowadays we can find solutions for three and more objects by using computers. In the case of superconductivity, the problem involves electrons and atomic nuclei, and one needs to explain the zero resistance by calculating how the electrons and nuclei dance with each other. This presents an even bigger problem than the classical many-body problems, and the BCS theory is, therefore, mathematically too complicated to be presented here.

However, there is a simple physical picture and analogy that explains superconductivity. Consider a membrane, such as a drum or a sheet of plastic. Consider further two heavy steel spheres that you put on this membrane. Then what you observe is that the steel spheres distort the membrane due to their weight by creating a slight depression on the membrane. This depression, in turn, causes the steel spheres to move close together as if they attract each other. A similar effect happens in superconductivity. The electrons distort the atomic nuclei of the metal, just as the membrane is distorted by steel spheres, because the negative charge of the electrons exerts a force on the positive protons of the nuclei. This leads to an effective attractive force between the electrons. Electrons normally repel each other. However, here they are immersed in the positive sea of the atomic nuclei, and together electrons and protons are neutral. The distortion, however, results in a net attraction between the electrons. The electrons form in this way something like a "goo" that sticks together. Such a goo may conduct electricity perfectly, meaning that the electrical resistance is zero. In this goo-like state, each of two electrons (a so-called Cooper pair) can be imagined as emitting and immediately reabsorbing lattice vibrations so that no net heat is generated: a really magical dance of electrons and lattice vibrations.

Superconducting materials do have many applications. One can build a coil made out of superconducting wire and run a current through it which, in turn, creates a magnetic field. That current will continue to flow when left alone; some have estimated for a hundred thousand years or more, because the resistance is really so close to zero. Such superconducting magnets, with their enormous possible magnetic fields, have numerous applications. The reader is encouraged to browse the Internet and look at the possibility of levitating trains with such strong magnets. The trains are not only levitated by the magnetic fields but can also be propelled by them. Record speeds of 361 miles per hour (581 km/h) have been achieved that way already.

Is this the solution to all of our problems with resistance? Can we also create chips that run at very low power? Can we supply cities without heating up the wires and losing 10–15% of the energy before it gets to the cities? Well, unfortunately not yet. The problem is that the superconducting goo is destroyed by high temperatures and works only at very low temperatures. Most superconductors need to be cooled down to the temperature of liquid helium to work. This means temperatures of about −270°C or an even more horrible number in Fahrenheit. The necessary cooling makes the applications of superconductivity very expensive and therefore often unpractical. There were recent advances, and materials have been found that superconduct over a 100°C (or Kelvin) higher. Unfortunately this is still very much in the negative temperature range: too much in the negative! Is it possible to increase the superconducting temperature to room temperature just by engineering design? The reason that this is not yet possible is that the many-body theory is too complicated to be solved, under general conditions, even for our largest computers. Such types of many-body theories are the holy grail of STEM researchers of the future. If we could solve all many-body problems of arbitrary complexity, then we could find solutions to all of our wishes in physics, chemistry, biology, and engineering. We could design optimal batteries and generate useful energy from sunlight, because these are all many-body problems. As far as I understand it, mankind is still a long way from being able to solve general many-body problems; there is a lot left to do for the future STEM experts.